PHYTOREMEDIATION AND BIOFORTIFICATION

Strategies for Sustainable Environmental and Health Management

PHYTOREMEDIATION AND BIOFORTIFICATION

Strategies for Sustainable Environmental and Health Management

Edited by

Nand K. Singh, PhD

Shadma Afzal, PhD

Tariq Aftab, PhD

AAP | APPLE ACADEMIC PRESS

First edition published 2024

Apple Academic Press Inc.
1265 Goldenrod Circle, NE,
Palm Bay, FL 32905 USA

760 Laurentian Drive, Unit 19,
Burlington, ON L7N 0A4, CANADA

CRC Press
2385 NW Executive Center Drive,
Suite 320, Boca Raton FL 33431

4 Park Square, Milton Park,
Abingdon, Oxon, OX14 4RN UK

Apple Academic Press exclusively co-publishes with CRC Press, an imprint of Taylor & Francis Group, LLC

Library and Archives Canada Cataloguing in Publication

CIP data on file with Canada Library and Archives

Library of Congress Cataloging-in-Publication Data

CIP data on file with US Library of Congress

ISBN: 978-1-77491-408-3 (hbk)
ISBN: 978-1-77491-409-0 (pbk)
ISBN: 978-1-00340-208-4 (ebk)

About the Editors

Nand K. Singh, PhD
Associate Professor, Department of Biotechnology,
Motilal Nehru National Institute of Technology Allahabad,
Prayagraj, Uttar Pradesh, India

Nand K. Singh, PhD, is an Associate Professor at the Department of Biotechnology, Motilal Nehru National Institute of Technology Allahabad, Prayagraj, Uttar Pradesh, India. After completing his doctorate in Biotechnology from the Indian Institute of Technology Roorkee, U.K., India, he was appointed as Assistant Professor in the Department of Biotechnology, Motilal Nehru National Institute of Technology, Allahabad, India. Dr. Singh has more than 16 years of teaching and research experience in genetics, genetic engineering, plant biotechnology, molecular biology, genomics, proteomics, molecular biology, recombinant DNA technology, metabolic engineering, and bioinformatics. He is the recipient of several fellowships at both national and international levels. Currently, he is involved with a number of national research projects with different government organizations.

So far, Dr. Singh has authored two books with national and international publishers. He also has to his credit more than 60 research publications in peer-reviewed international journals and 20 book chapters in edited volumes with international publishers. At present, Dr. Singh serves as an editorial board member and reviewer of several high-impact international scientific journals.

Shadma Afzal, PhD
Research Scholar, Motilal Nehru National Institute of Technology Allahabad,
Prayagraj, Uttar Pradesh, India

Shadma Afzal, PhD, is a Research Scholar at the Motilal Nehru National Institute of Technology Allahabad, Prayagraj, India. She received her postgraduate degree in Botany from the University of Allahabad, Prayagraj, India and is currently pursuing her PhD. She is a recipient of National Scholarship Programme of the Slovak Republic and worked as pre-doctoral visiting fellow at Faculty of Natural Sciences, Comenius University, Bratislava, Slovakia. She is an awardee of the CSIR-NET (2017 and 2018) in Life Science, UGC-NET (2017) in Environmental Science, and GATE in Life Science (2019). Her research interests include nanotechnology, plant biotechnology, physiology, and molecular biology. Her research includes more than 20 research articles in international journals of repute, and a contribution of 10 chapters to books. Findings of her research works have been published in the *Journal of Materials Science, Planta, Journal of Biotechnology, Environmental Science and Pollution Research, Plants, Journal of Plant Growth Regulation, Biotechnology Letters,* and *Environmental Pollution*. She is also a member of various societies.

Tariq Aftab, PhD
Department of Botany, Aligarh Muslim University, India

Tariq Aftab, PhD, is currently an Assistant Professor in the Department of Botany at Aligarh Muslim University, India. He is the recipient of a prestigious Leibniz-DAAD fellowship from Germany, a Raman Fellowship from the Government of India, and Young Scientist Awards from the State Government of Uttar Pradesh (India) and the Government of India. After completing his doctorate, he worked as a research fellow at the National Bureau of Plant Genetic Resources, New Delhi, and a post-doctorate fellow at Jamia Hamdard, New Delhi, India. Dr. Aftab also worked as visiting scientist at the Leibniz Institute of Plant Genetics and Crop Plant Research (IPK), Gatersleben, Germany, and in the Department of Plant Biology, Michigan State University, USA. He is a member of various scientific associations in India and abroad.

He has edited seven books with international publishers, including Elsevier Inc., Springer Nature, and CRC Press (Taylor & Francis Group), co-authored several book chapters, and published over 60 research papers in peer-reviewed international journals. His research interests include physiological, proteomic, and molecular studies on medicinal and aromatic plants. He received his PhD in the Department of Botany at Aligarh Muslim University, India.

Contents

Contributors

Majid Abdoli
Department of Plant Production and Genetics, Faculty of Agriculture, University of Maragheh, Maragheh, Iran; Nomads Affairs Organization of Iran (NAOI), Kermanshah, Iran

Shadma Afzal
Department of Biotechnology, Motilal Nehru National Institute of Technology, Allahabad, Prayagraj, Uttar Pradesh, India

Vladimir Aleksandrov
Institute of Plant Physiology and Genetics, Bulgarian Academy of Sciences, Sofia, Bulgaria

Natarajan Amaresan
C. G. Bhakta Institute of Biotechnology, Uka Tarsadia University, Maliba Campus, Bardoli, Surat, Gujarat, India

K. P. Raj Aswathi
Plant Physiology and Biochemistry Division, Department of Botany, University of Calicut, C.U. Campus P.O. Kerala, India

Balamurugan Balaji
Division of Molecular Biology and Biotechnology, ICAR–Indian Agricultural Research Institute, Pusa Campus, New Delhi, India

Hebert Jair Barrales-Cureño
Instituto Tecnológico de Estudios Superiores de Zamora, Carrera en Ingeniería en Innovación Agrícola Sustentable, Zamora, Michoacán, México

Michael Bell
The University of Queensland, School of Agriculture and Food Sciences; Queensland Alliance for Agriculture and Food Innovation, Gatton, Australia

B. K. A. Bellanthudawa
Department of Agricultural Engineering, Faculty of Agriculture, University of Ruhuna, Sri Lanka

Juliana Silva Rodrigues Cabral
Department of Agronomy, Universidade de Rio Verde (UniRV), Rio Verde, GO, Brazil

Debarati Chakraborty
International Crop Research Institute for Semi-Arid Tropics, India; ASDC Naturophilia Pvt. Ltd., India

Sapna Chandwani
C. G. Bhakta Institute of Biotechnology, Uka Tarsadia University, Maliba Campus, Bardoli, Surat, Gujarat, India

Zhong Xiang Cheah
The University of Queensland, School of Agriculture and Food Sciences, Gatton, Australia

Kricelle Mosquera Deamici
Faculty of Pharmacy, Federal University of Bahia, Salvador, Brazil

Elangovan Dharani
Division of Plant Physiology, ICAR–Indian Agricultural Research Institute, Pusa Campus, New Delhi, India

Noreen Fatima
Department of Plant Breeding and Genetics, University of Agriculture, Faisalabad, Pakistan

Luz María Basurto González
Instituto Tecnológico de Estudios Superiores de Zamora, Carrera en Ingeniería en Innovación Agrícola Sustentable, Zamora, Michoacán, México

Stephen Harper
The University of Queensland, School of Agriculture and Food Sciences, Gatton, Australia

Braulio Edgar Herrera-Cabrera
Colegio de Postgraduados, Campus Puebla, Puebla, México

Iqra Ijaz
Department of Plant Breeding and Genetics, University of Agriculture, Faisalabad, Pakistan

Edappayil Janeeshma
Department of Integrated Botany, MES-KVM College, Valanchery, Malappuram, Kerala, India

Faiza Khan
Department of Plant Breeding and Genetics, University of Agriculture, Faisalabad, Pakistan

Luis Germán López-Valdez
Universidad Autónoma Chapingo, Texcoco, Edo. de México

Gonzalo Guillermo Lucho-Constantino
Instituto Tecnológico Superior de Jesús Carranza, Veracruz, México

Fabiola Zaragoza Martínez
Centro de Investigación y Estudios Avanzados, CINVESTAV

Giselle Camargo Mendes
Department of Biotechnology, Instituto Federal de Santa Catarina (ISFC), Lages, SC, Brazil

Monalisha Mishra
Department of Biotechnology, Motilal Nehru National Institute of Technology, Allahabad, Prayagraj, Uttar Pradesh, India

Pratibha Mishra
Department of Botany, Ewing Christian College, University of Allahabad, Prayagraj, Uttar Pradesh, India

Sanjay K. Mishra
Department of Botany, Ewing Christian College, University of Allahabad, Prayagraj, Uttar Pradesh, India

Jorge Montiel-Montoya
Centro Interdisciplinario de Investigación para el Desarrollo Integral Regional, Instituto Politécnico Nacional, Guasave, Sinaloa, México

Caroline Müller
Department of Environmental Science and Technology, Universidade Federal da Fronteira Sul (UFFS), Erechim, RS, Brazil

Tim O'Hare
The University of Queensland, Queensland Alliance for Agriculture and Food Innovation, Gatton, Australia

Richa Parashar
Department of Biotechnology, Motilal Nehru National Institute of Technology, Allahabad, Prayagraj, Uttar Pradesh, India

C. Akshaya Prakash
Department of Botany, St. Joseph's College [Autonomous], Devagiri, Kozhikode, Kerala, India

Jos T. Puthur
Plant Physiology and Biochemistry Division, Department of Botany, University of Calicut, C.U. Campus P.O. Kerala, India

Disna Ratnasekera
Department of Agricultural Biology, Faculty of Agriculture, University of Ruhuna, Sri Lanka

César Reyes Reyes
Universidad Intercultural del Estado de Puebla, Huehuetla, Puebla, México

Rafael Salgado-Garciglia
Universidad Michoacana de San Nicolás de Hidalgo, Morelia, Michoacán, México

Salinda Sandamal
State Key Laboratory of Systematic and Evolutionary Botany, Institute of Botany, Chinese Academy of Sciences, Beijing, China; University of Chinese Academy of Sciences, Beijing, China

Nair G. Sarath
Department of Botany, Mar Athanasius College [Autonomous], Kothamangalam, Kerala, India

Mohammad Sarraf
Department of Horticultural Science, Faculty of Agriculture, Shahid Chamran University of Ahvaz, Ahvaz, Iran

Delse Parekkattil Sebastian
Department of Botany, St. Joseph's College [Autonomous], Devagiri, Kozhikode, Kerala, India

A. M. Shackira
Department of Botany, Sir Syed College, Taliparamba, Kannur, Kerala, India

Senthilkumar Shricharan
Division of Plant Physiology, ICAR–Indian Agricultural Research Institute, Pusa Campus, New Delhi, India

Nand K. Singh
Department of Biotechnology, Motilal Nehru National Institute of Technology, Allahabad, Prayagraj, Uttar Pradesh, India

Nivedita Singh
Phycology Laboratory, Department of Botany, University of Allahabad, Prayagraj, Uttar Pradesh, India

Saima Sohrab
Department of Botany, Ewing Christian College, University of Allahabad, Prayagraj, Uttar Pradesh, India

Houda Taimourya
Department of Production, Protection and Plant Biotechnology, Agronomy and Veterinary Institute Hassan II, Rabat, Morocco

Asanka Tennakoon
Department of Biosystems Technology, Faculty of Technology, Eastern University, Sri Lanka; Department of Agricultural Technology, Faculty of Technology, University of Colombo, Sri Lanka

Satyam Tiwari
Department of Botany, Ewing Christian College, University of Allahabad, Prayagraj, Uttar Pradesh, India

Vaishali Trivedi
Department of Botany, Ewing Christian College, University of Allahabad, Prayagraj, Uttar Pradesh, India

Muhammad Uzair
Department of Plant Breeding and Genetics, University of Agriculture, Faisalabad, Pakistan

W. L. N. Wasana
Department of Agronomy, Kansas State University, Manhattan, KS, 66506, USA

Jeshima Khan Yasin
Division of Genomic Resources, ICAR–National Bureau of Plant Genetic Resources, Pusa Campus, New Delhi, India

Abbreviations

AAN	anhydrous ammonium nitrate
AAS	atomic absorption spectroscopy
AC	amylase content
ACF	activated carbon fibers
Ag_2O	silver oxide
Al	aluminum
Alg	alginate
AMF	arbuscular mycorrhizal fungi
AN	acrylonitrile
AOB	ammonia-oxidizing bacteria
APX	ascorbate peroxidase
As	arsenic
AsA	ascorbate
AsA	ascorbic acid
ATP	adenosine triphosphate
ATS	algal turf scrubbers
Au	gold
B	boron
BCF	bioconcentration Factors
BET	Brunauer-Emmet-Teller
BMOs	beneficial microorganisms
$CaCO_3$	calcium carbonate
CAT	catalase
CBZ	climbazole
CCDs	carotenoid cleavage dioxygenases
Cd	cadmium
CDTA	cyclohexane-1,2-diaminetetraacetic acid
CGIAR	Consultative Group for International Agricultural Research
CH	chitin
CHCs	chlorinated hydrocarbons
CIMMYT	International Maize and Wheat Improvement Center
CL	cellulose
CNC-ALG	cellulose nanocrystal-alginate
CNTFs	carbon nanotube-fibers pyrograf

Cas9	CRISPR associated 9
CNTs	carbon nanotubes
CO2	carbon dioxide
CRISPR	clustered regulated interspaced short palindromic repeats
crRNA	CRISPR RNA
CS	chitosan
CS–ZnONPs	chitosan–ZnO nanoparticles
Cu	copper
CV	crystal violet
CWRs	crop wild relatives
DAP	diammonium phosphate
DCBs	dichlorobenzenes
DCF	diclofenac
DHA	dehydroascorbate
DHAR	dehydroascorbate reductase
DM	dry mass
DMA	deoxymugineic acid
DMAS	deoxymugineic acid synthase
DMPP	dimethylpyrazole phosphate
DNBP	diethyl 4-nitrobenzyl phosphonate
DSB	double-stranded break
DW	dry weight
DWF	dry weather flow
DXS	deoxy-d-xylulose-5-phosphate synthase
EDDHA	ethylene diamine-N-N'-bis
EDDS	ethylene diamine-N,N'-disuccinic acid
EDTA	ethylene diamine tetraacetic acid
ELSI	ethical, legal, and social issues
EPS	extracellular polysaccharides substances
EU	European Union
EWG	Environmental Working Group
FA	fatty acid
FAD2	fatty acid desaturase
Fe	iron
Fe^{3+}	ferric ions
$FeSO_4$	iron sulfate
FM	fish meal
FO	fish oil
FOB	ferrioxamine B

FRO	ferric-chelate reductase oxidase
FSSAI	Food Safety and Standards Authority of India
FTIR	Fourier Transform Infrared Spectroscopy
GABA	ɣ-aminobutyric acid
GAD	glutamate decarboxylase
GBS	genotyping by sequencing
GBSS	granule-bound starch synthase
GBSSI	encoding granule-bound starch synthase I
GE	genetically engineered
GE	genome editing
GGPP	geranylgeranyl diphosphate
GHS	greenhouse gas
GMA	glycidyl methacrylate
GMO	genetically modified organisms
GO	graphene oxide
GPC	grain protein content
GQDs	graphene quantum dots
gRNA	guide RNA
GRSP	glomalin-related soil protein
GSH	glutathione
GWAS	genome-wide association studies
H3BO3	boric acid
HA	humic acid
HDR	homology-directed repair
HE	homing endonucleases
HIV	human immunodeficiency virus
HS	humic substances
IAA	indole acetic acid
ICDS	integrated child development services
IITA	International Institute of Tropical Agriculture
IMNPs	iron-based magnetic nanoparticles
IPK1	inositol pentakisphosphate 2- kinase 1
IRT	iron-regulated transporters
IRT1	iron-regulated transporter gene
ISE	ion selective electrode
K	potassium
KCl	potassium chloride
KMC	Kolkata Municipal Corporation
KNP	keratin nanoparticles

KO	knockout
LCY	lycopene cyclases
LCYε	lycopene epsilon-cyclase
LMW	low molecular weight
MAL	multiple aleurone layer
MAs	mugineic acids
MB	methylene blue
MCB	monochlorobenzene
MDHA	monodehydroascorbate
MeNPs	metallic nanoparticles
MgO	magnesium oxide
MHLW	Ministry of Health, Labor, and Welfare
Mn	manganese
MNPs	magnetic nanoparticles
$MnSO_4$	manganese sulfate
MO	methyl orange
MOF	metal-organic framework
MONPs	metallic oxide nanoparticles
MPT	mycorrhizal inducible Pi transporters
MTs	metallothioneins
MUFA	monounsaturated fatty acids
MWCNTs	multiwall carbon nanotubes
N	nitrogen
NA	nicotianamine
NAAT	nicothiamine amino-transferase
NAS	nicothiamine synthase
NBPT	N-(n-butyl) thiophosphoric triamide
NBTs	new breeding techniques
NCC	nanocrystalline cellulose
NHEJ	nonhomologous end-joining
NIDDCP	National Iodine Deficiency Disorders Control Program
NIPI	National Iron Plus Initiative
NIs	nitrification inhibitors
NNMB	National Nutrition Monitoring Board
NPs	nanoparticles
NR	nitrate reductase
NRAMP	natural resistance-associated macrophage protein
nZVI	zero-valent iron
OAS-TL	O-acetylserine (thiol)lyase

OFs	organic fertilizers
OM	organic matter
OPT	oligopeptide transporters
P	phosphorus
PA	phytic acid
PA	proanthocyanidin
PAL	phenylalanine ammonia-lyase
PAM	protospacer adjacent motif
Pb	lead
PBDEs	polybrominated diphenyl ethers
PCBs	polychlorinated biphenyls
PCS	phytochelatin synthase
PCs	phytochelatins
Pd	Palladium
PGPB	plant growth-promoting bacteria
PGPM	plant growth-promoting microorganisms
PGPR	plant growth-promoting rhizobacteria
PhAs	pharmaceuticals
Phy A	phytochrome
POD	peroxidase
PUFA	poly-unsaturated fatty acids
PVA–CNF	polyvinyl alcohol/cellulose nanofibril
PVDF	polyvinylidene fluoride
PvFER	(P. vulgaris L.) FER
QPM	quality protein maize
QUT	Queensland University of Technology
RDA	recommended daily amount
RDI	recommended dietary intake
REFs	root endophytic fungi
RhB	rhodamine B
RISC	RNA-induced silencing complex
RNAi	RNA interference
ROS	reactive oxygen species
RS	resistant starch
S	sulfur
SB	speed breading
SBE	starch-branching enzymes
SBEII	encoding starch branching enzyme II
SDS	sodium dodecyl sulfate

Se	selenium
SEM	scanning electron microscopy
SFE	supercritical fluid extraction
SGR	stay green protein
sgRNA	short-guided RNA
SL	strigolactones
SOD	superoxide dismutase
SPIONs	superparamagnetic iron oxide nanoparticles
SSNs	sequence-specific nucleases
SWCNTs	single-wall carbon nanotubes
TaIPK1	inositol pentakisphosphate 2-kinase 1
TALEN	transcription activator-like effector nuclease
TALEs	transcription activator-like effectors
TCE	trichloroethene
TF	translocation factors
TOI	traits of interest
tracrRNA	trans-activating crRNA
UN	United Nations
USDA	United States Department of Agriculture
UV	ultraviolet
VAD	vitamin A deficiency
VIT	vacuolar iron transporter
VTC	vacuole transporter chaperones
WGS	whole-genome sequencing
WHO	World Health Organization
YSL	yellow stripe 1-like
ZFNs	zinc finger nuclease
Zn	zinc
ZnO	zinc oxide
$ZnSO_4$	zinc sulfate

Preface

For sustainable environmental management as a whole, the role of phytoremediation and biofortification as two sides of the same coin must be highlighted. Phytoremediation is a process in which biological organisms are used to eliminate or nullify an environmental pollutant by a metabolic course of action. The *biological* organisms consist of various types of plants, and the *remediation* refers to treating the situation. Over the past decade, this treatment has gained appreciation as a cost-efficient, non-invasive, complementary technology or substitute to engineering-based remediation processes. Biofortification, in contrast, is an agrarian process of enhancing the micronutrient content of a food crop through plant breeding, genetic modification, use of enriched fertilizers, or management of agricultural practices.

Malnutrition as a whole comprises diet-related chronic illnesses besides evident nutrient insufficiencies, which lead to disease and reduced physical and mental growth. Producing nutritious and safe foods adequately and sustainably has become an urgent goal of modern agriculture for the reason that about half of the global population suffers from the lack of micronutrients such as Fe, Cu, Zn, etc. However, strategies to enhance food fortification have not always been successful. Simultaneously, pollution and toxicity profiles of various soil contaminants, toxicity mechanisms, degradation, detoxification, and their phytoremediation approaches for ecological sustainability are also a matter of debate.

This book shows how phytoremediation and biofortification, although targeting different objectives, both technologies work on the same principle of the bio-extraction process that involves uptake, accretion, removal, and modification of micro-and macro-nutrient elements from the soil. Explicitly, this offers a comprehensive overview of phytoremediation and biofortification of Zn, Fe, Cu, and Cd, etc., and illustrates the emerging and innovative combination of these technologies.

The book comprises 16 chapters, most of them being review articles written by experts, highlighting a wide range of topics and discussing the integration of advanced phytoremediation and biofortification approaches for sustainable ecological and health management. We hope that this volume will serve as a prerequisite for all investigators who are working or have an interest in this particular field. In conclusion, this book will be helpful for general use by research students, teachers, scientists, and engineers, whether in academia, industry, or government and those who have an interest in environmental remediation and micronutrient fields.

We are extremely grateful to all our contributors for readily acknowledging our invitation for not only sharing their knowledge and research but also admirably integrating their expertise on scattered information from diverse fields in composing the chapters and enduring editorial suggestions to finally produce this venture. We also thank Apple Academic Press (CRC Press, Taylor & Francis Group) team for their generous cooperation at every stage of the book production.

Lastly, we extend our gratitude to all the well-wishers, research students, and authors' family members for their unconditional support, blessings, and encouragement in the compilation of this book.

—Nand K. Singh, PhD
Motilal Nehru National Institute of Technology, Allahabad, Uttar Pradesh, India

Shadma Afzal, PhD
Motilal Nehru National Institute of Technology, Allahabad, Uttar Pradesh, India

Tariq Aftab, PhD
Aligarh Muslim University, Uttar Pradesh, India

CHAPTER 1

Recent Advances and Challenges in Bioremediation and Biofortification

SATYAM TIWARI, SAIMA SOHRAB, PRATIBHA MISHRA, VAISHALI TRIVEDI, and SANJAY K. MISHRA

Department of Botany, Ewing Christian College, University of Allahabad, Prayagraj, Uttar Pradesh, India

ABSTRACT

Countries have been forced to increase crop yield due to the world's expanding population. One out of every three people on the planet is malnourished (according to the International Micronutrient Organization). One of the main reasons for the pollution of natural resources is contamination by heavy metals and many of the heavy metals that act as pollutants also act as trace elements for the human body. A plant used in bioremediation can be used as a source of trace elements as a biofortified food product. Plant-microbe interactions, plant breeding techniques, and genetic engineering processes have all been studied to improve plant ability to uptake heavy metals. In edible plants, the purpose of the biofortification process is to either increase trace element accumulation or bioavailability. Bacterial inoculation to wheat resulted in a remarkable increase in selenium accumulation as well as Ni, Mn, P, Mo, K, Fe, and Ca. *Rhodobacter sphaeroides* was used to bioremediate soils contaminated with Cadmium (Cd) and zinc (Zn), and it is a promising technique for heavy metal remediation. Transgenic plants having genes for reductase, transporters, and other enzymes are employed in genetic engineering procedures to create transgenic plants with an improved ability to acquire and utilize heavy metals. The plant wastes produced during the process of bioremediation of heavy metals such as Ca, Zn, and Se can be

Phytoremediation and Biofortification: Strategies for Sustainable Environmental and Health Management. Nand Kumar Singh, Shadma Afzal, & Tariq Aftab (Eds.)

used in biofortification as well as biofertilizer for soil with reduced levels of trace elements as well as fodder for dairy and meat animals to improve food quality. However, the challenge arises with the plant wastes produced by bioremediation of noxious heavy metals like Cd, Arsenic, and Mercury, which are difficult to get rid of. Biofortification and bioremediation are two interconnected aspects of achieving healthy and sustainable social growth, and more research is needed to investigate other options and solutions.

1.1 INTRODUCTION

Micronutrients, although needed in small quantities, are essential for the growth and development of the human body. Micronutrient deficiencies, sometimes known as "hidden hunger," have an impact on children's growth, immune systems, IQ, and brain development. Micronutrient deficiencies (such as iron, iodine, and vitamin A deficiency (VAD), as well as vitamin B12 deficiency) are a public health concern, with an estimated 2 billion individuals globally lacking one or more micronutrients [1]. According to the Food and Agriculture Organization report on the status of food security and nutrition in the world, it is estimated that 190.7 million people (14.5%) were malnourished in India in 2014–2016. Several schemes have been initiated over the years in India, such as ICDS (integrated child development services), Mid-Day Meal Program, NIPI (National Iron Plus Initiative), NIDDCP (National Iodine Deficiency Disorders Control Program), to improve the nutritional and health status of the population; however, much of the population continues to suffer from micronutrient deficiency, it is regarded as one of humanity's most serious global concerns. Plant breeding has traditionally been focused on increasing agronomic yields rather than nutritional quality, and other efforts to address the issue have generally focused on industrial fortification or pharmaceutical supplementation. Biofortification has been advocated as a long-term alternative to improve mineral nutrition, is a method that increases both mineral levels and their bioavailability in the edible component of staple crops, intending to improve their mineral nutritional properties at the source through agronomic intervention or genetic selection, and could be a solution to malnutrition or hidden hunger-reduction [2].

Another major risk to the environment as well as human health is toxic metal contamination which has increased quite a lot due to human activities. Studies have shown that there are certain bioremediating microbes and plants which are capable of removing the toxic metals from the environment and sequestering them in their bodies [3]. Such plants are able to grow and expand their roots even in the presence of heavy metal contamination of soil and also are more resistant to different pathologies (Paz-Alberto). The discovery of a new scientific approach to phytoremediation unveiled the noble concept of eliminating the heavy metals from contaminated soil and water by using those plants which naturally remove toxins from soil or water like Edenfern (brake fern) is used to make soil arsenic-free [4]. Some other examples of Phytoremediating plants are *Populus, Arundo, Salix*, etc. While biofortification is the process of increasing nutrient concentration in staple crops by simultaneous maintenance of agronomic traits like yield, pest resistance, etc. [5]. Thus, biofortification can be used as an armamentarium to combat the prevailing micronutrient deficiencies. Both the process of phytoremediation and biofortification follow the principle of accumulation of heavy metal/nutrients by plant uptake, translocation, bioavailability, and finally morphing in plant matrix [6–8]. Therefore, in this chapter, the two processes, i.e., phytoremediation and biofortification, are discussed by focusing specifically on their recent advances as well as challenges associated with them so that the clear understanding and concerting biofortification, phytoremediation, and bioenergy production in more advanced ways would lead to sustainable energy sufficiency and improve the degraded environmental remediation.

1.2 PHYTOREMEDIATION ADVANCES AND CHALLENGES

Contamination of soil, especially by heavy metals, is a grave environmental concern worldwide. Different types of contaminants, including fertilizers, pesticides, municipal wastes, heavy metals, flying ash, and wastewater irrigation, adversely affect the plants thriving therein [9]. Till date, around 53 heavy metal pollutants (having a density of more than 5 g/cm^3) have been reported. Heavy metals belong to two categories – essential (Co, Cr, Cu, Fe, Mn, Ni, and Zn) and non-essential (Pb, Cd, and Hg), the former being toxic when the excess amount is taken while the latter is toxic even in lower concentrations and causes the production of reactive oxygen species (ROS) which ultimately damages cells of the organisms by oxidative stress, and can even lead to death [10]. And thus, their entry into the food chain must be prevented to protect all other lifeforms. There already exist some soil remediation processes like soil washing, electrokinetic treatment, chemical reduction or oxidation, vitrification, etc., but they are complex and cost-intensive methods. In contrast to the conventional approaches, the most appropriate, efficient, less invasive, green, and gentle alternative technology is phytoremediation.

In the process of phytoremediation, natural or genetically modified plants (often fast-growing ones) uptake heavy metals, including pesticides, radionuclides, polynuclear aromatic hydrocarbons, and polychlorinated biphenyls (PCBs) from the environment and metabolize and store them as safer compounds, ultimately detoxifying the soil from hazardous compounds [11]. Hyper accumulative plants with high growth rate, biomass, yield, and toxic metal sequestration ability are employed for phytoremediation. There are some conventional approaches for phytoremediation, such as phytofiltration, phytostimulation, phytotransformation, phytostabilization, and phytoextraction [12]. However, with these classical approaches, certain limitations are associated, like soil conditions, rate, biomass production of the plant, solubilization of the heavy metals, etc. [13]. Therefore, recently advanced techniques have been developed which enhance the process of phytoremediation by overcoming the limitations and are discussed in subsections.

1.2.1 ADVANCES IN PHYTOREMEDIATION

1.2.1.1 BIOAUGMENTATION-ASSISTED PHYTOEXTRACTION METHOD

Bioaugmentation is one of the most fascinating strategies for the enhancement of the rate of biodegradation of contaminated sites by the introduction of a pregrown culture of microbes possessing required or desired features [14]. Bioaugmentation-assisted phytoextraction is a more advanced method where plants and microbes act synergistically and thus increase the bioavailability as well as uptake of the metals from the concerned site. Therefore, the selection of the appropriate microbes (which may be fungi or bacteria) and plants is of utmost importance in this technique. Furthermore, a number of hyperaccumulating and non-hyperaccumulating plants have been tested in this regard; some examples of non-hyperaccumulator plants include *Sorghum bicolor*, *Zea mays*, *Astragalus sinicus*, *Cannabis sativa*, *Hordeum vulgare*, *Salix dasyclados*, *Nicotiana tabacum*, *Prosopis juliflora*, *Trifolium pratense*, *Trifolium repens*, *Scirpus robustus*, *Vigna unguiculata* while plants like *Brassica juncea*, *Elsholtzia splendens*, *Solanum nigrum*, *Thlaspi praecox*,

Helianthus annuus are hyperaccumulators. In addition, plant growth-promoting rhizobacteria (PGPR) such as *Agrobacterium, Azotobacter, Bacillus, Pseudomonas, Alcaligenes, Arthrobacter, Serratia, Azospirillum, Burkholderia,* and *Rhizobium* also play a significant role in enhancing the biomass of the plant as well as metal accumulation [15]. Bacteria belonging to the bacillus species are reported to release some organic acids that increase Cd, Pd, and Zn bioavailability [16]. Studies have shown that *the Pseudomonas aeruginous*-Alfalfa (*Medicago sativa* L.) association helps in the enhanced mobility and availability of Cd, Ni, Cr, Zn, and Cu and thus encourages the growth and remediation process [17]. Apart from bacteria, symbiotic fungi, especially arbuscular mycorrhizal fungi (AMF), contribute to microbe-assisted phytoremediation.

1.2.1.2 HEAVY METAL CHELATION TECHNIQUE

Chelation of heavy metals for phytoremediation is an advanced and modified phytoextraction technique. The basic concept behind this technique is that chelating agents, when added to the soil, make soluble metal-chelate complexes through the process of desorption and dissolution and increase the feasibility of the phytoremediation process. In this procedure, synthetic chelating agents such as EDTA (ethylene diamine tetraacetic acid), DTPA (Pentetic acid or diethylene triamine penta acetic acid), EDDHA (Ethylene diamine-N-N'-bis), EDDS (ethylene diamine-N, N'-succinic acid), DOTA (1,4,7-tetra aza cyclo dodecane 1,4,7,10-tetra acetic acid), and natural, biodegradable or organic chelating agents such as Citric Acid, Nitrilotri-acetic acid, Gluconic acid, 2,3-Di Hydroxy Benzoic Acid are added to the fast-growing and high biomass producing plants and mobility of the heavy metals is thus increased in the soil [18–20]. The use of organic agents is often preferred over synthetic compounds as they show greater penetrating ability, stability, and non-toxic nature. It has also been seen that some plants produce natural chelating agents themselves, which help in the detoxification of the metals. Interestingly, the application of these chelating compounds to the non-hyperaccumulator plants shows an increase in their potential to accumulate metals [21]. The different steps and factors involved in the process of enhancement of chelating activity by increasing the heavy metal bioavailability and phytoextraction from the soil are described in the form of the flowchart (Figures 1.1 and 1.2) [22].

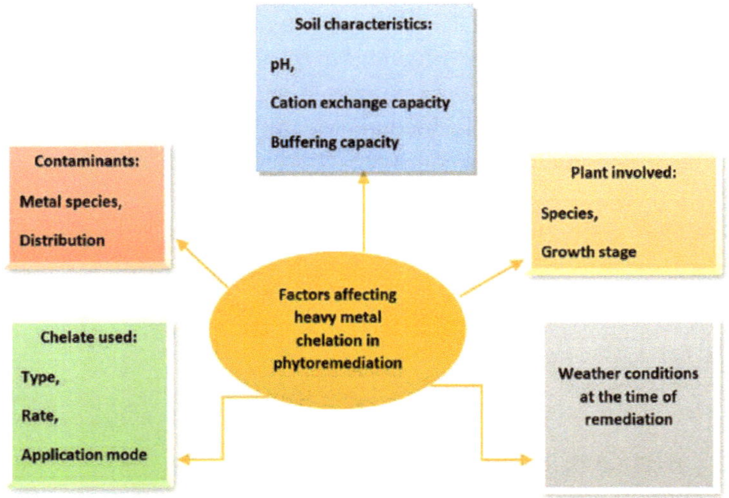

FIGURE 1.1 Factors affecting soil–metal–chelate–plant interactions and phytoremediation.
Sources: Ref. [22, 23]

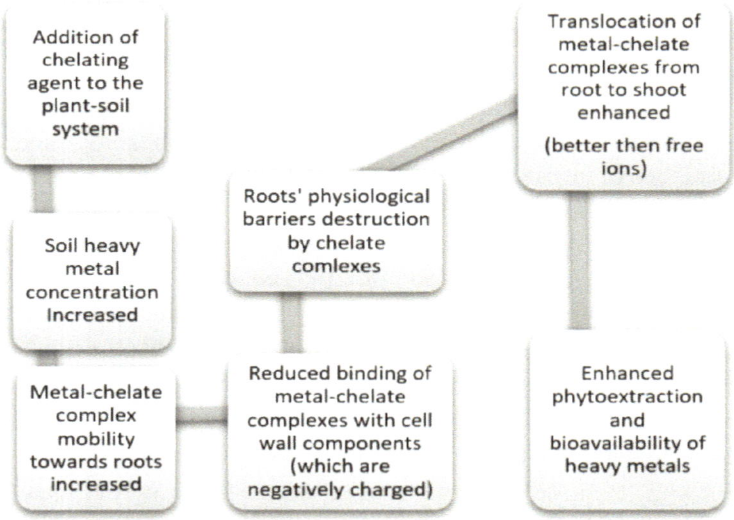

FIGURE 1.2 Enhancement of the bioavailability and phytoextraction of heavy metals from soil by the chelating agents [23].
Source: Ref. [23]

1.2.1.3 BIOLOGICAL SULFUR OXIDATION

This oxidation process is employed for remediating soil by using sulfur-oxidizing bacteria, for example, *Acidithio bacillus*, a chemolithotrophic gram-negative bacteria, and elemental sulfur, sulfides, and thiosulfate. This method is mainly an acidifying mechanism as there occurs a slow release of acids, particularly H_2SO_4 when sulfur is oxidized biologically using microbes [24]. The acidification of the soil conditions creates bioleaching of metals like Pb, Cu, Cd, Cr, Fe, and Zn [25].

1.2.1.4 APPLICATION OF ACIDIFIED COW DUNG

One of the most novel and interesting techniques is cow dung acidification and its application in phytoremediation. It involves two important methods, viz., bioaugmentation and then phytoextraction. Bioaugmentation is done using sulfur-oxidizing bacteria, which helps in the phytoextraction of heavy metals. Ashraf et al. [26] have developed this method in which sulfur and molasses solutions are used for cow dung acidification before they are subjected to use further. According to the studies, lead and Cd concentrations increased along with soil nutrient enrichment by the application of acidified cow dung [26].

1.2.1.5 PLANT GENETIC ENGINEERING

Plants with greater phytoremediation potential can be developed by applying advanced techniques of genetic engineering where the production of such genotypes is aimed that can possess better desirable traits along with higher tolerance, accumulation, and detoxification capabilities. For this purpose, genes involved in different stages of phytoremediation are actually taken from microbes and then transferred to the plants of interest. These genes can be extracted from fungi, bacteria, plants, or animals. The best-studied plants used for genetically modified phytoremediation plants are *Nicotiana tabacum* and *Arabidopsis thaliana* [27]. The plant metabolism, as well as phenotypic characteristics, both get altered because of the transgenic plants' genes overexpression and translocation, sequestration of heavy metals within plant cells gets improved. Other advantages include enhanced production of secretory enzymes responsible for metal removal in the exudates of roots [28].

1.2.2 CHALLENGES OF PHYTOREMEDIATION

Phytoremediation came into view as a technology that has clouded the previous physical and chemical techniques for soil recovery; notably, for its economic, commercial, and industrial benefits, they are being used in greenhouses and laboratory tests. To use the process at full scale, it is crucial that it is broadly field-tested to attain the best imaginable outcome [29, 30].

1.2.2.1 PHYSICAL RESTRICTIONS AND STRESSORS

Phytoremediation is growing and advancing very effectively, but still, there are few inevitable restrictions to technology. Various writings of failed and uncertain field trials put forward some insight into the most common challenges that have been endured [31–33]. A recently observed disadvantage of the process is that the pace of remediation is much sluggish in comparison to other processes. This may be because plant growth is hard to attain in highly contaminated soils, thereby decreasing the metabolism of contaminants [34–36]. On the other hand, there are stressors that disturb phytoremediation in the field, which were not observed during tests in laboratories or greenhouses. These include the change in precipitation, nutrients, temperature, and; herbivory, diseases causing microbes, competition by weeds that have been already adapted to the site [37]. Any of these biotic or abiotic stressors can decrease or halt plant development in the field, and this will adversely affect phytoremediation. Most of this information is from laboratory and greenhouse experiments. Other than environmental stressors in the field, various physical challenges also hinder the use of phytoremediation. For example, raising the moisture content of hydrophobic Petroleum Hydrocarbons-contaminated soils can be tricky once they have become dehydrated, which can halt germination and seedling growth. The problem also arises for phytoremediation when the polluted soil is farther than the rooting zone, e.g., the average root depth of herbaceous plants is 50 cm [38]. This creates the necessity of unearthing before phytoremediation. Trees show an average root depth of 3 m (much deeper roots in some plant species) which facilitates remediation to higher depths without unearthing [33, 39]. The challenge is it can be problematic to grow trees in contaminated soils, and once planted; they take many years to attain enough biomass for an adequate rate of phytoremediation [40]. There is also the problem of disposal if the roots and wood are accounted to be contaminated.

1.2.2.2 THE DULL GROWTH CYCLE OF THE PLANT USED FOR PHYTOREMEDIATION

The other major drawback of technology is the period of the lifecycle of most plants to attain full bloom decreases the process of contaminant uptake, as the plant uses all its energy for growth. Thus, the process becomes a slow process. Today plants with short life periods are being discovered to overcome this problem. Genetic improvements made in such plants can be really helpful [41, 42].

1.2.2.3 COMPLEXITY IN THE CONTAMINATED FIELD SITE

A challenge often encountered in the field is uneven dispersion of contaminants, with "hot spots" spread across the site [31, 33, 37, 43]. In controlled test environments, the soil is well mixed to attain an even matrix. This may not be the case in the field despite the site having been tilled before planting. The heterogeneity of contaminants causes data scatter, which will make it difficult for the statistical presentation of significant treatment aftermath for field trials. This is also challenging in points of regulatory standards as phytoremediation success is judged on a point-by-point basis rather than average of data from the whole site. If one point across a site fails to achieve the regulatory targets, the whole site goes down [33, 44]. Along with contaminant distribution, factors like soil pH, soil structure, microbial activity, root structure, the organic composition of the soil, and moisture content often exhibit variability across a given site, and can vary over time [37, 45], e.g., water added to hydrophobic Petroleum Hydrocarbon-polluted soil may show uneven distribution, causing uneven moisture content throughout the site [31, 46]. This can affect microbial and plant growth, and so the remediation capacity. One way to check the problem of uneven dispersion is to step up the number of samples investigated per treatment plot. However, because many field studies have limitations, including resources and time, the required large-scale sampling is often not a practical option [37]. Agronomic methods, like the addition of nutrients and organic matter (OM) and tilling, are commonly used before planting [47, 48]. These methods are generally applied to improve soil composition and texture that facilitate microbial and/or plant growth. They cause a change in oxygen content and soil pH, which will alter the bioavailability, and hence the breakdown of contaminants even in bulk soil. Nutrients in the non rhizosphere region, such as phosphorus and nitrogen, will undoubtedly affect the domestic microbial populations, in addition to contaminant utilizers [48]. Aeration of site soil enhances the soil oxygen amount, which enhances oxidative degradation of various organic contaminants [35, 36]. Photo-oxidation of formerly existing organic contaminants can also occur.

1.2.2.4 CONTAMINANT ACCUMULATION IN PLANT TISSUES

Phytoextraction may cause the accumulation of non-degraded contaminants in plant organs, which may become the main reason for the ecosystem to get exposed to the contaminants; else, the plants may require harvesting reaping to extract the contaminant. Another process called phytovolatilization is the reason for removing toxic elements in their gaseous form from the exposed surface of plants' bodies, thereby causing an increase in airborne contamination [42].

1.2.2.5 PROBLEMS IN CONVINCING AND ACCEPTANCE BY REGULATORY BODIES

The fundamental necessity of government regulators is to assure the safety of the community and to protect the environment. They should be convinced that Bioaccumulation of hazardous contaminants in plants is an efficient and safe way for remediation. Arrangements should be made for the deportation of contaminated plant parts and remain as part of a restorative plan in case the plants do not break down the contaminants. Another consideration is the probable ecological risks of introducing and bringing non-native microbial and plant species into the field [39, 49, 50]. These species may disperse from the contaminated site, deposing or hybridizing with domestic species. Phytoremediation of organic contaminants generally takes place in the region of fine roots that have high turnover. These roots typically do not survive a full flourishing season, and the deteriorating root tissue is changed into bulk soil while the process of humification [48, 51]. OM may also be added to a field site from other remains from the plants used for remediation, animal manure, peat moss, biosolids, and woodchips/sawdust, which may be added to the soil before planting. This is required, as it enhances the soil quality at field sites that commonly consist of greatly degraded soil. Hydrocarbons can also be added by the indigenous microbial populations. Regrettably, the addition of OM from such sources can baffle sample analyses due to structural similarities with the compounds that are being remediated.

1.2.2.6 LESS TOLERANCE AGAINST HIGH CONTAMINANTS LEVEL IN PLANTS

Phytotoxicity is also one of the leading causes of the decline of remediation, as very high amounts of contaminants in the plant body will not permit its growth or survival, so the remediation is seen to be compelling only at the lower amount of contaminant. This problem has been solved by genetically modifying plants by inserting genes for resistance to heavy metal aggregation, thereby making the remediation work more swiftly [41, 42].

1.2.2.7 USE OF GENETICALLY MODIFIED ORGANISMS (GMOS) IN THE FIELD

Although encouraging results have been observed by using genetically modified organisms (GMOs) in the greenhouse and laboratory, regulated in situ tests have prevented any considerable amassment of field data [38]. If GMOs are being used without sufficient regulation systems, there may be a possibility for introduced genetic material to be introduced into indigenous populations. As well as, recombinant microbial strains that consist of antibiotic resistance genes from the cloning processes cannot be exposed to the environment [52]. There is also a problem of low public acceptance of GMOs [54]. Even if problems of public acceptance and regulatory restrictions can be overcome, there are problems that are to hinder the use of GMOs for remediation at the field sites. Genetically modified microbes and plants often fail to compete with the native plants and microbes at the field site [54, 55]. Breakdown of organic contaminants frequently requires the collaborative action of diverse enzymes, and it is practically not possible to introduce all genes required to degrade an organic contaminant into a single microbial or plant genome [54]. It can be problematic to consistently maintain even a single gene in recombinant organisms, and the introduced trait is often lost [56, 57]. In plants, transgene silencing, through the mechanisms like cytosine methylation, makes use of the technology intrinsically unpredictable.

1.3 RECENT ADVANCES AND CHALLENGES IN BIOFORTIFICATION

1.3.1 RECENT ADVANCES IN BIOFORTIFICATION

Malnutrition is becoming a more critical concern as the world's population grows. In 2030, the world's population is expected to hit 8.3 billion, up from 7.8 billion today. Malnutrition affects 792.5 million people worldwide, according to UNFAO, with the developing world bearing the brunt of the burden. The majority of people in impoverished countries either go hungry or eat nutrient-deficient food. Hunger claims the lives of 24,000 people every day around the world. "Hidden hunger" affects almost a third of the population [1]. In their diets, they are deficient in one or even more critical micro- or macronutrients such as zinc, iron, selenium, iodine, folic acid, lysine, vitamin D, vitamin A, vitamin B12, vitamin C, and others. These vitamins and nutrients are vital for the human body's growth, development, and proper functioning. Short-term deficiency in these micronutrients somehow does not cause significant harm, but long-term deficiency can result in a variety of disorders such as Anemia (Fe deficiency), beriberi (Vit B deficiency), pellagra (niacin deficiency), rickets (Vit D deficiency), and scurvy (Vit C deficiency) can all be lethal if these micronutrients are deficient for long periods of time. Fe deficiency is the major cause of anemia, affecting 50–60% of the world's population [59]. Beriberi was responsible for half of the mortality of children in rice-consuming countries around the world in the early 1920s. To combat malnutrition, a variety of direct and indirect methods are utilized to boost the concentration of important nutrients in grain and other plant parts. Malnutrition is treated by dietary diversity, artificial supplementation, and Biofortification. Biofortification, on the other hand, is a cost-effective, realistic, and long-term solution to the nutritional shortage. Biofortified crops provide all of the important vitamins, minerals, and fatty acids (FAs) that humans require to meet their nutritional needs. Furthermore, biofortified crops are environmentally favorable since they boost soil quality and nutrient absorption, and they are also sustainable [3].

Crop Biofortification can be accomplished in a variety of ways, including traditional methods and new breeding techniques (NBTs). Mineral fertilization via foliar or soil application, microbe-mediated increased nutrient uptake, and conventional plant crossing to achieve the optimal combination of genes for balanced nutrient uptake and bioavailability are all examples of traditional Biofortification methods now those Gene suppressing, genetic manipulation, overexpression, and gene transfer from some of the other species for maintenance of balanced and nutritious compositions in mutated plants have been used by NBTs [3] (Figure 1.3).

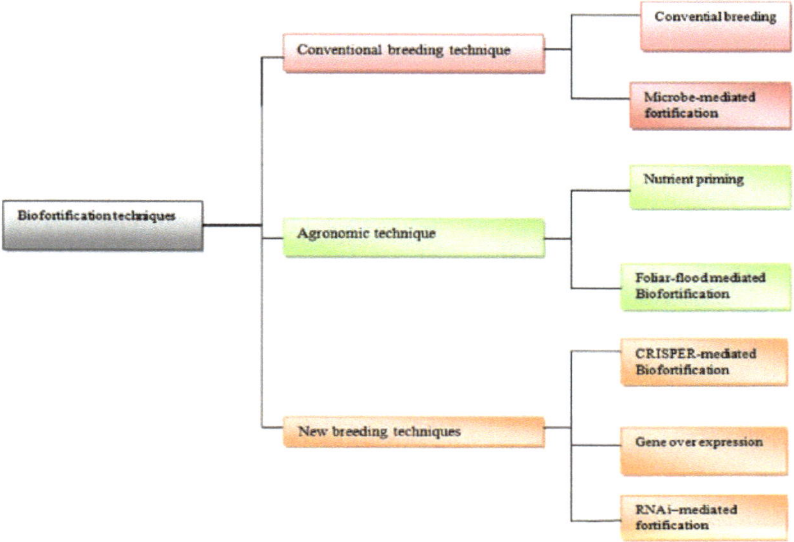

FIGURE 1.3 Different conventional and advanced techniques are involved in biofortification.
Source: Ref. [3]

Some NBTs, such as Transgenic breeding, RNAi interference, and genome editing (GE), are promising to biofortified crops by facilitating the generation of unique genetic variation which does not exist within the gene pool [3].

1.3.1.1 TRANSGENIC BREEDING INVOLVED IN BIOFORTIFICATION

Biofortification by transgenic genetics is a cost-effective, efficient, and long-term solution to hunger. When there is limited genetic variety in the gene pool and wild relatives' transgenic crops are formed. Transgenic breeding aids biofortification by introducing genes that either reduce anti-nutrient molecules or boost the accumulation and bioavailability of micronutrients [61]. The importation and overexpression of genes related to micronutrients, including zinc, iron, vitamin A, and proteins, in cereal grains could be used to address micronutrient deficiencies. For example, ectopic overexpression of the TaFer1-A and OsNAS2 genes raised iron content in wheat grains. Similarly, overexpression of TaVIT increased the iron content of wheat grain, and Overexpression of the Ama1 gene, which is based on Amaranthus albumin, compensates critical proteins and amino acid shortage such as tyr, lys, cys, and met. Apart from this, silencing the phyA and TaABC13 transporter genes decreased phytic acid (PA) production and boosted iron and zinc bioavailability in plants, but overexpression of the phytochrome (Phy A) gene increased phytase synthesis and anti-nutrient activity [60]. To prevent obesity and overnutrition, the SBEIIA gene was silenced, which raised the proportion toward less edible amylose starch. Carotene desaturase CrtI, CrtB, and PSY gene overexpression in bacteria also boosted vitamin A concentration, as well as CrtB or Crtl, likewise resulting in enhanced pro-vitamin A accumulation [3].

Few pulses, such as common beans, soybeans, chickpeas, and lupines, have been modified to overcome malnutrition as a result of transgenic breeding. The goal of transgenic breeding for biofortification of pulses is to increase important amino acids, Fe and Zn fortification, and minimize anti-nutrient chemicals [62]. Overexpression of heterologous proteins rich with sulfur-rich amino acids was used to overcome a sulfur-rich amino acid deficiency [3]. For nutritional enhancement of cysteine amino acid, maize-based 27 kDa-zein, a cysteine-rich protein, has been produced and overexpressed in various crop varieties. Similarly, overexpression of S-rich proteins increased methionine levels in carbon bean and lupin. After transformation, the rice OASA1D transgene increased free tryptophan production in adzuki beans. To improve iron bioavailability in chickpeas, the Glycine max ferritin and chickpea NAS2 genes have been inserted and overexpressed, and Carotene desaturase as well as bacterial PSY, bkt1, crtW, and crtB genes overexpression increased pro-vitamin A accumulation in soybeans [63]. To diminish α-linolenic levels in soybean, researchers used siRNA-mediated knockout (KO) to silence the -3 FAD3 gene. Additionally, the 16-desaturase gene has been overexpressed to promote the conversion of α-linolenic acid to its stable form [3].

1.3.1.2 RNAI INTERFERENCE TECHNIQUE INVOLVED IN BIOFORTIFICATION

RNAi is a sequence-specific gene regulatory method involving a (ds-RNA) molecule that causes a particular genetic transcription or translation to be inhibited. Ever since its invention RNAi has opened up a whole new world of possibilities for agricultural crops. This is a more accurate, reliable, and efficient technique than the antisense technique. Through biofortification and bio-elimination, RNAi gives a platform for the inclusion of various biotic and abiotic tolerance as well as the supply of high-quality meals [64]. It is commonly used to improve the nutritional quality of crops and to remove allergies and pollutants from food. Because of its potential to chelate micronutrients, thus limiting the bioavailability of key nutrients, PA is regarded as a prominent anti-nutrient in grains. RNAi-based interruption of carotenoid cleavage dioxygenases (CCDs) genes, which degrade carotenoids and increased carotenoid concentration in rice. RNAi was used to disrupt three genes named OsCCD1, OsCCD4a, and OsCCD4b. As a result, when mutants were compared to the wild kind, they produced more carotenoids [66].

1.3.1.3 GENOME EDITING (CLUSTERED REGULARLY INTERSPACED SHORT PALINDROMIC REPEATS (CRISPR) SYSTEMS) INVOLVED IN BIOFORTIFICATION

Plant genome editing (GE) uses sequence-specific nucleases (SSNs) to produce transgene-free plants via stably inherited and targeted gene modification in the desired crop. For plant GE, SSNs like as TALENs, ZFNs, and the CRISPR-Cas system are utilized. Cas9/13, RNA-guided DNA endonucleases directed by a short-guided RNA (sgRNA) are used in CRISPR GE to create a complex at the target site for targeted gene editing [67]. Emerging molecular biology trends have advanced our understanding of plant-specialized metabolites in terms of structure, function, and biosynthesis pathways. Transcription activator-like effector nuclease, zinc finger nuclease (ZFNs), and CRISPR systems have potential applications in genome and epigenome engineering since they are powerful genome-editing tools. CRISPR/Cas9 is being utilized in plants to investigate the activities of a variety of genes and gene families [68]. CRISPR technology is continually improving and now includes possibilities for creating KOs, precise changes, multiplex GE, and activating and suppression of target genes among other genetic manipulations [69].

Pulses There are currently no examples in the literature that illustrate the role of GE in pulse Biofortification. However, GE for iron and zinc biofortification has the potential to increase iron and zinc uptake in all pulse crops by manipulating iron-regulated transporters (IRT), ferric-chelate reductase oxidase (FRO), zinc-regulated transporters natural resistance-associated macrophage protein (NRAMP), and IRT like protein (ZIP) [3]. Numerous studies show the use of CRISPR-based GE in rice, and a few research has indicated using GE to improve biotic and abiotic challenges for rice crop development. The rice genome has a lot of potential PAM sites [70]. In the near future, CRISPR technology could be used to address any desired traits in the rice genome [71]. For the first time in any crop plant, researchers used both protoplast and particle-bombarded rice calli systems to demonstrate sequence-specific CRISPR/Cas9 mediated genomic alteration of genes, phytoene desaturase (OsPDS), betaine aldehyde dehydrogenase

(OsBADH2), and mitogen-activated protein kinase (OsMPK2), which have been involved in regulating responses to different environmental stresses stimuli. Xie & Yang [69] developed two rice GE vectors, pRGE3 and pRGE6, to demonstrate an RNA-guided GE technique. OsMPK5, a negative regulator for abiotic stress stressors in rice, was chosen for targeted mutagenesis and testing in rice protoplasts utilizing three gRNAs. An even more precise gRNA design method resulted in a low level of off-targets [70] (Table 1.1).

TABLE 1.1 List of Biofortified Crops, Target Genes, and Techniques Used in Biofortification

Crops	Target Genes	Used Techniques	Stress/Trait	References
Rice	OsERF922	CRISPER technique	Blast resistance.	[18]
Chickpea	GmFER, NAS2	Transgenic breeding and genome editing.	Increased iron bioavailability.	[72]
Bread wheat	TaMLO-A1, TaMLO-B1, and TaMLOD1	CRISPER technique	Powdery mildew disease resistance.	[18]
Rice	OsVIT2 gene	Genome editing	Increased iron bioavailability.	[73]
Maize	ARGOS8	CRISPER technique	Increased grain yield under drought conditions.	[74]
Soybean	GmPDS11, GmPDS18	CRISPER technique	Enhance carotenoid biosynthesis.	[74]
Tomato	Rin	CRISPER technique	Fruit ripening	[75]
Tobacco	NtSTF1, NtPDS	CRISPER technique	Effect on pigmentation and leaf morphology.	[75]
Wheat	[inositol pentakisphosphate kinase (TaIPK1)]	RNA I technique	Increased micronutrients (phosphate, zinc, and iron) contents.	[76]
Rice	OsCCD1, OsCCD4a, and OsCCD4b	RNA I technique	Enhanced carotenoid biosynthesis.	[65]
Sorghum	PSY-1, CRT-I, PMI, LPA-1	Transgenic breeding	Enhanced vitamin A content.	[77]

1.3.1.4 GENOME-WIDE ASSOCIATION STUDIES (GWAS) USED IN BIOFORTIFICATION

Developmental biology requires a thorough understanding of the genetics of complex phenotypes. Plant scientists have long been fascinated by the phenotypic variation in different genetic backgrounds, which has led to the development of several association studies. Genome-wide association studies (GWAS) were made possible by advances in genome sequencing technologies, the creation of high-density genotyping arrays, and high-throughput phenotyping platforms [78]. GWAS begins with genotyping using various next-generation sequencing technologies, such as genotyping by sequencing (GBS) or whole-genome sequencing (WGS), and then continues with thorough phenotyping in a variety of contexts. The genotyping data is trimmed to produce relevant SNPs, which are then exposed to GWAS for each trait independently in order to develop SNP-trait associations. As a result, candidate genes for specific features can be predicted [79].

1.3.1.5 THE IMPLICATIONS OF SPEED BREEDING ON BIOFORTIFIED CROPS ADVANCEMENT

The University of Queensland proposed this strategy in 2003 as a blend of many methods to speed up wheat breeding. The objective behind this strategy is to alter the environment in stimulating early flowering and shorten the generation time [80]. It supports a broad range of germplasm and does not really necessitate *in vitro* culture or traveling across the nation to find a suitable climate for numerous generations in a year, like double haploid, as well as shuttle breeding [80].

Plant growth is influenced by a variety of internal and external elements, including photoperiod, light intensity, temperature, and light quality. SB (speed breading) alters these natural processes by hijacking various biological systems in plants in order to boost generation speed. For the creation of effective SB platforms for a specific crop, knowledge of these essential processes is required [80]. Photoperiod is important in the shift to reproductive development because it detects any changes in the external environment that photoreceptors detect. As a result of synchronizing with changes in external stimuli, a species' reproductive success increases [80].

1.3.2 CHALLENGES OR LIMITING FACTORS IN BIOFORTIFICATION

Biofortification is a major tool to solve the riddle of hidden hunger in a large sector of society. Even though several advances have been made in this field, several factors are there which hinder the proper implementation of the tools, techniques, and programs related to biofortification. The limiting factors of biofortification are as in subsections.

1.3.2.1 REDUCTION OF ANTINUTRIENT ACCUMULATION

Antinutrient factors like tannins and phytate, which are naturally found in seeds, grain, and other edible parts of the plant, restrict the proper absorption of essential nutrients like iron, calcium, zinc, etc. [81]. Hence, breeding such biofortified crop varieties should be done which are having restricted or low-concentration antinutrient factors. These antinutrient substances impair the absorption of essential nutrient factors present in food by the gut, like phytic acid (phytate) restricts the absorption of iron in the human gut by chelating iron (Table 1.2) [82]. PA is the derivative of phosphate which is found in plant seeds and plays a major role in its metabolism [83]. It has also been observed that there is a lack of awareness in society that tea, coffee, and cocoa should not be consumed with meals if the poor iron status is suspected. Polyphenols in these beverages inhibit the absorption of nonheme iron.

1.3.2.2 ENHANCED CONCENTRATION OF PROMOTER SUBSTANCES

Nutrient assimilation from biofortified crops gets accelerated by the presence of promoter substances. Promoter substances are the organic compounds like Amino acids (methionine, lysine, and cysteine), which accelerate nutrient absorption in the gut. For example, fruits and vegetables contain vitamin C and organic acids (e.g., citric acid) that keep iron in a reduced form, increases absorption of nonheme iron when consumed in the same meal [84].

TABLE 1.2 List of Elements, Their Antinutrients, and Promoters [3, 85]

SL. No.	Element	Antinutrients	Promoters
1.	Calcium (Ca)	Oxalate, Phylate, Tannins, Fiber	Inuline
2.	Magnesium (Mg)	Phylate	–
3.	Iron (Fe)	Oxalate, Phylate, Tannins, Fiber, hemaglutinins	B-carotene, cysteine, histidine, lysine, methionine, fumarate, malate, citrate
4.	Zinc (Zn)	Phylate, Tannins, Fiber, hemaglutinins	Cysteine, histidine, lysine, methionine, fumarate, malate, citrate, Palmitic acid, riboflavin, ascorbate
5.	Iodine	Goitrogens	Selenium

Sources: Ref. [3, 85]

1.3.2.3 POST-HARVEST MANAGEMENT

The impact of different post-harvest processes on the micronutrient content of biofortified crops varies from one biofortified crop to another depending upon the type of post-harvest process and specific nutrient which has been biofortified. It is so because there is a specific region of deposition of that particular nutrient, like the leaching of water-soluble vitamins occurring during the post-harvest process of soaking [85]. Some of the major post-harvest processes and essential nutrient loss due to them are mentioned in Table 1.3.

1.3.2.4 AVAILABILITY OF GENETIC DIVERSITY AND ITS EXPLORATION

There is a lack of genetic diversity in the biofortified crops which are grown through conventional breeding, like improving Se contents in wheat and improving oleic and linoleic acid in the soybean [92]. Transgenic breeding operations can be done to introduce traits of interest (TOI) in the desired crop by exploiting the genetic diversity of crop wild relatives (CWRs) with high nutritional value having TOI to be incorporated in the desired crop can be used. Exploration of new genetic pathways for the introduction of TOI will enhance the micronutrient loading by adsorption, translocation, and remobilization. The discovery of new genetic pathways for the introduction of essential nutrients in the specific part of plant-like in the aleurone layer of seed or its endosperm, along with the discovery of pathways for the elimination of antinutrients and introduction promoters helps in nutrient assimilation in the gut of a human. All these can be achieved by further studies of the genetic diversity of essential nutrients in desired crop and its CWRs like pulse biofortification [93].

1.3.2.5 NUTRIENT MOBILIZATION, TRANSLOCATION, AND BIOAVAILABILITY

Uptake of essential micronutrients occurs from the vicinity of the rhizosphere, which is then translocated to seeds through xylem and phloem loading for the bioavailability of essential nutrients (Figure 1.4) [94].

TABLE 1.3 Mechanical Operation and Chemical Effect of Essential Micronutrients

SL. No.	Essential Micronutrient	Mechanical/ Chemical Operation	Effect
1.	Carotenoids [85–87]	Drying	Reduction of carotenoids (depending upon exposed temperature and genotype of the plant).
		Storage	Loss of carotenoids (depending upon storage condition, species, and genotype of the plant.)
		Milling	Increases bioavailability due to reduced particle size.
		Blanching	Enhanced retention due to inactivation of peroxidases and release of carotenoids from carotenoid-protein complexes.
		Fermentation	Removal of inhibitors and accumulation of nutritional promoters.
		Nixtamalization	Reduction in carotenoid level.
		Heating	Increases bioavailability
2.	Iron and Zinc [88–91]	Dehulling	Reduction of inhibitors and loss of minerals (loss approx. 50% iron and a variable amount of zinc.)
		Milling	Enhanced bioavailability.
		Soaking	Mineral leaching
		Fermentation	Enhance absorption by degrading phytate.
		Nixtamalization	It can cause a reduction in iron absorption or improve the absorption of iron and zinc by reduction of phytates.
		Heating	Enhanced mineral absorption.
3.	Water-soluble vitamins [86, 88, 90]	Drying	Loss of vitamins
		Storage	No effect except the long-stage storage of vitamins B and C.
		Dehulling	Loss of certain vitamin B
		Milling	–
		Soaking	Leaching of vitamins
		Blanching	Inactivation and loss of vitamin C.
		Fermentation	Increased bioavailability of vitamin B.
		Nixtamalization	Reduce in content of vitamin B but increase in its bioavailability.
		Cooking	Loss of vitamins due to leaching or heat sensitivity (B1, B2, and B3).

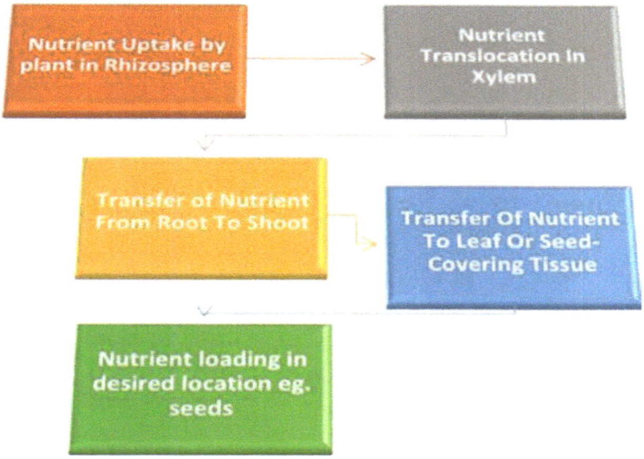

FIGURE 1.4 Diagram showing the general translocation of nutrients from the rhizosphere to the desired location, like in seeds.

1.3.2.6 COMMERCIALIZATION AND BIOSAFETY ASSESSMENT

Commercialization of biofortified crops should be done to promote consumer acceptability. Apart from its proper regulation of biofortified crops by biosafety assessment agencies and genetically modified crops regulatory bodies with nominal charges irrespective of current charges, which are very high and equivalent to development charges.

1.4 CONCLUSION

Environmental contamination and micronutrient malnutrition are the most challenging and serious problems faced by the world today and demand concerted actions to be taken. Collective approaches from different fields like nutritional sciences, environmental sciences, biotechnology along basic plant biology must be done. For phytoremediation, challenges arising due to various factors like low uptake and slow metabolism of metal contaminants, less tolerance against high heavy metal concentration, the safety regulation standards for plants and soil-microbe system, and other components of the environment must be addressed by encouraging more effective techniques like plant genetic engineering, microbe assisted bioaugmentation, heavy metal chelation, etc., synergistically with the conventional methods. Enhancing the bioavailability of metals or trace elements and their translocation and sequestration in the plants' grain tissues should be targeted for biofortification as they enrich the soil by nutrient absorption in an environmentally friendly and sustainable manner. Biofortified crops rich in important minerals, vitamins, and FAs should be promoted to combat hidden hunger in developing and underdeveloped parts of the world. Conventional plant breeding and microbe-assisted breeding should be checked for their shortcomings and replaced by new and advanced techniques that operate at molecular levels, such as RNAi-mediated biofortification, gene overexpression, and CRISPR-mediated biofortification. Evaluation of any of these potential technologies should be done under real conditions so that it can raise the environmental, nutritional, as well as economic status of human beings.

KEYWORDS

- **biofortification**
- **bioremediation**
- **genetic engineering**
- **heavy metals**
- **plant-microbe interaction**
- **transgenic plants**

REFERENCES

1. Sedeek, K. E., Mahas, A., & Mahfouz, M., (2019). Plant genome engineering for targeted improvement of crop traits. *Front Plant Sci., h10*, 114. doi: 10.3389/fpls.2019.00114.
2. Shahzad, R., Jamil, S., Ahmad, S., Nisar, A., Khan, S., Amina, Z., Kanwal, S., et al., (2021). Biofortification of cereals and pulses using new breeding techniques: Current and future perspectives. *Frontiers in Nutrition, 8*, 721728. https://doi.org/10.3389/fnut.2021.721728.
3. Dixit, R., Wasiullah, E., Malaviya, D., Pandiyan, K., Singh, U., Sahu, A., Shukla, R., et al., (2015). Bioremediation of heavy metals from soil and aquatic environment: An overview of principles and criteria of fundamental processes. *Sustainability, 7*(2), 2189.
4. Beans, C., (2017). Core concept: Phytoremediation advances in the lab but lags in the field. *Proceedings of the National Academy of Sciences, 114*(29), 7475–7477. doi: 10.1073/pnas.1707883114.

5. Rouached, H., (2013). Recent developments in plant zinc homeostasis and the path toward improved biofortification and phytoremediation programs. *Plant Signal. Behav., 8*, e22681. doi: 10.4161/psb.22681.
6. Zhao, F. J., & McGrath, S. P., (2009). Biofortification and phytoremediation. *Curr. Opin. Plant Biol., 12*, 373–380. doi: 10.1016/j.pbi.2009.04.005.
7. Zhu, Y. G., Pilon-Smits, E. A. H., Zhao, F. J., Williams, P. N., & Meharg, A. A., (2009). Selenium in higher plants: Understanding mechanisms for biofortification and phytoremediation. *Trends Plant Sci., 14,* 436–442. doi: 10.1016/j.tplants.2009.06.006.
8. Bañuelos, G. S., Arroyo, I., Pickering, I. J., Yang, S. I., & Freeman, J. L., (2015). Selenium biofortification of broccoli and carrots grown in soil amended with Se-enriched hyperaccumulator *Stanleya pinnata. Food Chem., 166*, 603–608. doi: 10.1016/j.foodchem.2014.06.071.
9. Ashraf, S., Ali, Q., Zahir, Z. A., Ashraf, S., & Asghar, H. N., (2019). Phytoremediation: An environmentally sustainable way for the reclamation of heavy metal polluted soils. *Ecotoxicology and Environmental Safety, 174*, 714–727.
10. Sumiahadi, A., & Acar, R., (2018). A review of phytoremediation technology: Heavy metals uptake by plants. *IOP Conference Series: Earth and Environmental Science, 142*, 012023. doi: 10.1088/1755-1315/142/1/012023.
11. Khalid, S., Shahid, M., Niazi, N. K., Murtaza, B., Bibi, I., & Dumat, C., (2017). A comparison of technologies for remediation of heavy metal contaminated soils. *Journal of Geochemical Exploration, 182*, 247–268.
12. Ramanjaneyulu, A. V., Neelima, T. L., Madhavi, A., & Ramprakash, T., (2017). Phytoremediation: An overview. In: Humberto, R. M., Ashok, G. R., Thakur, K., & Sarkar, N. C., (eds.), *Applied Botany* (pp. 42–84). New York: American Academic Press.
13. Antoniadis, V., Golia, E. E., Shaheen, S. M., & Rinklebe, J., (2017). Bioavailability and health risk assessment of potentially toxic elements in thriasio plain, near Athens, Greece. *Environmental Geochemistry and Health, 39*(2), 319–330.
14. Goswami, M., Chakraborty, P., Mukherjee, K., et al., (2018). Bioaugmentation and biostimulation: A potential strategy for environmental remediation. *J Microbiol Exp., 6*(5), 223–231. doi: 10.15406/jmen.2018.06.00219.
15. Glick, B. R., (2003). Phytoremediation: Synergistic use of plants and bacteria to clean up the environment. *Biotechnol. Adv., 21*, 383–393.
16. Chen, B. D., Li, X. L., Tao, H. Q., Christie, P., & Wong, M. H., (2003). The role of arbuscular mycorrhiza in zinc uptake by red clover growing in a calcareous soil spiked with various quantities of zinc. *Chemosphere, 50*, 839e846.
17. Wang, Y., Song, J., Zhao, W., He, X., Chen, J., & Xiao, M., (2011). In situ degradation of phenol and promotion of plant growth in contaminated environments by a single *Pseudomonas aeruginosa* strain. *Journal of Hazardous Materials, 192*, 354–360.
18. Ali, S. Y., & Chaudhury, S., (2016). EDTA-enhanced phytoextraction by *Tagetes* sp. and effect on bioconcentration and translocation of heavy metals. *Environmental Processes, 3*(4), 735–746.
19. Kaushik, P., (2015). Use of natural organic and synthetic chelating agents for efficient phytoremediation. *International Journal of Enhanced Research in Science Technology & Engineering, 4*(6), 99–101.
20. Shinta, Y. C., Zaman, B., & Sumiyati, S., (2021). Citric acid and EDTA as chelating agents in phytoremediation of heavy metal in polluted soil: A review. *IOP Conf. Series: Earth and Environmental Science, 896,* 012023. doi: 10.1088/1755-1315/896/1/012023.

21. Wiszniewska, A., Hanus-Fajerska, E., Muszynska, E., & Ciarkowska, K., (2016). Natural organic amendments for improved phytoremediation of polluted soils: A review of recent progress. *Pedosphere, 26*(1), 1–12.
22. Pinto, A. P., De Varennes, A., Fonseca, R., & Teixeira, D. M., (2015). Phytoremediation of soils contaminated with heavy metals: Techniques and strategies. In: Ansari, A. A., Gill, S. S., Gill, R., Lanza, G. R., & Newman, L., (eds.), *Phytoremediation: Management of Environmental Contaminants* (Vol. 1, pp. 133–158). Switzerland: Springer International Publishing.
23. Evangelou, M. W. H., Ebel, M., & Schaeffer, A., (2007). Chelate-assisted phytoextraction of heavy metals from soil. Effect, mechanism, toxicity, and fate of chelating agents. *Chemosphere, 68*, 989–1003.
24. Vamerali, T., Bandiera, M., & Mosca, G., (2010). Field crops for phytoremediation of metal-contaminated land: A review. *Environ. Chem. Lett., 8*, 1–17. doi: 10.1007/s10311-009-0268-0.
25. Gupta, V. V., Zhao, C., Degryse, F., & McLaughlin, M. J., (2017). Abundance and diversity of sulfur-oxidizing bacteria and their role in oxidizing elemental sulfur in cropping soils. *Biology and Fertility of Soils, 53*(2), 159–169.
26. Ashraf, S., Zahir, Z. A., Asghar, H. N., & Asghar, M., (2018). Isolation, screening, and identification of lead and cadmium-resistant sulfur-oxidizing bacteria. *Pakistan Journal of Agricultural Sciences, 55*(2), 349–359.
27. Maestri, E., & Marmiroli, N., (2011). Transgenic plants for phytoremediation. *International Journal of Phytoremediation, 13*(1), 264–279. doi: 10.1080/15226514.2011.568549.
28. Ibañez, S. G., Paisio, C. E., Wevar, O. A. L., Talano, M. A., González, P. S., & Medina, M. I., (2014). *Overview and New Insights of Genetically Engineered Plants for Improving Phytoremediation,* 99–113. https://doi.org/10.1007/978-3-319-10395-2_8.
29. Campos, V., Merino, I., Casado, R., & Gómez, L., (2008). Review of phytoremediation of organic pollutants. Span. *J. Agric. Res., 6*, 38–47.
30. Kulakow, P. A., & Pidlisnyuk, V. V., (2009). *Application of Phytotechnologies for Cleanup of Industrial, Agricultural, and Wastewater Contamination.* Springer Verlag.
31. Ferro, A. M., Rock, S. A., Kennedy, J., Herrick, J. J., & Turner, D. L., (1999). Phytoremediation of soils contaminated with wood preservatives: Greenhouse and field evaluations. *Int. J. Phytorem.,* 289–306.
32. Lalande, T. L., Skipper, H. D., Wolf, D. C., Reynolds, C. M., Freedman, D. L., Pinkerton, B. W., Hartel, P. G., & Grimes, L. W., (2003). Phytoremediation of pyrene in a Cecil soil under field conditions. *Int. J. Phytorem., 5*, 1–12.
33. U.S., (2008). Environmental Protection Agency website. *Remediation, Phytotechnology Project Profiles Database*. http://www.cluin.org/products/phyto/search/phyto_search.cfm (accessed on 22 May 2023).
34. Chaudhry, Q., Blom-Zandstra, M., Gupta, S., & Joner, E. J., (2005). Utilizing the synergy between plants and rhizosphere microorganisms to enhance the breakdown of organic pollutants in the environment. *Environ. Sci. Pollut. Res., 12*, 34.
35. Huang, X. D., El-Alawi, Y. S., Penrose, D., Glick, B. R., & Greenberg, B. M., (2004). Responses of three grass species to creosote during phytoremediation. *Environ. Pollut.,* 453–463.
36. Huang, X. D., El-Alawi, Y. S., Gurska, J., Glick, B. R., & Greenberg, B. M., (2005). A multi-process phytoremediation system for decontamination of persistent total petroleum hydrocarbons (TPHs) from soils. *Microchem. J., 81*, 139–147.

37. Nedunuri, K. V., Govindaraju, R. S., Banks, M. K., Schwab, A. P., & Chen, Z., (2000). Evaluation of phytoremediation for field-scale degradation of total petroleum hydrocarbons. *J. Environ. Eng., 126*, 483–490.
38. Pilon-Smits, E. A. H., & Freeman, J. L., (2006). Environmental cleanup using plants: Biotechnological advances and ecological considerations. *Front. Ecol. Environ.*, 203–210.
39. Pilon-Smits, E., (2005). Phytoremediation. *Annu. Rev. Plant Biol.*, 15–39.
40. Schoenmuth, B. W., & Pestemer, W., (2004). Dendroremediation of trinitrotoluene (TNT)— Part 2: Fate of radio-labeled TNT in trees. *Environ. Sci. Pollut. Res.,* 331–339.
41. Russell, K., (2005). The use and effectiveness of phytoremediation to treat persistent organic pollutants. *Environ. Careers Organ.* (for US EPA).
42. Kozdrój, J., & Van, E. J. D., (2001). Structural diversity of microbial communities in arable soils of a heavily industrialized area determined by PCR-DGGE fingerprinting and FAME profiling. *Appl. Soil Ecol., **2001***, 31–42.
43. Van, D. P., Caballero, A., Paz, J. A., Gonza´lez-Pe´rez, M. M., Oliva, J. M., & Ramos, J. L., (2007). Bioremediation of 2,4,6-trinitrotoluene under field conditions. *Environ. Sci. Technol., 41,* 1378–1383.
44. Alberta Environment, (2007). *Environmental Policy Branch, Alberta Tier 1 Soil and Groundwater Remediation Guidelines.* https://open.alberta.ca/dataset/842becf6-dc0c-4cc7-8b29-e3f383133ddc/resource/2b1d5eb8-258d-49ce-92ce-197253600a01/download/161552-2007-12.pdf (accessed on 22 June 2023).
45. Corwin, D. L., Hopmans, J., & De Rooij, G. H., (2006). From field-to-landscape-scale vadose zone processes: Scale issues, modeling, and monitoring. *Vadose Zone J.*, 129–139.
46. Leahy, J. G., & Colwell, R. R., (1990). Microbial degradation of hydrocarbons in the environment. *Microbiol. Rev., 54*, 305–315.
47. White, Jr. P. M., Wolf, D. C., Thoma, G. C., & Reynolds, C. M., (2006). Phytoremediation of alkylated polycyclic aromatic hydrocarbons in crude oil-contaminated soil. *Water Air Soil Pollut.,* 207–220.
48. Thoma, G. J., Lam, T. B., & Wolf, D. C., (2003). A mathematical model of phytoremediation for petroleum contaminated soil: Sensitivity analysis. *Int. J. Phytorem.,* 125–136.
49. Gressel, J., & Al-Ahmad, H., (2005). Assessing and managing biological risks of plants used for bioremediation, including risks of transgene flow. *Z. Naturforsch [C],* 154–165.
50. Ronchel, M. C., & Ramos, J. L., (2001). Dual system to reinforce biological containment of recombinant bacteria designed for rhizoremediation. *Appl. Environ. Microbiol., 67*, 2649–2656.
51. Leigh, M. B., Fletcher, J. S., Fu, X., & Schmitz, F. J., (2002). Root turnover: An important source of microbial substrates in rhizosphere remediation of recalcitrant contaminants, *Environ. Sci. Technol., 36*, 1579–1583.
52. Villacieros, M., Whelan, C., Mackova, M., Molgaard, J., Sa´nchez-Contreras, M., Lloret, J., Aguirre De, C. D., et al., (2005). Polychlorinated biphenyl rhizoremediation by *Pseudomonas fluorescens* F113 derivatives, using a *Sinorhizobium meliloti* nod system to drive bph gene expression. *Appl. Environ. Microbiol., 71*, 2687–2694.
53. INRA(Europe)—ECOSA, Eurobarometer 52.1, (2000). The Europeans and Biotechnology Report. https://vdocuments.net/the-europeans-and-biotechnology-european-521-the-europeans-and-biotechnology.html?page=1 (accessed on 22 June 2023).

54. Macek, T., Mackova, M., & Kas, J., (2000). Exploitation of plants for the removal of organics in environmental remediation. *Biotechnol. Adv., 18*, 23–34.
55. Gilbertson, A. W., Fitch, M. W., Burken, J. G., & Wood, T. K., (2007). Transport and survival of GFP-tagged root-colonizing microbes: Implications for photodegradation. *Eur. J. Soil Biol.,* 224–232.
56. Newman, L. A., & Reynolds, C. M., (2005). Bacteria and phytoremediation: New uses for endophytic bacteria in plants. *Trends Biotechnol., 23*, 6–8.
57. Jussila, M. M., Zhao, J., Suominen, L., & Lindstro¨m, K., (2007). TOL plasmid transfer during bacterial conjugation in vitro and rhizoremediation of oil compounds *in vivo. Environ. Pollut., 146*, 510–524.
58. Mahadea, D., Adamczewska, E., Ratajczak, A. E., et al., (2021). Iron deficiency anemia in inflammatory bowel diseases: A narrative review. *Nutrients, 13*(11). doi: 10.3390/nu13114008.
59. Borg, S., Brinch-Pedersen, H., Tauris, B., Madsen, L. H., Darbani, B., Noeparvar, S., et al., (2012). Wheat ferritins: Improving the iron content of the wheat grain. *J. Cereal Sci., 56*, 204–213. doi: 10.1016/j.jcs.2012.03.005.
60. Malik, K. A., & Maqbool, A., (2020). Transgenic crops for biofortification. *Front Sustain. Food Syst., 4*, 182. doi: 10.3389/fsufs.2020.571402.
61. Tan, G. Z., DasBhowmik, S. S., Hoang, T. M., Karbaschi, M. R., Johnson, A. A., Williams, B., et al., (2017). Finger on the pulse: Pumping iron into chickpea. *Front Plant Sci., 8*, 1755. doi: 10.3389/fpls.2017.01755.
62. Kaur, S., Kumari, A., Singh, P., Kaur, L., Sharma, N., & Garg, M., (2020). Biofortification in pulses. In: Tilak, R. S., Rupesh, D., & Humira, S., (eds.), *Advances in Agri-Food Biotechnology* (pp. 85–103). New York, NY: Springer.
63. Rajam, M., (2020). RNA silencing technology: A boon for crop improvement. *J. Biosci., 45*, 1–5. doi: 10.1007/s12038-020-00082-x 127.
64. Ko, M. R., Song, M. H., Kim, J. K., Baek, S. A., You, M. K., Lim, S. H., et al., (2018). RNAi-mediated suppression of three carotenoid-cleavage dioxygenase genes, OsCCD1, 4a, and 4b, increases carotenoid content in rice. *J Exp Bot., 69*, 5105–5016. doi: 10.1093/jxb/ery300 130.
65. Yang, Q. Q., Yu, W. H., Wu, H. Y., Zhang, C. Q., Sun, S. S. M., & Liu, Q. Q., (2020). Lysine biofortification in rice by modulating feedback inhibition of aspartate kinase and dihydrodipicolinate synthase. *Plant Biotechnol. J.,* 19490–19501. doi: 10.1111/pbi.13478.
66. Ahmad, S., Shahzad, R., Jamil, S., Tabassum, J., Chaudhary, M. A. M., Atif, R. M., et al., (2021). Regulatory aspects, risk assessment, and toxicity associated with RNAi and CRISPR methods. In: Abd-Elsalam, K., & Ki-Taek, L., (eds.), *CRISPR and RNAi Systems* (pp. 687–721). Amsterdam: Elsevier.
67. Bhambhani, S., Kondhare, K. R., & Giri, A. P., (2021). Advanced genome editing strategies for manipulation of plant specialized metabolites pertaining to biofortification. *Phytochem. Rev.* https://doi.org/10.1007/s11101-021-09749-1.
68. Arora, L., & Narula, A., (2017). Gene editing and crop improvement using CRISPR-Cas9 system. *Frontiers in Plant Science, 8*, 1932. https://doi.org/10.3389/fpls.2017.01932.
69. Xie, K., & Yang, Y., (2013). RNA-guided genome editing in plants using a CRISPR-Cas system. *Mol. Plant, 6*, 1975–1983. doi: 10.1093/mp/sst119.

70. Shan, Q., Wang, Y., Li, J., Zhang, Y., Chen, K., Liang, Z., et al., (2013). Targeted genome modification of crop plants using a CRISPR-Cas system. *Nat. Biotechnol., 31*, 686–688. doi: 10.1038/nbt.2650c.
71. Tan, G. Z., DasBhowmik, S. S., Hoang, T. M., Karbaschi, M. R., Johnson, A. A., Williams, B., et al., (2017). Finger on the pulse: Pumping iron into chickpea. *Front Plant Sci., 8*, 1755. doi: 10.3389/fpls.2017.01755.
72. Ludwig, Y., & Slamet-Loedin, I. H., (2019). Genetic biofortification to enrich rice and wheat grain iron: From genes to product. *Front Plant Sci., 10*, 833. doi: 10.3389/fpls.2019.00833.
73. Shi, J., Gao, H., Wang, H., Lafitte, H. R., Archibald, R. L., Yang, M., et al., (2017). ARGOS8 variants generated by CRISPR-Cas9 improve maize grain yield under field drought stress conditions. *Plant Biotechnol. J., 15*, 207–216. doi: 10.1111/ pbi.12603.
74. Ito, Y., Nishizawa-Yokoi, A., Endo, M., Mikami, M., & Toki, S., (2015). CRISPR/Cas9-mediated mutagenesis of the RIN locus that regulates tomato fruit ripening. *Biochem. Biophys. Res. Commun., 467*, 76–82. doi: 10.1016/j.bbrc. 2015.09.117.
75. Aggarwal, S., Kumar, A., Bhati, K. K., Kaur, G., Shukla, V., Tiwari, S., et al., (2018). RNAi-mediated downregulation of inositol pentakis phosphate kinase (IPK1) in wheat grains decreases phytic acid levels and increases Fe and Zn accumulation. *Front Plant Sci., 9*, 259. doi: 10.3389/fpls.2018. 00259.
76. Lipkie, T. E., De Moura, F. F., Zhao, Z. Y., Albertsen, M. C., Che, P., Glassman, K., et al., (2013). Bioaccessibility of carotenoids from transgenic provitamin A biofortified sorghum. *J Agric. Food Chem., 61*, 5764–5771. doi: 10.1021/jf305361s.
77. Tibbs, C. L., Zhang, Z., & Yu, J., (2021). Status and prospects of genome-wide association studies in plants. *Plant Genome, 14*, e20077. doi: 10.1002/tpg2.20077.
78. Xiao, Y., Liu, H., Wu, L., Warburton, M., & Yan, J., (2017). Genome-wide association studies in maize: Praise and stargaze. *Mol. Plant, 10*, 359–374. doi: 10.1016/j.molp.2016.12.008.
79. Bhatta, M., Sandro, P., Smith, M. R., Delaney, O., Voss-Fels, K. P., Gutierrez, L., et al., (2021). Need for speed, manipulating plant growth to accelerate breeding cycles. *Curr. Opin. Plant Biol., 60*, 101986. doi: 10.1016/j.pbi.2020.101986.
80. Mendoza, C., (2002). Effect of genetically modified low phytic acid plants on mineral absorption. *Int. J. Food Sci. Technol., 37*, 759–767.
81. Jin, F., Frohman, C., Thannhauser, T. W., Welch, R. M., & Glahn, R. P., (2009). Effects on ascorbic acid, phytic acid, and tannic acid on iron availability from reconstituted ferritin measured by an in vitro digestion-Caco-2 cell model. *Br. J. Nutr., 101*, 972–981.
82. Murgia, I., Arosio, P., Tarantino, D., & Soave, C., (2012). Biofortification for combating hidden hunger for iron. *Trends Plant Sci., 17*, 47–55.
83. Garg, M., Sharma, N., Sharma, S., Kapoor, P., Kumar, A., Chunduri, V., & Arora, P., (2018). Biofortified crops generated by breeding, agronomy, and transgenic approaches are improving the lives of millions of people around the world. *Front. Nutr., 5*, 12. doi: 10.3389/fnut.2018.00012.
84. Díaz-Gómez, J., Twyman, R. M., Zhu, C., Farré, G., Serrano, J. C. E., Portero-Otin, M., Muñoz, P., et al., (2017). Biofortification of crops with nutrients: Factors affecting utilization and storage. *Current Opinion in Biotechnology, 44*, 115–123. doi: 10.1016/j.copbio.2016.12.002.
85. Ball, G. F. M., (2006). Vitamins in foods. *Analysis, Bioavailability, and Stability*. CRC Press.

86. La Frano, M. R., De Moura, F. F., Boy, E., Lo¨ nnerdal, B., & Burri, B. J., (2014). Bioavailability of iron, zinc, and provitamin A carotenoids in biofortified staple crops. *Nutr. Rev., 72*, 289–307.
87. Suri, D. J., & Tanumihardjo, S. A., (2016). Effects of different processing methods on the micronutrient and phytochemical contents of maize: From A to Z. *Compr. Rev. Food Sci. Food Saf., 15*, 912–926.
88. Raes, K., Knockaert, D., Struijs, K., & Van, C. J., (2014). Role of processing on bioaccessibility of minerals: Influence of localization of minerals and anti-nutritional factors in the plant. *Trends Food Sci Technol., 37*, 32–41.
89. Gibson, R. S., (2007). The role of diet- and host-related factors in nutrient bioavailability and thus in nutrient-based dietary requirement estimates. *Food Nutr Bull., 28*, S77–S100.
90. Pereira, E. J., Carvalho, L. M. J., Dellamora-Ortiz, G. M., Cardoso, F. S. N., & Carvalho, J. L. V., (2016). Effect of different home-cooking methods on the bioaccessibility of zinc and iron in conventionally bred cowpea (*Vigna unguiculata* L. Walp) consumed in Brazil. *Food Nutr. Res., 60*, 29082.
91. Yeom, W. W., Kim, H. J., Lee, K. R., Cho, H. S., Kim, J. Y., Jung, H. W., & Chung, Y. S., (2020). Increased production of α-linolenic acid in soybean seeds by overexpression of *Lesquerella* FAD3-1. *Frontiers in Plant Science,* 10.
92. Waters, B. M., & Sankaran, R. P., (2011). Moving micronutrients from the soil to the seeds: Genes and physiological processes from a biofortification perspective. *Plant Science, 180*, 562–574.
93. Giehl, R. F. H., Meda, A. R., & Wiren, N. V., (2009). Moving up, down, and everywhere: Signaling of micronutrients in plants. *Curr. Opin. Plant Biol., 12*, 320–327.

CHAPTER 2

Nanoremediation: A New and Emerging Technology

VLADIMIR ALEKSANDROV

Institute of Plant Physiology and Genetics, Bulgarian Academy of Sciences, Sofia, Bulgaria

ABSTRACT

In recent decades, nanoparticles (NPs) have begun to be used not only in industry but also in agriculture. The size (0.1–100 nm) of NPs determines their different chemical, physical, and biological properties, distinguishing them from large-scale particles. These new features of the particles allow to use them for remediation of contaminated soils and waters. Nano-remediation technologies are new and fast growing. The base advantages of these methods for remediation are high efficiency and rapidness, but they are relatively expensive technologies. Organic contaminants and metals from soils and waters can be easily removed using NPs. Latterly, a new type of contaminant is emerged – nano- and micro-plastic particles. Contamination of environment with nano- and microplastic is a huge problem for ecosystem of the Earth. Remediation of the plastic contaminant with NPs is a new process but with very god prospects. Also, the ecological impact of nanomaterials on the environment is still poorly studied. There are many types of NPs but only some of them are used for the processes of removal of contaminants. The most used nanomaterials for remediation are carbon particles, especially carbon nanotubes (CNTs), zero-valent iron (nZVI), metallic (MeNPs) and magnetic nanoparticles (MNPs). CNTs are mostly used to remove organic pollutants due to their absorption properties. nZVI NPs are high reactive, and this is the main reason for their widespread use in remediation of soil

Phytoremediation and Biofortification: Strategies for Sustainable Environmental and Health Management.
Nand Kumar Singh, Shadma Afzal, & Tariq Aftab (Eds.)

and ground water. MNPs have unique properties for magnetic separation and absorption of metal ions and are therefore used in the purification of soil and groundwater contamination with metals. Other particles such as chitosan (CS) NPs are increasingly being used in the process of remediation. The advantages of CS NPs are their nontoxicity and biodegradability. In this chapter, I will discuss the principles and methods of nanoremediation, the types of nanomaterials used for remediation, as well as their environmental impact.

2.1 INTRODUCTION

The rapid technical and industrial development of mankind leads to environmental pollution. The contamination of groundwater and soils is a huge problem for our civilization and the removing of contaminants becoming increasingly urgent. There are many methods for remediation of pollutants such as physical remediation [1], chemical remediation, bioremediation [2]. Nanoremediation is a way to purify polluted waters and soils and it is a method between chemical, physical, and biological remediation techniques. The properties of nanoparticles (NPs) depend mainly on their physical structure, their chemical composition and their shape. For some NPs the ratio between surface area and volume is very high and this is the main reason for some of their unusual properties [3]. Due of that the outer layer of a nanoparticle have different chemistry then the core. In practice, very often the surface compositions of NPs determine their applications. NPs easily aggregate and precipitate, leading to loss of their properties. To prevent these negative effects, NPs are often pre-treated. Therefore, most NPs are splitted in two or three layers – surface, shell, and core. To the surface of a nanoparticle may be attached some metal jones, organic molecules, surfactants or polymers. In this way, the surface of the particle becomes functional with the desired properties. The second layer of the nanoparticle is called the shell and has a chemical structure different from that of the core. However, the core of the nanoparticle is its most important part, and its chemical composition determines the basic properties of the particle. Also, the morphology of NPs has impact of their properties such as a catalytic activity [4, 5], optical [6], electrical [7], and magnetic [8] features. The most common shapes of NPs are spherical and wire as carbon nanotubes (CNTs), but there exist NPs with different shape such as cube [9, 10], prism [11], hexagon [12], octagon [13]. Due of their unique properties some NPs are used for remediation processes.

The use of carbon nanomaterials such as nanotubes, graphene, graphene oxide (GO), fullerenes, and activated carbon is not limited to medicine, electronics, and agriculture, but is also widely used in remediation of groundwater and soils from organic and inorganic substances. Fullerenes are carbon nanomaterials and they have closed-caged pentagonal and hexagonal ring structure [14]. In remediation processes, fullerenes are used to break down organic pollutants in soil and water but can also be used very successfully to destroy pathogens in water [15, 16]. There exist three types of CNTs – single-wall carbon nanotubes (SWCNTs), multiwall carbon nanotubes (MWCNTs) and carbon nanotube-fibers pyrograf (CNTFs) [17]. SWCNT are one-dimensional (1-D) nanomaterials whose walls are about 1 nm thick. CNTs have a large surface and tubular structure, which makes them suitable for use in pollution control and remediation of contaminated waters and soils [18, 19]. Graphene is single-layered graphite with honeycomb lattice. The structure of this nanomaterial provides it many properties, which determines its application in pollution remediation [20].

nZVI NPs are often used to remove organic contaminants due of their reductant properties. The size of particles is very important for their mobility and low toxicity [21].

TiO_2 NPs have applications in different environmental remediation processes such as organic pollutions removing [22], removing of cyanide [23] and metals from wastewater.

The nZVI and TiO_2 NPs fall into the category of metallic nanoparticles (MeNPs). Other MeNPs, which have application in soil and water remediation are ZnO, CuO, SnO_2, silver, and so on [24].

The MNPs due of their magnetic properties, demonstrate capabilities for remediation of waters from different contaminants. MNPs have the ability to be held in a magnetic field due to their properties. This quality allows these particles to be used repeatedly for several cycles [25].

Biopolymers such as chitosan (CS), starch, dextran, and cellulose (CL) are biologically synthesized macromolecules and are often used for the synthesis of nanomaterials, as the biopolymeric NPs are unique in composition and have different physiological properties [26–28]. However, the biggest advantage of biopolymer nanomaterials is their biodegradability and non-toxicity. Restoring the environment using biopolymer nanomaterials involves removing and reducing various types of pollutants.

The nanomaterials are increasingly used in processes related to the purification of the environment (especially surface water and soil) from contamination with organic materials, heavy metals or waste products in oil refining processes. The NPs have unique physicochemical and biological properties due to their large surface area per unit volume as well as their nanosizes.

This chapter gives an overview of the different types of NPs that are used for remediation processes in soils and waters as well as their impact on the environment. However, the focus of the discussions will be on the latest methods and research for remediation of groundwater and soils.

The applications of nanomaterials as nanotubes, NPs, nanowires are critical reviewed based on their chemical composition, size, and structure.

2.2 CARBON NANOMATERIALS

Carbon nanomaterials can be divided into several categories – fullerene, CNTs, GO, and graphene quantum dots (GQDs). According to their dimension, carbon nanomaterials are divided into three groups – zero-dimensional (D0), one-dimensional (D1) and two-dimensional (D2). In the processes of remediation of contaminated water from zero-dimensional carbon nanomaterials, fullerene is most often used. One-dimensional carbon nanomaterials used in water and soil purification are nanofibers and nanotubes, while graphene is the most widely used two-dimensional carbon nanomaterial [29].

2.2.1 ZERO-DIMENSIONAL CARBON NANOMATERIALS FOR NANOREMEDIATION

Fullerenes are carbon allotropes, and their molecules consists of carbon atoms connected by covalent bonds with fused rings of five to seven atoms. The shape of the fullerene is shaped like a closed mesh. The most important fullerenes are C60 and C70. The molecule of C60 fullerene is composed of 60 carbon atoms grouped as 12 pentagons and 20 hexagons and molecule of C70 consist of 12 pentagons and 25 hexagons [30]. Due to the chemical composition of fullerene molecules C60 and C70, fullerenes are very stable, but with increasing number of carbon atoms the stability of these molecules decrease. The density of fullerenes is (1.65 g/cc), and fullerenes are stable up to a temperature of 1,000°C. Fullerenes strongly attract electrons, which means that they can easily react with nucleophiles [31]. The fullerene molecules can undergo various chemical reactions such as reduction, oxidation, hydrogenation, halogenation, nucleophilic reactions and so on [30]. The physical properties of fullerene nanotubes, especially their small diameters, suggests that membranes made from these materials might be highly selective. The fullerenes have antimicrobial properties, and they can be used for disinfection of wastewater [32]. The Fullerenes have strong ROS-producing properties, and this is used to break down organic compounds in water [33]. The sorption capacity of fullerenes is not good, but they are highly effective in removing organometallic contaminants [34]. Despite promising properties of fullerenes there exist evidence of the cytotoxicity of these nanomaterials. On the other hand, fullerenes have a toxic effect on cells due to a change in their structure when they interact with water. Toxicity is manifested in cellular uptake and intracellular catalytic activity [35]. This will be a problem for the use of fullerenes for remediation of tap water and groundwater.

2.2.2 ONE-DIMENSIONAL CARBON NANOMATERIALS FOR NANOREMEDIATION

According to the number of layers of carbon atoms in the walls of nanotubes, CNTs are divided into single-walled carbon nanotubes (SWCNTs) and multiwalled carbon nanotubes (MWCNTs) [17]. Due to their unique properties, CNTs have a large number of applications such as solar cells, nanomedicine, nanoelectronics, optics, sensors, and others, but they can also have important applications for the environment [36]. The diameter of CNTs can be from 1 nm to several nanometers. The structure of CNTs make them a perfect tool for cleaning wastewater and soils from heavy metals by adsorption [37, 38]. Absorption affinity of CNTs to heavy metal ions depends on various factors such as ionic radius of heavy metals, the metal electronegativity, which mean that that metals with higher electronegativity have stronger adsorption on the negatively charged CNT surface, reduction potential of heavy metal ions and according to the theory of absorption processes, the affinity of the sorbent decreases with increasing reduction potential. The pH of the solution plays a major role in the absorption and when the pH in the medium decreases, the adsorption of cations also decreases. The change in temperature also affects the absorption of heavy metal ions by CNTs. As the temperature increases, the rate of diffusion of heavy metal ions increases, which leads to increased levels of adsorption [39]. The magnetic multiwall composite of CNTs and iron oxide increases its efficiency for adsorption of As(III) and As(V) in high pH environments [40]. The absorption properties of CNTs can be increased by modifying their outer walls with polymeric functional groups and groups such as -OH, C = O and -COOH [41].

SWCNTs are widely used for remediation processes alone or as nanocomposites. The adsorption of heavy metals by the CNTs depends on the interactions between available functional groups on the surface of CNTs and particular contaminants rather than the size of nanotubes [36]. SWCNTs nanocomposites have strong absorption abilities and Alijani et al. demonstrated that the nanocomposite synthesized by them SWCNTs-CoS has greeted abilities to absorb mercury for short time period – over 7 minutes. The rate of mercury absorption for this nanohybrid was more than 99.56%, but absorption rate only for SWCNTs were 45.39% [42].

Due of their physicochemical properties MWCNTs are more widely used for water and soil purification than SWCNTs.

Zirconia-modified MWCNTs have high absorption rate of As(III) and As(V) than common MWCNTs [43]. However, functionalized MWCNTs have higher absorption rate than common CNTs. Iannazzo et al. developed dendrimer-functionalized multi-walled CNTs to remove heavy metal ions from wastewater, with тriazole dendrimers that were built directly in the nanotubes [44]. MWCNTs with amino and thiol-functionalized groups have been used successfully to remove cadmium(II) and mercury(II) from highly polluted waters [45]. Except to removing heavy metal ions from wastewater, functionalized CNTs have applications for cleaning organic pollutants from contaminated water [46, 47] as well as pesticides [48] and dyes [49].

Microbiological pollutants are a big problem for humanity, and this is the reason why we are constantly looking for effective methods to remove them from the environment. Due to the specific properties of CNTs they have high-affinity interactions with biological contaminants, which are expressed in high absorption efficiency, inactivation of viruses and bacteria and high antimicrobial potential [50, 51]. Functionalized CNTs with different chemical groups can destroy cell walls of pathogens. The strong interaction between modified CNTs and bacteria often lead to detrimental effects on the metabolism and morphology of microorganisms. CNTs easily penetrate the cell walls of microorganisms, and this is the main reason for their high ability to inactivate organisms [52].

There exist many other modifications of CNTs used for remediation processes. Compositions between MWCNTs and magnetic particles are very promising for removing of toxic metal ions from wastewater [53–55]. In addition, combination between CNTs and polymers showed positive effect on the removing of toxins.

Nanoplastics and microplastics are some of the world's biggest polluters of the world's oceans and beaches. These pollutants from water, through marine animals, enter the food chain and terrestrial animals, and from there into humans [56]. The effects of micro- and nanoplastics in mammals have not been fully elucidated, but recent findings in this area suggest that they may be toxic [57]. As the natural degradation of nanoplastics and microplastics takes decades and their degradation releases toxins from polymers and additives, it is necessary to seek and apply effective methods for their safe degradation. Nanomaterials are a promising method for the safe degradation of micro- and nanoplastics. Spiral CNTs with high level nitrogen additives and encapsulated metal NPs were designed. Magnetic nanohybrids have been used for peroxymonosulfate activation to generate highly oxidative radicals for the decomposition of microplastics under hydrothermal conditions. The results of the toxicity tests show that the organic intermediates are environmentally friendly [58].

Materials based on CNTs are increasingly used for remediation of water and soil due of their unique properties, such as high surface area and abundant pore structures, π-π interaction, high negative charge density, and hydrophilicity.

2.2.3 TWO-DIMENSIONAL CARBON NANOMATERIALS FOR NANOREMEDIATION

Graphene and GO are a major group of graphite-like products, and perhaps the most important of them. Monolayer graphene exhibits unique physical properties such as anomalous half integer quantum Hall effect, super-ion thermal conductivity, extremely high charge carrier mobility, large conductivity, significant surface area, and high mechanical and thermal stability, which make this material applicable for many fields such as electronics, active sensors, microelectronics batteries, supercapacitor electrode materials, quicker DNA sequencing, fluorescence bioimaging, photocatalysis, gene, and bioactive delivery, water treatment, and biosensing [59]. Also, graphene is extremely hydrophobic material, and it forms large aggregates in solutions. The oxygen molecules in GO improves hydrophilicity of these carbon nanomaterials, which permits its utilization in natural conditions and water. GO nanomaterials have a very good sorption capacity due to high their solubility in water and they have a large surface on which there are oxygen-containing functional groups [60, 61]. These properties of GO allow this nanomaterial to be widely used in the removal of toxins, dyes, heavy metals, antibiotics, and other contaminants from wastewater and soil.

Pure graphene tends to form aggregates in solutions, and this significantly reduces its available absorption area, but when graphene is modified with functional groups such as oxygen and sulfonate, its ability to aggregate in solution decreases. Graphene functionalization methods can be divided into two major groups: covalent and non-covalent [62]. Hydrogenated graphene at moderate temperatures may lose its hydrogen groups and this may limit its application in practice [63]. However, graphene modified with fluorine tends to bind carbon more strongly than hydrogen. Fluorinated graphene is more mechanically stable, more thermally stable and has better insulating properties than hydrogenated grapheme [64]. Functionalized graphene with oxygen groups increases the adsorption of anionic organic molecules and this absorption is proportional of surface area of graphene. Also, functionalized graphene with oxygen groups can binding metal ions and positively charged organic molecules via electrostatic interaction and coordination [65]. Non-covalent functionalization of graphene occurs through a π-π bond between graphene and the modifier. There are three base mechanisms of non-covalent interactions that take place – van der Waals, hydrophobic and electrostatic interactions. Non-covalent functionalization of graphene with a pyrimidine derivative can stabilize composites of graphene and metal ions [66]. In addition, graphene nanocomposite materials can also be non-covalently bound to metal NPs (e.g., Au, Pt, Ag, Ni, Ru) through various physicochemical processes such as in situ reduction, thermal evaporation, electrical deposition, photochemical, and solvent-free bulk synthesis [67].

Due to the properties of graphene-based materials, they are used for remediation of wastewater and contaminated soils [68, 69].

Shahzadi et al. proposed metal-organic framework (MOF) bedded with GO for removing heavy metals and organic pollutants from wastewater. The porous material MOF is constructed of FeBT (MIL-100; BT = 1,3,5-benzenetriboxylate), which consists of Fe_3O triangular clusters. Between them there are nanofibers of CL and GO NPs. In this study, they showed that the nanocomposite used successfully removes Pb^{2+} and Hg^{2+} ions and gamma-HCH and PCB44 from water [70].

Electrostatic interactions, π-π stacking and hydrogen bonding interactions between N-bis (2-hydroxyethyl) glycine (Bicine) modified GO composite and contaminants make this nanocomposite an ideal material for adsorption and separation of phenols, REE3+ (Y^{3+}, Nd^{3+}, Eu^{3+}) and organic dyes from aqueous solution [71].

Functionalized single-layer GO nanosheets have demonstrated high efficiency in removing methylene blue (MB) from sands [72]. Also, composite with GO and hollow mesoporous silica was used for the removal of methyl orange (MO) and MB dyes [73].

Tian et al. constructed nanocomposite between GO and magnetite polydopamine for absorption of amphetamine-type stimulants from urban wastewater [74].

Mura et al. constructed a functionalized GO nanocomposite with FeO-Fe (VI) having paramagnetic properties, which showed approximately 99% efficiency in removing rhodamine B (RhB), diethyl 4-nitrobenzyl phosphonate (DNBP) and diclofenac (DCF) from aqueous samples. When the process of remediation of the aquatic environment from the contaminants is completed, the NPs are removed by a magnet [75].

Moreover, graphene-based nanomaterials can be used to modify the surface of membranes for better filtration and GO nanosheets were applied as a material in the surface modification of membranes. There exist several methods for modification of membranes with graphene-based nanomaterials such as coating, grafting, blending, and chemical [76].

Graphene and its modified forms have vast application in industry and ecology due to the unique properties of this carbon nanomaterial. As the rate of environmental pollution increases, the use of nanomaterials, in particular graphene and graphene-based nanomaterials, will increase.

2.3 METALLIC NANOPARTICLES (MENPS)

2.3.1 METALLIC OXIDE NANOPARTICLES (MONPS)

Metallic nanoparticles (MeNPs) such as TiO_2, ZnO, CuO, SnO_2, AgO, Al_2O_3, MgO, ZrO_2 fall into the class of metallic oxide nanoparticles (MONPs). The metal oxide NPs have unique physicochemical properties that allow them to be used in industry, pharmacy, agriculture, medicine, electronics, ecology. Many metal NPs are supermagnets, they are very good catalysts for chemical reactions, they have nonlinear optical properties, high ductility at high temperatures [77].

Titanium dioxide or Titania is a semiconductor with photocatalytic properties and has many applications such as photocatalytic hydrogen evolution, removal of organic pollutants, synthesis of polymers, drug delivery and so on [78]. Titanium dioxide is known in four different polymorphic forms – rutile, anatase, brookite, and TiO_2(B). The properties of the phases are different, due to differences in their crystal structure, shape, and dimension and the most stable form is rutile, while the other phases are metastable [79]. However, when exposed to high temperatures, all metastable Titania forms can be transformed into rutile.

TiO_2 is widely used in water remediation processes due to its chemical and physical properties, low cost and corrosion resistance [80].

Nanocomposite of TiO_2 NPs and high-performance polyphenyl sulfone mixed matrix membranes have been used for ultrafiltration of water with a very high degree of water purification (80%) [81].

TiO_2 was used successfully for degradation of MB dye and fungi *Fusarium equiseti* from drinking water. The results obtained during the experiments showed 86% MB degradation at 4-h reaction time and 50% inhibition of the *F. equiseti* after 14 days [82].

Through the process of photocatalysis, pollutants in wastewater can be destroyed. TiO_2 NPs have high photocatalytic activity, they are chemically stable, they are resistant to corrosion, and they are non-toxic. These properties of this type of NPs allow them to be used in the decomposition of dyes, pesticides, and other environmental pollutants. Diazinon is an organophosphate insecticide that is toxic and should not accumulate in soils and water reservoirs. Titanium NPs doped with iron in the presence of ultraviolet (UV) rays have been proven to be an effective tool of destroying diazinon [83]. Vatanpour et al. used a hybrid system combining UV/TiO_2 NPs and polyvinylidene fluoride (PVDF) membrane to degradation of RhB. The dye RhB was degraded with high efficiency up to 95% [84]. Also, TiO_2 NPs were used to degrade toluene exposed to sunlight with 83.8% efficiency [22].

The pharmaceuticals (PhAs) constitute have been identified as a serious polluter of the environment, which requires the search for methods for their removal from contaminated soils and waters. Nanotechnology is a promising tool for removing PhAs from the environment and Das et al. used successfully TiO_2 photocatalyst-based process for treatment of a PhAs in pharmaceutical wastewater. Their method can reduce the toxicity level of wastewater to 68.2% in 1 hour [85].

Zinc oxide (ZnO) NPs have a porous nanostructure and a high BET area, which makes them an excellent absorbent for heavy metals. But modified forms of ZnO (nano-plates, nano-sheet microspheres, and hierarchical ZnO nano-rods) have a higher capacity to remove heavy metals than the unmodified form of ZnO [86].

ZnO nano-plates for removal of Cu(II) from wastewater were used with high efficiency at pH=4 [87]. Single-crystalline mesoporous ZnO nanosheets with a thickness of about 10 nm show excellent photocatalytic efficiency and have been used to degrade MB in solutions [88]. In other study mesoporous hierarchical ZnO nanorods (ZnOs) with surface areas of 15.75 m^2 g^{-1} and pore volumes of 0.038 cm^3g^{-1} was used as adsorbents for the removal of Pb(II) and Cd(II) from waters [89].

CuO has very good semiconducting property and this property make the CuO NPs useful in gas sensors, batteries, catalysis, solar energy conversion, high-temperature superconductors, photocatalysis, and in antimicrobial and antifouling applications [90].

Copper oxide has been used successfully as a photocatalyst to decolorize various dyes, such as MB, acid yellow 23 (AY 23) and reactive black 5 (RB 5). The results showed that the highest decolorization achieved was 67.8% and 66.3% for RB-5 dye and AY-23 dye, while 43.5% for MB dye from aqueous solution at 5 h illumination [91]. In their research Mahar et al. used cupric oxide/polyvinyl alcohol nanocomposite (CuO/PVA) for the degradation of RhB dye in contaminated water solution. They have achieved extremely high efficiency in the decomposition of the dye – more than 98% [92].

CuO NPs have been used to absorb heavy metals from wastewater with efficiency. However, more research is needed to properly evaluate the effectiveness of these NPs.

Mercury (Hg) is metal with high toxicity, and this make it one of the most hazard metals for the environment. Hg(II) has a high priority for removal from the environment and NPs are a promising method to achieve this goal. The absorption of Hg(II) from aqueous solution was successfully performed in the presence of CuO NPs as absorbent. The temperature for the optima rate of absorption of mercury by CuO NPs are equal 278 K and optimal pH is 9 [93]. In addition, Mahmoud et al. showed that CuO NPs can be used effectively to remove lead, nickel, and cadmium from contaminated water [94].

Magnesium oxide (MgO) NPs can be used as an absorbent of various toxic gases. MgO NPs can be used as an absorbent of various toxic gases and to remove organic dyes, heavy metals and chemical toxins from polluted waters. MgO has great metal adsorption capacity, nontoxicity, high specific surface area, thermally stable, noncorrosive, and environmentally friendly, as well as pH-independent adsorption [95]. Uranium ions in the aqueous medium have been removed with high efficiency by MgO NPs [96].

Taourati et al. reported highly efficient degradation (80%) of the dye Alizarin red S for a short period of time (20 minutes) by MgO NPs [97]. The dye Acid red 73 (AR73) is used in textile industry and it is toxic water pollutant. MgO NPs have been used to remove acid red 73 dye from wastewater. The authors found that the complete removal of AR73 was observed with MgO NPs of 0.8 g/L, initial pH of 5 and reaction time of 60 min [98].

The agarose-MgO nanocomposite was used successfully for removal of iron from aqueous solution with 91% efficiency at 0.2 g of the nanocomposite [99]. MgO NPs have also been used to degrade diazinon in aqueous solution with efficiency of 99.46 % obtained at pH 7 [100].

In recent years, MgO NPs are increasingly used in processes related to remediation of soils and waters due to their properties.

Other NPs, such as SnO_2, Ag_2O, Al_2O_3 are beginning to find increasing use in the remediation of soils and water. Silver oxide (Ag_2O) NPs are spherical with the particle size between 20 and 80 nm with a surface area between 10 and 50 m^2/g. These NPs are also available in coated and dispersed and they have high purity. Completely degradation of MO was observed by Ag_2O under UV or visible light irradiation for a period of 120 s [101]. Peikam & Jalali studied the extent to which Al_2O_3 NPs are able to remove zinc (Zn), cadmium (Cd) and nickel (Ni) from soils [102].

2.3.2 ZERO-VALENT IRON NANOPARTICLES (NZVI)

Zero-valence iron (ZVI) is a form of iron with a standard redox potential (E_0 = –0.44 V), which makes it an effective reducing agent. The production of ZVI is relatively cheap and is a non-toxic metal. Directional transfer of electrons from nZVI to oxidized organic pollutants is a mechanism expressed in a series of chemical reactions in which ZVI can degrade and oxidize organic compounds [103]. In the presence of dissolved oxygen ZVI transfers $2e^-$ to O_2 to produce H_2O_2.

$$Fe^0 + O_2 + 2H^+ \rightarrow Fe^{2+} + H_2O_2$$

Next the hydrogen peroxide can be reduced to water by new electron transfer from ZVI.

$$Fe^0 + H_2O_2 + 2H^+ \rightarrow Fe^{2+} + 2H_2O$$

The combination between Fe_2 and H_2O_2 known as Fenton reaction produce hydroxyl radicals $^{\bullet}OH$.

$$Fe^{2+} + H_2O_2 \rightarrow Fe^{3+} + {}^{\bullet}OH + OH^-$$

Nano-ZVI (nZVI) has a larger surface area and higher reactivity than ZVI and this makes them preferred in remediation processes. However, nZVIs have limited efficiency because they easy form aggregates in aqueous media and it is difficult to separate them from the treated solution. These problems can be solved if the NZVI participates in nanocomposites with porous materials such as zeolite, carbon, kaolinite.

The nZVI as a composite material with other metals can be improve its reactivity. Palladium (Pd) extremely enhanced reactive properties of nZVI, but this application is limited because of its high cost. Instead of Pd, Cu, and Ni are often used as doped metals for nZVI [104]. Bimetallic nZVI particles are often used to increase the efficiency of removing organic matter (OM) from wastewater and soil.

The use of nZVI for the decomposition of chlorinated hydrocarbons is one of the most common methods for removing these substances from water and soil. Both trichloroethene (TCE) and polychlorinated biphenyls (PCBs) are carcinogens and their removing from environment is very important. It has been demonstrated that Fe/Pd BNPs significantly increase the degradation rates of TCE [105]. He and Zhao prepared Fe/Pd composition and successfully removed 98% TCE and transform over 80% of PCBs in contaminated water. Both TCE and PCBs are carcinogens and their removing from environment is very important [106]. Apart from the purification of soils and waters from TCE, nZVI have also been successfully used to remove of chlorobenzenes (monochlorobenzene (MCB), dichlorobenzenes (DCBs), and 1,2,4-trichlorobenzene (124TCB)) [107].

Polybrominated diphenyl ethers (PBDEs) are flame retardants that are toxic to the environment and therefore need to be removed. In their study, Wang et al. used n-ZVI/Pd bimetallic system to rapidly debromination of PBDEs to diphenyl ether [108].

A large group of dyes are resistant to light, water, and oxidizing chemicals and this leads to their accumulation in the environment. Due to the azo bonds (-N=N-) in their chemical structure, azo dyes are resistant to degradation. In addition, these dyes are known to be highly toxic and carcinogenic, making them a major threat to the nature. Many methods have been developed to remove azo dyes from wastewater and soils, but most of them are inefficient and expensive. Luo et al. used bimetallic n-ZVI/Pd to degrade azo dyes Orange II with high efficiency (98% of Orange II was removed) [109]. Another group of researchers showed that n-ZVI/Pd are effective to degradation of MO, which is one of the most common and extensively used acidic/anionic dyes in textiles, printing, cosmetics, electroplating, mining, leather, pharmaceutical, tanning paper industries [110].

Fe/Ni nZVI nanoparticles (NPs) are used successfully for degradation of flame retardant decabromodiphenyl ether (BDE209) in contaminated soils. The results were very promising, as bimetallic NPs of Fe/Ni were removed and degraded to 72% efficiency at ambient temperature [111]. In another study, chloroform-contaminated water was treated with Fe/Ni nZVI composite maintained with activated carbon fibers (ACF). The results showed that Fe/Ni bimetallic NPs with ACF support had a 92.1% efficiency in removing chloroform from water [112].

Very often in nature pollutants are just a mixture of organic compounds and heavy metals and removing them from the environment is a big challenge. A study by Lynn et al. demonstrates that Fe/Ni NPs can remove a mixture of triclosan (TCS) and copper (Cu(II)) with highly efficiency – 75.8 and 44.1% respectively. On the other hand, the efficiency was slightly higher when the contaminants are not in the mixture but were removed independently. In this case TCS was removed with 85.8% and Cu(II) was removed with 52.5% efficiency. This effect is due to competitive relationship between the two contaminants and Fe/Ni NPs [113].

The dyes are also successfully removed from the aqueous solution by Fe/Ni bimetallic NPs. The dye Acid Red 14 was removed with 92.57% efficiency using Fe/Ni NPs [114] and Sarvari et al. used Fe/Ni bimetallic NPs to remove MO from contaminated water. Their results showed that 60 mg of Fe/Ni NPs at ambient temperature can remove 99.5% of 50 mg/L dye solution within 15 min [115].

Other nZVI have also been studied for remediation of wastewater and soils. The Fe/Cu bimetallic NPs were successfully applied to remediation of hexavalent chromium (Cr IV) at a concentration of 120 mg/kg in contaminated soil. The results show that the reduction of Cr IV reaches 99% at pH 5 [116]. Cr IV was also removed from wastewater by Fe/Cu NPs with 84.0% efficiency [117]. Ou et al. used Fe/Al NPs for the reduction of Cr IV to Cr III in groundwater [118].

Sulfidated Zerovalent Iron NPs S-nZVI are often used in the last 10 years for remediation of wastewater and soils from chlorinated hydrocarbons (CHCs) mainly TCE, heavy metals, organic dyes and PhAs. Organic contaminants are predominantly transformed by S-nZVI through reduction while the removal of metals and metalloids occurs through sorption and co-precipitation, and reduction simultaneously [119].

Different co-contaminants can influence S-nZVI degradation of TCE. Wang et al. observed that the removal of TCE in the presence of various co-pollutants (Cr^{6+}, Cd^{2+}, and NO^{3-}) under acidic and alkaline conditions decreases significantly with increasing concentration of pollutants [120].

The nZVIs are widely used for remediation of wastewater and soils due to its low cost and physicochemical properties.

2.3.3 MAGNETIC NANOPARTICLES (MNPS)

Magnetic nanoparticles (MNPs) are nanomaterials composed of metals with paramagnetic, ferromagnetic, or superparamagnetic properties. The metals that have these properties are cobalt, nickel, and iron and their MNPs are prepared from a mixture of the metals and their oxides. Iron-based magnetic nanoparticles (IMNPs) are zero-sized NPs and have very good mechanical, optical, electrical properties, and if an external magnetic field is applied, IMNPs form clusters with useful properties for manipulation and separation. Coating the surfaces of IMNPs with organic, inorganic or doped hybrid polymers is done to increase the resistance of these type of NPs that are unstable in air or aqueous medium, oxidize, and tend to form aggregates [121]. Fe_2O_3 and Fe_3O_4 are the iron oxides which are the base of IMNPs, because they are superparamagnetic materials and can be magnetized in external low-intensity magnetic field. IMNPs have strong adsorption and reduction properties, and this makes them very appropriate for remediation processes. On the other hand, as an example, nZVIs are very good reductants and are able to degrade organic contaminants, toxins, and detoxify heavy metals. For that reason, to improve the properties of IMNPs, they are used as composite particles with other nanomaterials. The advantages of IMNOs and their composites for remediation are low cost, nontoxicity, and effectiveness for accumulation and transformation of contaminants [121].

Maghemite NPs (pristine ɣ-Fe_2O_3) have high colloidal stability and specific binding properties. These properties of maghemites are the reason why these NPs are used in the process of purification of contaminated water. Hu et al. used maghemite NPs for the removal and recovery of Cr(VI) from wastewater. They observed maximum adsorption rate at pH 2.5 [122]. Other researcher has used ɣ-Fe_2O_3 NPs to adsorb of AsV and AsIII from aqueous solutions with high efficiency [123]. Because of hydrophobic properties of INMPs they form aggregates in aqueous solutions which lowers adsorption efficiency. To prevent the formation of aggregates in aqueous solutions, nanocomposites are produced between MNPs and various organic or inorganic materials that have porous structures. Silicon-based materials are hydrophilic, good adsorbents, chemically, and physically stable and inexpensive, and this makes them very suitable for forming nanocomposites with MNPs. Silica-coated magnetic nanocomposites were used for adsorption of Pb^{2+} ions from aqueous media with good efficiency [124]. In another study, Kaur et al. also used silica-coated MNPs to adsorb Cd(II) from wastewater [125].

A three-component nanocomposite of Fe_3O_4 was used successfully as a potential adsorbent for uranium(VI) and also provide a simple, fast separation method for removal of uranium(VI) ions from aqueous solution [126].

IMNPs nanocomposites are applied to remove of dyes from wastewater with high rate of success. Fe_3O_4/SiO_2/glycidyl methacrylate/acrylonitrile nanocomposites (Fe_3O_4/SiO_2/GMA/AN) were used for adsorption of basic violet 7 dye. The nanocomposites were synthesized by γ-irradiation copolymerization of glycidyl methacrylate (GMA) and acrylonitrile (AN) of different ratios onto the surface of Fe_3O_4/SiO_2 NPs. The maximum level of dye adsorption was 90.65% and was reached at pH 7 in 2 hours [127]. Hernández et al. showed that MNPs can be regenerate and reuse with high efficiency up to 99% for degradation of MB in wastewater [128].

Graphene-Fe_3O_4 nanocomposite was used as the adsorbent for the extraction of some triazine herbicides (atrazine, prometon, propazine, and prometryn) in water [129].

Sulfonamides are antimicrobials in human and veterinary medicine, and they have high resistance to biological degradation. When released into the environment, sulfonamides can cause the development of antibacterial resistance among microorganisms and therefore they must be destroyed or removed quickly when they enter in the nature. Humic acid (HA) coated Fe_3O_4 MNPs (Fe_3O_4/HA)were used in wastewater to remove sulfathiazole. Degradation of sulfathiazole depends on temperature and is preferred in acidic solution. The catalytic rate increases with Fe_3O_4/HA dose and H_2O_2 concentration. When 3 g of L^{-1} Fe_3O_4/HA and 0.39 M H_2O_2 were introduced into the aqueous solution, most of the sulfathiazole degraded within 1 hour and more than 90% of the total organic carbon was removed within 6 hours [130].

Sarcletti et al. developed core-shell superparamagnetic iron oxide nanoparticles (SPIONs) that attract micro- and nanoplastics and turn them into larger agglomerates which then can be removed from water by applying an external magnetic field. The results showed that the SPIONs are able to link and agglomerate to oppositely charged nanoplastics of different chemical structure and size with high efficiency, but more investigation is needed [131].

The composites of MNPs, as well as the particles themselves, are very effective for environmental pollution remediation. Their high reducing activity and catalytic activity, low cost and non-toxicity make them one of the most preferred nanomaterials used for purification of polluted waters and soils.

Biologically synthesized metal NPs have very good abilities to purify wastewater from heavy metals. Biosynthesized metallic nanoparticles (MeNPs) are low cost and environment friendly, and this make them ideal for remediation of contaminated soils and waters. Most of the used MeNPs for the remediation processes in recent years are biologically synthesized.

2.4 BIOPOLIMERIC NANOPARTICLES (NPS)

Proteins and polysaccharide NPs form a class of biopolymeric NPs. The choice of biopolymer NPs for a given activity depends on several factors such as the size and morphology of the NPs, the surface charge and permeability of the NPs, biodegradability, and toxicity. Although biopolymeric NPs are used primarily for medical and pharmacological purposes, there have been increasing studies in recent years in which these NPs have been used for remediation of water and soils.

2.4.1 PROTEIN NANOPARTICLES (NPS)

The protein NPs are biodegradable and chemical stable, less immunogenic and non-toxic, they are easy to manufacture. The surface of protein NPs can be easily modified due to their chemical structure. Also, protein NPs can form covalent bonds with drugs [132]. Albumin is a protein found in blood plasma as well as in plants, which is biodegradable, biocompatible, and less-immunogenic protein. It is involved in the transport of hormones, drugs, and more. As well as in the regulation of osmotic pressure and neutralization of free radicals [133]. These properties also determine the main application of albumin NPs as carriers of various drugs to diseased cells, tissues or organs. However, there are some studies in which albumin NPs have been used to treat contaminated water. Hyunji at el. used serum albumin nanoparticle for adsorbents to remove amoxicillin antibiotics from water. The maximum adsorption capacity measured by spectrophotometry was found to be 12.4 micrograms of amoxicillin per milligram of albumin nanoparticle [158].

Collagens are a large group of proteins in animals with simple, repetitive sequence of amino acids that defines their unique properties. This structure of amino acids sequence lead to the formation of large triple helix fibers. In addition, collagens are biodegradable and biocompatible, and these properties make them suitable for use as biomaterials in medicine. Because collagen is easily modified, it is appropriate for the production of composite

NPs. Modifications often involve the incorporation of other proteins into collagen NPs such as elastin, fibronectin, and glycosaminoglycans, which lead to improved biodegradability of NPs or other physicochemical properties. Collagen NPs and their modification are actively used to deliver drugs to cells, but there are studies in which these NPs are used to remediate contaminated water. Shalaby at el. manufactured magnetic collagen nanocomposite to remove crystal violet (CV) dye from industrial waters. The synthesis of this nanocomposite is carried out by the method of co-precipitation and mixing of iron NPs with collagen. They reach an adsorption capacity of 92% at 100 ppm CV, while cytotoxicity is reduced to 15% [134]. Collagen based magnetic nanocomposite was successfully used to absorption of oil from aqueous solution by magnetic tracking. The oil absorbed magnetic nanobiocomposite has been converted into bi-functional carbon materials that can be used for a variety of applications [135].

Keratin is intermediate filament-forming protein with specific physicochemical properties. This protein consists of 19 amino acids that are linked together by disulfide bonds, forming a ladder structure with high mechanical strength. Also, keratin is biodegradable and biocompatible material, making it an excellent choice for drug delivery. Due to the huge amount of keratin waste that is discarded each year, some scientists are beginning to do research on the use of keratin nanoparticles (KNP) or keratin nanocomposites in the remediation processes of wastewater and contaminated soils.

Mousavi et al. synthesized KNP from chicken feathers that were used to remove Cu(II). They conducted experiments at pH 5 and 25°C and the results showed that KNP is a promising method for purifying wastewater from heavy metals [136]. In other study Mousavi used magnetic composite of KNP for remove of Zn(II) from aqueous solution with high efficiency at pH in range of 4 to 6 and 25°C [137].

The study of the effect of the use of protein NPs for the remediation of contaminated water and soils is still in its infancy. These studies show that these NPs will be increasingly used because they are biodegradable and non-toxic.

2.4.2 POLYSACCHARIDE NANOPARTICLES (NPS)

Polysaccharides are a major class of hydrophilic polymers with natural origin and biocompatibility. Due to the properties of polysaccharides such as biodegradability, biocompatibility, and low toxicity, polysaccharide-based

NPs are mainly used in medicine, cosmetics, and food industry, but there are many studies that apply these NPs in remediation processes.

Chitosan (CS) is an important derivative of chitin, and it is a linear polysaccharide obtained from the partial N-deacetylation of chitin with a deacetylation degree of over 40% under alkaline conditions and elevated temperature [138]. Chitin (CH) is insoluble in aqueous solutions, but CS dissolves well in water, which makes CS more useful for the production of NPs. Due of the functional reactive groups (primary/secondary amines and hydroxyl groups) and their physicochemical properties such as biocompatibility, non-toxicity, biodegradability, adsorption capacity, mucoadhesion, antioxidant, antimicrobial, and anticancer activities, CS NPs are widely used in medicine to delivering drug to cells and for tissue engineering. Also, CS NPs and nanocomposites are used in agriculture as potential nanofertilizers, as well as in the cosmetics industry as cosmetic ingredients. However, in recent years CS NPs have found application for wastewater remediation.

Almutairi et al. extracted CS from shrimp shells and alginate (Alg) from brown algae (*Sargassum linifolium*) and then synthesized CS/alginate nanocomposite. CS/Alg nanocomposites had a mean diameter of 311.2 nm and were negatively charged (–23.2 mV). Iron NPs with a 21.4 nm mean diameter were then added to this nanocomposite. The nanocomposite CS/Alg/Fe were used to remove Cr^{6+} from wastewater with high efficiency nearly 100% at pH 5, adsorbant dose = 5 g/L, contact time = 200 min and initial ion concentration = 100 ppm [27].

In another study Dubey et al. prepared CS/Alg for adsorption of Hg(II) ions from aqueous solutions. The maximum adsorption of the nanocomposite for Hg(II) was found to be 217.39 mg/g at 30°C [139].

Alsharari et al. used CS NPs from fungal CS for absorption of Pb^{2+} and Cu^{2+} at concentration range of 100–300 ppm, from aqueous solution and soil matrix. They achieved a high level of heavy metal remediation at high concentrations of NPs in wastewater. Three concentrations of heavy metals – 100 ppm, 200 ppm and 300 ppm – were used for remediation of soils contaminated with Pb^{2+} and Cu^{2+}. The Application of CS NPs, at concentrations of 0.5%, resulted in a 98.6% reduction in Pb^{2+} concentrations for 100 ppm, 87.2% for 200 ppm and 94.8% for 300 ppm. With regard to soils contaminated with Cu^{2+}, the highest percentages of reduction, i.e., 97.3%, 90.4% and 88.8% were achieved by applying NPs with a concentration of 0.5% in the soil with a pollutant concentration of 100 ppm, 200 ppm and 300 ppm, respectively [140].

CS NPs have been used to purify aqueous solutions of dyes. The NPs showed high adsorption rates and dye binding capacity (300–1,050 mg g^{-1}) and the sorption rate decreasing with increasing dye molecule size [141].

Tunable Schiff's base-cross-linked CS-glutaraldehyde nanocomposite with TiO_2 nanoparticles (CS-GLA/TNC) with different ratios (25% and 50%) was used for the removal of reactive red 120 dye (RR 120). CS-GLA/TNC with 25% TiO_2 exhibited the best adsorptive property and this nanocomposite had adsorption capacity 103.1 mg/g at 29.85°C [142].

Chitosan–ZnO nanoparticles (CS–ZnONPs) composite was synthesized to remove of permethrin pesticide from water. The highest adsorption (99% removal of the pesticide from the solution) was achieved at pH 7 [143].

CL has linear structure which is formed by glucose units (>10,000) linked by β-(1-4)-glycosidic bonds. The molecular structure of CL defines its properties such as strength, hydrophilicity, biodegradable, biocompatible as well as low toxicity. The exceptional strength of CL is due to the hydrogen bonds that add crystal fiber structures to the CL. CL is mostly extracted from plants, but algae and bacteria can also be a source of CL [144]. There are two main types of CL NPs and nanocomposites – CL nanofibers and nanocrystals, which are based on the difference between CL structures.

CL NPs are being used more and more successfully in remediation of wastewaters from heavy metals. Sun et al. synthesized magnetic silica NPs and then coated them with CL. This CL-based magnetic nanomaterial was used to remove Cr (VI) from wastewater and a maximum adsorption capacity of 171.5 mg/g was achieved at 25°C. and pH 2 [145].

Amino modified nanostructured microfibrillated CL was prepared for remove of Ni(II), Cu(II), and Cd(II) ions from aqueous solutions. The maximum removal capacities of the modified CL nanocomposite adsorbent for Ni(II), Cu(II), and Cd(II) ions were 2.734, 3.150, and 4.195 mmol/g, respectively [146].

Nemr et al. have been designed and synthesized from sugarcane magnetic CL nanocomposites for adsorption of Cu(II), Cd(II), and Pb(II) ions from contaminated waters. They prepared three types of nanocomposites – the first nanocomposite being prepared as MNPs (Fe_3O_4) are coated with CL dissolved in aqueous solution (NFC1), in the second type of nanocomposite, the magnetic particles are coated with CL through the co-precipitation (NFC2) and the third type of nanocomposite is obtained by crosslinking CL and Fe_3O_4 with formaldehyde (NFC3). They found that NFC1 was best for removing Pb(II) and Cu(II) (3,150 and 2,150 μmol/g), and NFC2 was a good nanocomposite for removing Cd(II) 800 μmol g^{-1} [147]. In addition, contaminated with uranium(IV) aqueous solutions were successfully treated with CL nanocomposites at pH 5.5 [148].

CL nanocomposites are applicable also for removing of dyes from polluted waters. Anionic dyes were removed from aqueous solutions by amino-functionalized nanocrystalline cellulose (ANCC), which was prepared by processes of extraction of nanocrystalline cellulose (NCC) from fully bleached hardwood kraft pulp by sulfuric acid hydrolysis, sodium periodate oxidation of NCC to yield the corresponding C-2/C-3 dialdehyde nanocellulose and grafting with ethylenediamine to obtain ANCC through a reductive amination treatment [149]. Mohamed et al. used cellulose nanocrystal-alginate (CNC-ALG) hydrogel beads in a fixed bed column to adsorb of MB. They reached maximum capacity of adsorption of 255.5 mg/g [150]. Polyvinyl alcohol/cellulose nanofibril (PVA–CNF) aerogel nanocomposites loaded with ZnO particles were used for adsorption of organic dye in aqueous solution with maximum adsorption capacity of 24.53 mg/g [151].

Starch is polysaccharide which is composed of two different polysaccharides amylose and amylopectin. Amylose is a linear polymer linked by α-d-(1-4) glycosidic bonds while amylopectin is a branched chain polymer formed by α-d-(1-4) glycosidic bonds in a linear chain and α-d-(1-6) branched connections. Due to hydrogen bonds in its molecular structure, starch is insoluble in water at low temperatures. However, when water is heated to temperatures higher than 52°C, a process known as gelatinization begins. Gelatinization is an endothermic process in which the hydrogen bonds in amylose are destroyed and then under the influence of water amylose hydrates and swells and starch granules finally ruptured and the amylose is released out in the form of a starch paste. Also, starch polysaccharide is chemically stable, biocompatible, and biodegradable, which makes it useful in medical applications and food packaging [152].

Starch-stabilized zero valent iron NPs for soil remediation of a Pb-acid battery were synthesized by a reduction process. The researchers found that concentrations of nZVI-Starch NPs greater than 1,000 mg/kg of soil would be needed to completely immobilize Pb in extremely polluted soils [153]. Wang et al. used sodium carboxymethyl CL stabilized nanoscale zero-valent iron (CMC-stabilized nZVI) nanocomposite for remediation of contaminated soils with Cr(VI).

The results showed that 0.09 g/L^{-1} of CMC-stabilized nZVI NPs significantly reduced the toxicity characteristic of Cr by 82% [154].

Starch nanocomposites are also used as an adsorbent of wastewater contaminants. Ahamad et al. fabricated magnetic nanocomposite based on the starch and salicylaldehyde resin embedded with magnetic Fe_3O_4 NPs. The maximum adsorption capacity was found to be 265.4 and 247.2 mg/g for Pb(II) and Cd(II), respectively at pH 6 and 25°C [155].

Gad et al. prepared corn starch nanoparticles by pullulanase to remove crystalline violet dye from aqueous solution. They achieved optimal results at pH 8.5, 50°C for 120 min [156]. The removal of MB from wastewater was successfully achieved with modified starch nanocrystals, which were obtained by grafting an epichlorohydrin-dimethylamine copolymer onto oxidized starch nanocrystals. The results show that the maximum adsorption capacity of this nanocomposite is 169.49 mg g^{-1} [157]. Starch nanocomposite was also used to degradation of CHCs in water. The researchers developed palladized iron (Fe–Pd) nanoparticles with water-soluble starch as a stabilizer. This starch nanocomposite remains suspended in water for a long time compared to starch-free nanoparticles. The results showed that at a dose of 0.1 g L^{-1}, the starched particles were able to destroy 98% of TCE and 80% of PCBs within 1 h [106].

Biopolymeric nanoparticles due of their biodegradability, biocompatibility, and low toxicity are widely used in medicine. These nanoparticles are also used in remediation processes, but more research is needed. There are many challenges for the further development of biopolymer nanoparticles and nanocomposites, but their promising physicochemical properties are a guarantee that research will continue.

2.5 CONCLUSION

This chapter summarizes the application of different types of nanomaterials in remediation processes. Carbon nanoparticles and nanocomposites are divided into zero, one-, and two-dimensional particles depending on their size, shape, and structure. They are widely used in medicine, food industry and cosmetics, but in recent years have found applications in the processes of remove of contaminants from wastewater and soils.

The properties of metal nanoparticles, as well as their relatively inexpensive production, make these nanomaterials suitable for the remediation processes of contaminated water and soils. MONPs are class of metallic nanomaterials that are used frequently to remove contaminants from waters because of their unique catalytic properties. TiO_2, ZnO, CuO, SnO_2, AgO, Al_2O_3, MgO nanoparticles and their nanocomposites are most widely used in the remediation of water.

nZVI nanoparticles and MNPs are two other types of metal nanoparticles that are also used in the decontamination of contaminants in water and soil. nZVI nanoparticles have a low cost and properties that make them very suitable for removing heavy metals and dyes from wastewater.

Magnetic nanoparticles have very good mechanical, optical, electrical properties. When an external magnetic field is applied, the MNPs are easily manipulated and separated, and these properties make them useful for removing contaminants from contaminated media.

Biopolymeric nanoparticles due to their biodegradability and biocompatibility are mainly used in medicine, food industry and cosmetics, but in recent years, biopolymeric nanocomposites are increasingly used in remediation of water and soils.

Nanoremediation is a promising technology that will be increasingly used, but which still faces many problems to solve such as cost reduction, research on the toxicity of these particles and more.

KEYWORDS

- **carbon nanomaterials**
- **contamination**
- **metallic nanoparticles**
- **nanocomposites**
- **nanoparticles**
- **nanoremediation**
- **zero-dimensional carbon nanomaterials**

REFERENCES

1. Agency, U. S. E. P., (2006). *In Situ Treatment Technologies for Contaminated Soil in Engineering Forum.* U.S. Environmental Protection Agency.
2. Rafael, G. L., José, M. B., & Carlos, G. (2020). Biological methods of polluted soil remediation for an effective economically-optimal recovery of soil health and ecosystem services. *Journal of Environmental Science and Public Health, 4*, 112–133.
3. Christian, P., Von, D. K. F., Baalousha, M., & Hofmann, T., (2008). Nanoparticles: Structure, properties, preparation and behavior in environmental media. *Ecotoxicology, 17*(5), 326–343.

4. Wang, Z. L., Ahmad, T. S., & El-Sayed, M. A., (1997). Steps, ledges and kinks on the surfaces of platinum nanoparticles of different shapes. *Surface Science, 380*(2), 302–310.
5. Lee, H., Habas, S. E., Kweskin, S., Butcher, D., Somorjai, G. A., & Yang, P., (2006). Morphological control of catalytically active platinum nanocrystals. *Angew Chem. Int. Ed. Engl., 45*(46), 7824–7828.
6. Millstone, J. E., Hurst, S. J., Métraux, G. S., Cutler, J. I., & Mirkin, C. A., (2009). Colloidal gold and silver triangular nanoprisms. *Small, 5*(6), 646–664.
7. Smalley, R. E., & Yakobson, B. I., (1998). The future of the fullerenes. *Solid State Communications, 107*(11), 597–606.
8. Essajai, R., Benhouria, Y., Rachadi, A., Qjani, M., Mzerd, A., & Hassanain, N., (2019). Shape-dependent structural and magnetic properties of Fe nanoparticles studied through simulation methods. *RSC Advances, 9*(38), 22057–22063.
9. Mami, Y., Syusaku, K., & Mikio, M., (2005). Synthesis and size control of Pt nanocubes with high selectivity using the additive effect of NaI. *Chemistry Letters, 34*(7), 1050, 1051.
10. Clark, B. D., Jacobson, C. R., Lou, M., Renard, D., Wu, G., Bursi, L., Ali, A. S., et al., (2019). Aluminum nanocubes have sharp corners. *ACS Nano, 13*(8), 9682–9691.
11. Frank, A. J., Cathcart, N., Maly, K. E., & Kitaev, V., (2010). Synthesis of silver nanoprisms with variable size and investigation of their optical properties: A first-year undergraduate experiment exploring plasmonic nanoparticles. *Journal of Chemical Education, 87*(10), 1098–1101.
12. Theiss, S., Voggel, M., Schlötter, M., Sutter, S., Stöckl, M. T., & Polarz, S., (2019). Tolerance in superstructures formed from high-quality colloidal ZnO nanoparticles with hexagonal cross-section. *CrystEngComm, 21*(34), 5137–5144.
13. He, C., & Wang, W., (2020). 2D octagon-structure carbon and its polarization resolved Raman spectra. *Nanomaterials, 10*(11), 2252.
14. Kroto, H. W., Heath, J. R., O'Brien, S. C., Curl, R. F., & Smalley, R. E., (1985). C60: Buckminsterfullerene. *Nature, 318*(6042), 162, 163.
15. Brunet, L., Lyon, D. Y., Hotze, E. M., Alvarez, P. J. J., & Wiesner, M. R., (2009). Comparative photoactivity and antibacterial properties of C60 fullerenes and titanium dioxide nanoparticles. *Environmental Science & Technology, 43*(12), 4355–4360.
16. Bakry, R., Vallant, R. M., Najam-ul-Haq, M., Rainer, M., Szabo, Z., Huck, C. W., & Bonn, G. K., (2007). Medicinal applications of fullerenes. *Int. J. Nanomedicine, 2*(4), 639–649.
17. Mousa, M. S., (2018). Comparison between single-walled CNT, multi-walled CNT, and carbon nanotube-fiber pyrograf III. *IOP Conference Series: Materials Science and Engineering, 305*, 012025.
18. Sarkar, B., Mandal, S., Tsang, Y. F., Kumar, P., Kim, K. H., & Ok, Y. S., (2018). Designer carbon nanotubes for contaminant removal in water and wastewater: A critical review. *Science of The Total Environment, 612*, 561–581.
19. Polizu, S., Savadogo, O., Poulin, P., & Yahia, L., (2006). Applications of carbon nanotubes-based biomaterials in biomedical nanotechnology. *J. Nanosci. Nanotechnol., 6*(7), 1883–1904.
20. Zhao, L., Yang, S. T., Yilihamu, A., & Wu, D., (2019). Advances in the applications of graphene adsorbents: From water treatment to soil remediation. *Reviews in Inorganic Chemistry, 39*(1), 47–76.

21. Galdames, A., Ruiz-Rubio, L., Orueta, M., Sánchez-Arzalluz, M., & Vilas-Vilela, J. L., (2020). Zero-valent iron nanoparticles for soil and groundwater remediation. *International Journal of Environmental Research and Public Health, 17*(16), 5817.
22. Ahmad, S., Nazir, A., & Hussain, T., (2017). Photocatalytic degradation of organics by using nanocrystalline titania. *Journal of Chemistry, 2017*, 7408972.
23. Ijadpanah-Saravy, H., Safari, M., Khodadadi-Darban, A., & Rezaei, A., (2014). Synthesis of titanium dioxide nanoparticles for photocatalytic degradation of cyanide in wastewater. *Analytical Letters, 47*(10), 1772–1782.
24. Gupta, A., Tandon, M., & Kaur, A., (2020). Role of metallic nanoparticles in water remediation with special emphasis on sustainable synthesis: A review. *Nanotechnology for Environmental Engineering, 5*(3), 27.
25. Govan, J., (2020). Recent advances in magnetic nanoparticles and nanocomposites for the remediation of water resources. *Magnetochemistry, 6*(4), 49.
26. Farshchi, E., Pirsa, S., Roufegarinejad, L., Alizadeh, M., & Rezazad, M., (2019). Photocatalytic/biodegradable film based on carboxymethyl cellulose, modified by gelatin and TiO_2-Ag nanoparticles. *Carbohydrate Polymers, 216*, 189–196.
27. Almutairi, F. M., El Rabey, H. A., Alalawy, A. I., Salama, A. A. M., Tayel, A. A., Mohammed, G. M., Aljohani, M. M., et al., (2021). Application of chitosan/alginate nanocomposite incorporated with phycosynthesized iron nanoparticles for efficient remediation of chromium. *Polymers, 13*(15), 2481.
28. Kolangare, I. M., Isloor, A. M., Karim, Z. A., Kulal, A., Ismail, A. F., Inamuddin, & Asiri, A. M., (2019). Antibiofouling hollow-fiber membranes for dye rejection by embedding chitosan and silver-loaded chitosan nanoparticles. *Environmental Chemistry Letters, 17*(1), 581–587.
29. Mostofizadeh, A., Li, Y., Song, B., & Huang, Y., (2011). Synthesis, properties, and applications of low-dimensional carbon-related nanomaterials. *Journal of Nanomaterials, 2011*, 685081.
30. Taylor, R., & Walton, D. R. M., (1993). The chemistry of fullerenes. *Nature, 363*(6431), 685–693.
31. Mohan, G. M., & Ravindra, S. R., (2015). Fullerenes: Chemistry and its applications. *Mini-Reviews in Organic Chemistry, 12*(4), 355–366.
32. Chae, S. R., Hotze, E. M., & Wiesner, M. R., (2014). Chapter 21 – possible applications of fullerene nanomaterials in water treatment and reuse. In: Street, A., Sustich, R., Duncan, J., & Savage, N., (eds.), *Nanotechnology Applications for Clean Water* (2nd edn., pp 329–338). William Andrew Publishing: Oxford.
33. Bottero, J. Y., Rose, J., & Wiesner, M. R., (2006). Nanotechnologies: Tools for sustainability in a new wave of water treatment processes. *Integrated Environmental Assessment and Management, 2*(4), 391–395.
34. Ballesteros, E., Gallego, M., & Valcárcel, M., (2000). Analytical potential of fullerene as adsorbent for organic and organometallic compounds from aqueous solutions. *Journal of Chromatography A, 869*(1), 101–110.
35. Zhang, Q., Cui, Y., Gu, C., & Zhang, C., (2020). Potential concerns in fullerene application to water treatment related to transformation, cellular uptake and intracellular catalysis. *Science of The Total Environment, 728*, 138754.
36. Aslam, M. M. A., Kuo, H. W., Den, W., Usman, M., Sultan, M., & Ashraf, H., (2021). Functionalized carbon nanotubes (CNTs) for water and wastewater treatment: Preparation to application. *Sustainability, 13*(10), 5717.

37. Gupta, V. K., Agarwal, S., & Saleh, T. A., (2011). Chromium removal by combining the magnetic properties of iron oxide with adsorption properties of carbon nanotubes. *Water Research, 45*(6), 2207–2212.
38. Burakov, A. E., Galunin, E. V., Burakova, I. V., Kucherova, A. E., Agarwal, S., Tkachev, A. G., & Gupta, V. K., (2018). Adsorption of heavy metals on conventional and nanostructured materials for wastewater treatment purposes: A review. *Ecotoxicology and Environmental Safety, 148*, 702–712.
39. Mallakpour, S., & Khadem, E., (2019). 8 – carbon nanotubes for heavy metals removal. In: Kyzas, G. Z., & Mitropoulos, A. C., (eds.), *Composite Nanoadsorbents* (pp. 181–210). Elsevier.
40. Chen, B., Zhu, Z., Ma, J., Yang, M., Hong, J., Hu, X., Qiu, Y., & Chen, J., (2014). One-pot, solid-phase synthesis of magnetic multiwalled carbon nanotube/iron oxide composites and their application in arsenic removal. *Journal of Colloid and Interface Science, 434*, 9–17.
41. Abdel, S. E. T., Abou El-Nour, K. M., Awad, A. A., & Orabi, A. S., (2020). Carbon nanotubes modified with 5,7-dinitro-8-quinolinol as potentially applicable tool for efficient removal of industrial wastewater pollutants. *Arabian Journal of Chemistry, 13*(1), 109–119.
42. Alijani, H., & Shariatinia, Z., (2018). Synthesis of high growth rate SWCNTs and their magnetite cobalt sulfide nanohybrid as super-adsorbent for mercury removal. *Chemical Engineering Research and Design, 129*, 132–149.
43. Ntim, S. A., & Mitra, S., (2012). Adsorption of arsenic on multiwall carbon nanotube-zirconia nanohybrid for potential drinking water purification. *Journal of Colloid and Interface Science, 375*(1), 154–159.
44. Iannazzo, D., Pistone, A., Ziccarelli, I., Espro, C., Galvagno, S., Giofré, S. V., Romeo, R., et al., (2017). Removal of heavy metal ions from wastewaters using dendrimer-functionalized multi-walled carbon nanotubes. *Environmental Science and Pollution Research, 24*(17), 14735–14747.
45. Hadavifar, M., Bahramifar, N., Younesi, H., Rastakhiz, M., Li, Q., Yu, J., & Eftekhari, E., (2016). Removal of mercury(II) and cadmium(II) ions from synthetic wastewater by a newly synthesized amino and thiolated multi-walled carbon nanotubes. *Journal of the Taiwan Institute of Chemical Engineers, 67*, 397–405.
46. Zare, K., Gupta, V. K., Moradi, O., Makhlouf, A. S. H., Sillanpää, M., Nadagouda, M. N., Sadegh, H., et al., (2015). A comparative study on the basis of adsorption capacity between CNTs and activated carbon as adsorbents for removal of noxious synthetic dyes: A review. *Journal of Nanostructure in Chemistry, 5*(2), 227–236.
47. Engel, M., & Chefetz, B., (2016). Adsorption and desorption of dissolved organic matter by carbon nanotubes: Effects of solution chemistry. *Environmental Pollution, 213*, 90–98.
48. Deng, J., Shao, Y., Gao, N., Deng, Y., Tan, C., Zhou, S., & Hu, X., (2012). Multiwalled carbon nanotubes as adsorbents for removal of herbicide diuron from aqueous solution. *Chemical Engineering Journal, 193, 194*, 339–347.
49. Shinde, S. G., Patil, M. P., Kim, G. D., & Shrivastava, V. S., (2020). Ni, C, N, S multi-doped ZrO_2 decorated on multi-walled carbon nanotubes for effective solar induced degradation of anionic dye. *Journal of Environmental Chemical Engineering, 8*(3), 103769.

50. Brady-Estévez, A. S., Kang, S., & Elimelech, M., (2008). A single-walled-carbon-nanotube filter for removal of viral and bacterial pathogens. *Small, 4*(4), 481–484.
51. Kovach, K., Sabaraya, I. V., Patel, P., Kirisits, M. J., Saleh, N. B., & Gordon, V. D., (2020). Suspended multiwalled, acid-functionalized carbon nanotubes promote aggregation of the opportunistic pathogen *Pseudomonas aeruginosa. PLoS One, 15*(7), e0236599.
52. Kang, S., Pinault, M., Pfefferle, L. D., & Elimelech, M., (2007). Single-walled carbon nanotubes exhibit strong antimicrobial activity. *Langmuir, 23*(17), 8670–8673.
53. Taghizadeh, M., Asgharinezhad, A. A., Samkhaniany, N., Tadjarodi, A., Abbaszadeh, A., & Pooladi, M., (2014). Solid phase extraction of heavy metal ions based on a novel functionalized magnetic multi-walled carbon nanotube composite with the aid of experimental design methodology. *Microchimica Acta, 181*(5), 597–605.
54. Zhao, N., Bian, Y. W., Dong, X. Y., Gao, X., & Zhao, L. S., (2021). Magnetic solid-phase extraction based on multi-walled carbon nanotubes combined ferroferric oxide nanoparticles for the determination of five heavy metal ions in water samples by inductively coupled plasma mass spectrometry. *Water Science and Technology, 84*(6), 1417–1427.
55. Islam, A., Chauhan, A., Javed, H., Rais, S., & Ahmad, I., (2021). Magnetic carbon nanotubes-silica binary composite for effective Pb(II) sequestration from industrial effluents: Multivariate process optimization. *Clean-Soil Air Water, 49*(8).
56. Moore, C. J., (2008). Synthetic polymers in the marine environment: A rapidly increasing, long-term threat. *Environmental Research, 108*(2), 131–139.
57. Yong, C. Q. Y., Valiyaveettil, S., & Tang, B. L., (2020). Toxicity of microplastics and nanoplastics in mammalian systems. *International Journal of Environmental Research and Public Health, 17*(5), 1509.
58. Kang, J., Zhou, L., Duan, X., Sun, H., Ao, Z., & Wang, S., (2019). Degradation of cosmetic microplastics via functionalized carbon nanosprings. *Matter, 1*(3), 745–758.
59. Yang, G., Li, L., Lee, W. B., & Ng, M. C., (2018). Structure of graphene and its disorders: A review. *Science and Technology of Advanced Materials, 19*(1), 613–648.
60. Sitko, R., Turek, E., Zawisza, B., Malicka, E., Talik, E., Heimann, J., Gagor, A., et al., (2013). Adsorption of divalent metal ions from aqueous solutions using graphene oxide. *Dalton Transactions, 42*(16), 5682–5689.
61. Li, D., Müller, M. B., Gilje, S., Kaner, R. B., & Wallace, G. G., (2008). Processable aqueous dispersions of graphene nanosheets. *Nature Nanotechnology, 3*(2), 101–105.
62. Georgakilas, V., Otyepka, M., Bourlinos, A. B., Chandra, V., Kim, N., Kemp, K. C., Hobza, P., et al., (2012). Functionalization of graphene: Covalent and non-covalent approaches, derivatives and applications. *Chemical Reviews, 112*(11), 6156–6214.
63. Morse, J. R., Zugell, D. A., Patterson, E., Baldwin, J. W., & Willauer, H. D., (2021). Hydrogenated graphene: Important material properties regarding its application for hydrogen storage. *Journal of Power Sources, 494*, 229734.
64. Batzill, M., (2012). The surface science of graphene: Metal interfaces, CVD synthesis, nanoribbons, chemical modifications, and defects. *Surface Science Reports, 67*(3), 83–115.
65. Chen, Y., Chen, L., Bai, H., & Li, L., (2013). Graphene oxide–chitosan composite hydrogels as broad-spectrum adsorbents for water purification. *Journal of Materials Chemistry A, 1*(6), 1992–2001.

66. Arranz-Mascarós, P., Godino-Salido, M. L., López-Garzón, R., García-Gallarín, C., Chamorro-Mena, I., López-Garzón, F. J., Fernández-García, E., & Gutiérrez-Valero, M. D., (2020). Non-covalent functionalization of graphene to tune its band gap and stabilize metal nanoparticles on its surface. *ACS Omega, 5*(30), 18849–18861.
67. Ali, I., Basheer, A. A., Mbianda, X. Y., Burakov, A., Galunin, E., Burakova, I., Mkrtchyan, E., et al., (2019). Graphene based adsorbents for remediation of noxious pollutants from wastewater. *Environment International, 127*, 160–180.
68. Mahmood, A. N. A., William, I. M., Cengiz, S. O., Stanislaw, M., & Juana, L. G., (2016). Fabrication of graphene-based porous materials and their applications in environmental fields. In: *Graphene Science Handbook* (1st edn., p. 492). CRC Press.
69. Baragaño, D., Forján, R., Welte, L., & Gallego, J. L. R., (2020). Nanoremediation of As and metals polluted soils by means of graphene oxide nanoparticles. *Scientific Reports, 10*(1), 1896.
70. Shahzadi, I., Khan, Z. H., Akram, W., Khan, W. U., Ahmad, A., Yasin, N. A., & Yujie, L., (2022). Heavy metal and organic pollutants removal from water using bilayered polydopamine composite of sandwiched graphene nanosheets: One solution for two obstacles. *Separation and Purification Technology, 280*, 119711.
71. Ren, H., Cao, Z. F., Chen, Y. Y., Jiang, X. Y., & Yu, J. G., (2021). Graphene oxide-Bicine composite as a novel adsorbent for removal of various contaminants from aqueous solutions. *Journal of Environmental Chemical Engineering, 9*(6), 106769.
72. Beryani, A., Bianco, C., Casasso, A., Sethi, R., & Tosco, T., (2022). Exploring the potential of graphene oxide nanosheets for porous media decontamination from cationic dyes. *Journal of Hazardous Materials, 424*, 127468.
73. Sari, Y. M., (2022). Graphene oxide/hollow mesoporous silica composite for selective adsorption of methylene blue. *Microporous and Mesoporous Materials, 330*, 111570.
74. Tian, Y., Cao, S., Xi, C., Su, H., & Chen, Z., (2021). Designing polydopamine nanohybrid based on template-mediated for effectively remove amphetamine-type stimulants in sewage: Performance and mechanism. *Journal of Environmental Chemical Engineering, 9*(5), 105870.
75. Mura, S., Jiang, Y., Vassalini, I., Gianoncelli, A., Alessandri, I., Granozzi, G., Calvillo, L., et al., (2018). Graphene oxide/iron oxide nanocomposites for water remediation. *ACS Applied Nano Materials, 1*(12), 6724–6732.
76. Januário, E. F. D., Vidovix, T. B., Beluci, N. D. C. L., Paixão, R. M., Silva, L. H. B. R. D., Homem, N. C., Bergamasco, R., & Vieira, A. M. S., (2021). Advanced graphene oxide-based membranes as a potential alternative for dyes removal: A review. *Science of The Total Environment, 789*, 147957.
77. Chavali, M. S., & Nikolova, M. P., (2019). Metal oxide nanoparticles and their applications in nanotechnology. *SN Applied Sciences, 1*(6), 607.
78. Alsheheri, S. Z., (2021). Nanocomposites containing titanium dioxide for environmental remediation. *Des. Monomers Polym., 24*(1), 22–45.
79. Banfield, J. F., Veblen, D. R., & Smith, D. J., (1991). The identification of naturally occurring TiO_2 (B) by structure determination using high-resolution electron microscopy, image simulation, and distance-least-squares refinement. *American Mineralogist, 76*(3, 4), 343–353.
80. Theron, J., Walker, J. A., & Cloete, T. E., (2008). Nanotechnology and water treatment: Applications and emerging opportunities. *Critical Reviews in Microbiology, 34*(1), 43–69.

81. Vani, B., Shivakumar, M., Kalyani, S., & Sridhar, S., (2021). TiO2 nanoparticles incorporated high-performance polyphenyl sulfone mixed matrix membranes for ultrafiltration of domestic greywater. *Iranian Polymer Journal, 30*(9), 917–934.
82. Krishnan, S., Zulkapli, N. S., Din, M. F. B. M., Majid, Z. A., Nasrullah, M., & Sairan, F. M., (2021). Photocatalytic degradation of methylene blue dye and fungi *Fusarium equiseti* using titanium dioxide recovered from drinking water treatment sludge. *Biomass Conversion and Biorefinery.*
83. Baneshi, Rezaei, S., Sadat, A., Mousavizadeh, A., Barafrashtehpour, M., & Hekmatmanesh, H., (2012). *In Investigation of Photocatalytic Degradation of Diazinon Using Titanium Dioxide (TiO_2) Nanoparticles Doped with Iron in the Presence of Ultraviolet Rays from the Aqueous Solution.*
84. Vatanpour, V., Karami, A., & Sheydaei, M., (2017). Central composite design optimization of rhodamine B degradation using TiO_2 nanoparticles/UV/PVDF process in continuous submerged membrane photoreactor. *Chemical Engineering and Processing, 116*, 68–75.
85. Das, R., Sarkar, S., Chakraborty, S., Choi, H., & Bhattacharjee, C., (2014). Remediation of antiseptic components in wastewater by photocatalysis using TiO_2 nanoparticles. *Industrial & Engineering Chemistry Research, 53*(8), 3012–3020.
86. Anjum, M., Miandad, R., Waqas, M., Gehany, F., & Barakat, M. A., (2019). Remediation of wastewater using various nano-materials. *Arabian Journal of Chemistry, 12*(8), 4897–4919.
87. Wang, X., Cai, W., Lin, Y., Wang, G., & Liang, C., (2010). Mass production of micro/nanostructured porous ZnO plates and their strong structurally enhanced and selective adsorption performance for environmental remediation. *Journal of Materials Chemistry, 20*(39), 8582–8590.
88. Dong, J. Y., Lin, C. H., Hsu, Y. J., Lu, S. Y., & Wong, D. S. H., (2012). Single-crystalline mesoporous ZnO nanosheets prepared with a green antisolvent method exhibiting excellent photocatalytic efficiencies. *CrystEngComm, 14*(14), 4732–4737.
89. Kumar, K. Y., Muralidhara, H. B., Nayaka, Y. A., Balasubramanyam, J., & Hanumanthappa, H., (2013). Hierarchically assembled mesoporous ZnO nanorods for the removal of lead and cadmium by using differential pulse anodic stripping voltammetric method. *Powder Technology, 239*, 208–216.
90. Katwal, R., Kaur, H., Sharma, G., Naushad, M., & Pathania, D., (2015). Electrochemical synthesized copper oxide nanoparticles for enhanced photocatalytic and antimicrobial activity. *Journal of Industrial and Engineering Chemistry, 31*, 173–184.
91. Ramesh, M., (2021). CuO as efficient photo catalyst for photocatalytic decoloration of wastewater containing azo dyes. *Water Practice and Technology, 16*(4), 1078–1090.
92. Mahar, Z. A., Shar, G. Q., Balouch, A., Pato, A. H., & Shaikh, A. R., (2021). Effective and viable photocatalytic degradation of rhodamine B dye in aqueous media using CuO/PVA nanocomposites. *New Journal of Chemistry, 45*(36), 16500–16510.
93. Fakhri, A., (2015). Investigation of mercury (II) adsorption from aqueous solution onto copper oxide nanoparticles: Optimization using response surface methodology. *Process Safety and Environmental Protection, 93*, 1–8.
94. Mahmoud, A. E. D., Al-Qahtani, K. M., Alflaij, S. O., Al-Qahtani, S. F., & Alsamhan, F. A., (2021). Green copper oxide nanoparticles for lead, nickel, and cadmium removal from contaminated water. *Scientific Reports, 11*(1), 12547.

95. Mahdavi, S., Jalali, M., & Afkhami, A., (2013). Heavy metals removal from aqueous solutions using TiO_2, MgO, and Al_2O_3 nanoparticles. *Chemical Engineering Communications, 200*(3), 448–470.
96. Camtakan, Z., Erenturk, S., & Yusan, S., (2012). Magnesium oxide nanoparticles: Preparation, characterization, and uranium sorption properties. *Environmental Progress & Sustainable Energy, 31*(4), 536–543.
97. Taourati, R., Khaddor, M., Laghzal, A., & El Kasmi, A., (2020). Facile one-step synthesis of highly efficient single oxide nanoparticles for photocatalytic application. *Scientific African, 8*, e00305.
98. Jorfi, S., Barzegar, G., Ahmadi, M., Darvishi, C. S. R., Alah, J. H. N., Takdastan, A., Saeedi, R., & Abtahi, M., (2016). Enhanced coagulation-photocatalytic treatment of acid red 73 dye and real textile wastewater using UVA/synthesized MgO nanoparticles. *Journal of Environmental Management, 177*, 111–118.
99. Mai, N. X. D., Le, T. A. C., Doan, T. L. H., Park, S., Park, K. H., & Phan, B. T., (2020). Agarose@MgO composite tablet for heavy metal removal from acid sulfate water. *Journal of Electronic Materials, 49*(3), 1857–1863.
100. Ahmadifard, T., Heydari, R., Tarrahi, M. J., & Khorramabadi, G. S., (2019). Photocatalytic degradation of diazinon in aqueous solutions using immobilized MgO nanoparticles on concrete. *International Journal of Chemical Reactor Engineering, 17*(9).
101. Jiang, W., Wang, X., Wu, Z., Yue, X., Yuan, S., Lu, H., & Liang, B., (2015). Silver oxide as superb and stable photocatalyst under visible and near-infrared light irradiation and its photocatalytic mechanism. *Industrial & Engineering Chemistry Research, 54*(3), 832–841.
102. Naderi, P. E., & Jalali, M., (2019). Application of three nanoparticles (Al_2O_3, SiO_2 and TiO_2) for metal-contaminated soil remediation (measuring and modeling). *International Journal of Environmental Science and Technology, 16*(11), 7207–7220.
103. Fu, F., Dionysiou, D. D., & Liu, H., (2014). The use of zero-valent iron for groundwater remediation and wastewater treatment: A review. *Journal of Hazardous Materials, 267*, 194–205.
104. Wu, Y., Yue, Q., Gao, Y., Ren, Z., & Gao, B., (2018). Performance of bimetallic nanoscale zero-valent iron particles for removal of oxytetracycline. *Journal of Environmental Sciences, 69*, 173–182.
105. Wang, C. B., & Zhang, W. X., (1997). Synthesizing nanoscale iron particles for rapid and complete dechlorination of TCE and PCBs. *Environmental Science & Technology, 31*(7), 2154–2156.
106. He, F., & Zhao, D., (2005). Preparation and characterization of a new class of starch-stabilized bimetallic nanoparticles for degradation of chlorinated hydrocarbons in water. *Environmental Science & Technology, 39*(9), 3314–3320.
107. Zhu, B. W., & Lim, T. T., (2007). Catalytic reduction of chlorobenzenes with Pd/Fe nanoparticles: Reactive sites, catalyst stability, particle aging, and regeneration. *Environmental Science & Technology, 41*(21), 7523–7529.
108. Wang, R., Lu, G., Lin, H., Huang, K., Tang, T., Xue, X., Yang, X., Yin, H., & Dang, Z., (2017). Relative roles of H-atom transfer and electron transfer in the debromination of polybrominated diphenyl ethers by palladized nanoscale zerovalent iron. *Environmental Pollution, 222*, 331–337.

109. Luo, F., Yang, D., Chen, Z., Megharaj, M., & Naidu, R., (2016). One-step green synthesis of bimetallic Fe/Pd nanoparticles used to degrade orange II. *Journal of Hazardous Materials, 303*, 145–153.
110. Kubendiran, H., Hui, D., Pulimi, M., Chandrasekaran, N., Murthy, P. S., & Mukherjee, A., (2021). Removal of methyl orange from aqueous solution using SRB supported bio-Pd/Fe NPs. *Environmental Nanotechnology, Monitoring & Management, 16*, 100561.
111. Xie, Y., Fang, Z., Cheng, W., Tsang, P. E., & Zhao, D., (2014). Remediation of polybrominated diphenyl ethers in soil using Ni/Fe bimetallic nanoparticles: Influencing factors, kinetics and mechanism. *Science of The Total Environment, 485, 486*, 363–370.
112. Jin, X., Chen, H., & Yang, Q. (2019). Removal of chloroform by Fe/Ni nanoparticles supported on activated carbon fibers. *Environmental Engineering Science, 36*(6), 681–689.
113. Lin, Y., Jin, X., Owens, G., & Chen, Z., (2019). Simultaneous removal of mixed contaminants triclosan and copper by green synthesized bimetallic iron/nickel nanoparticles. *Science of The Total Environment, 695*, 133878.
114. Niksefat, M., & Ayati, B., (2016). Reduction and reuse of Fe–Ni bimetallic nanoparticles oxide and evaluating its ability in acid red 14 removal. *Environmental Progress & Sustainable Energy, 35*(6), 1646–1656.
115. Sarvari, H., Goharshadi, E. K., Samiee, S., & Ashraf, N., (2018). Removal of methyl orange from aqueous solutions by ferromagnetic Fe/Ni nanoparticles. *Physical Chemistry Research, 6*(2), 433–446.
116. Zhu, F., Li, L., Ma, S., & Shang, Z., (2016). Effect factors, kinetics and thermodynamics of remediation in the chromium contaminated soils by nanoscale zero valent Fe/Cu bimetallic particles. *Chemical Engineering Journal, 302*, 663–669.
117. Ye, J., Wang, Y., Xu, Q., Wu, H., Tong, J., & Shi, J., (2021). Removal of hexavalent chromium from wastewater by Cu/Fe bimetallic nanoparticles. *Scientific Reports, 11*(1), 10848.
118. Ou, J. H., Sheu, Y. T., Tsang, D. C. W., Sun, Y. J., & Kao, C. M., (2020). Application of iron/aluminum bimetallic nanoparticle system for chromium-contaminated groundwater remediation. *Chemosphere, 256*, 127158.
119. Garcia, A. N., Zhang, Y., Ghoshal, S., He, F., & O'Carroll, D. M., (2021). Recent advances in sulfidated zerovalent iron for contaminant transformation. *Environmental Science & Technology, 55*(13), 8464–8483.
120. Wang, B., Dong, H., Li, L., Wang, Y., Ning, Q., Tang, L., & Zeng, G., (2020). Influence of different co-contaminants on trichloroethylene removal by sulfide-modified nanoscale zero-valent iron. *Chemical Engineering Journal, 381*, 122773.
121. Zhou, Q., Li, J., Wang, M., & Zhao, D., (2016). Iron-based magnetic nanomaterials and their environmental applications. *Critical Reviews in Environmental Science and Technology, 46*(8), 783–826.
122. Hu, J., Chen, G., & Lo, I. M. C., (2005). Removal and recovery of Cr(VI) from wastewater by maghemite nanoparticles. *Water Research, 39*(18), 4528–4536.
123. Molinari, S., Magro, M., Baratella, D., Salviulo, G., Ugolotti, J., Filip, J., Petr, M., et al., (2020). Smart synthetic maghemite nanoparticles with unique surface properties encode binding specificity toward AsIII. *Science of The Total Environment, 741*, 140175.

124. Nicola, R., Costişor, O., Ciopec, M., Negrea, A., Lazău, R., Ianăşi, C., Picioruş, E. M., et al., (2020). Silica-coated magnetic nanocomposites for Pb^{2+} removal from aqueous solution. *Applied Sciences, 10*(8), 2726.
125. Kaur, S., Kaur, K., Kaur, A., & Singh, A. J., (2020). SiO_2@Fe_3O_4-SB nano-hybrids as nanosorbent for preconcentration of cadmium ions from environmental water samples. *Journal of Environmental Chemical Engineering, 8*(2), 103448.
126. Zhang, X., Wang, J., Li, R., Dai, Q., Gao, R., Liu, Q., & Zhang, M., (2013). Preparation of Fe_3O_4@C@layered double hydroxide composite for magnetic separation of uranium. *Industrial & Engineering Chemistry Research, 52*(30), 10152–10159.
127. Gad, Y. H., & Elbarbary, A. M., (2021). Radiation synthesis of Fe_3O_4/SiO_2/glycidyl methacrylate/acrylonitrile nanocomposite for adsorption of basic violet 7 dye: Kinetic, isotherm, and thermodynamic study. *Applied Organometallic Chemistry, 35*(7), e6258.
128. Hernández, L., Augusto, P. A., Castelo-Grande, T., & Barbosa, D., (2021). Regeneration and reuse of magnetic particles for contaminant degradation in water. *Journal of Environmental Management, 285*, 112155.
129. Zhao, G., Song, S., Wang, C., Wu, Q., & Wang, Z., (2011). Determination of triazine herbicides in environmental water samples by high-performance liquid chromatography using graphene-coated magnetic nanoparticles as adsorbent. *Analytica Chimica Acta, 708*(1), 155–159.
130. Niu, H., Zhang, D., Zhang, S., Zhang, X., Meng, Z., & Cai, Y., (2011). Humic acid coated Fe_3O_4 magnetic nanoparticles as highly efficient Fenton-like catalyst for complete mineralization of sulfathiazole. *Journal of Hazardous Materials, 190*(1), 559–565.
131. Sarcletti, M., Park, H., Wirth, J., Englisch, S., Eigen, A., Drobek, D., Vivod, D., et al., (2021). The remediation of nano-/microplastics from water. *Materials Today, 48*, 38–46.
132. Sundar, S., Kundu, J., & Kundu, S. C., (2010). Biopolymeric nanoparticles. *Science and Technology of Advanced Materials, 11*(1), 014104.
133. Miller, A., & Jedrzejczak, W. W., (2001). Albumin--biological functions and clinical significance. *Postepy Hig. Med. Dosw., 55*(1), 17–36.
134. Shalaby, M., Farag, S., Haddad, A., Hamouda, D., Hassan, A., & Saeed, H., (2020). Collagen polymer and magnetic collagen nanocomposite recycled from waste to reduce polluted water toxicity. *Polymers and Polymer Composites, 29*(9), 1515–1527.
135. Thanikaivelan, P., Narayanan, N. T., Pradhan, B. K., & Ajayan, P. M., (2012). Collagen based magnetic nanocomposites for oil removal applications. *Scientific Reports, 2*, 230.
136. Mousavi, S. Z., Manteghian, M., Shojaosadati, S. A., & Pahlavanzadeh, H., (2018). Keratin nanoparticles: Synthesis and application for Cu(II) removal. *Advances in Environmental Technology, 4*(2), 83–93.
137. Mousavi, S. Z., Manteghian, M., & Ahmadpoor, F., (2020). Performance of keratin nanoparticle and its magnetic nanocomposite for Zn(II) removal from its aqueous solution. *Advances in Environmental Technology, 6*(1), 37–45.
138. Singla, A. K., & Chawla, M., (2010). Chitosan: Some pharmaceutical and biological aspects – an update. *Journal of Pharmacy and Pharmacology, 53*(8), 1047–1067.
139. Dubey, R., Bajpai, J., & Bajpai, A. K., (2016). Chitosan-alginate nanoparticles (CANPs) as potential nanosorbent for removal of Hg (II) ions. *Environmental Nanotechnology, Monitoring & Management, 6*, 32–44.
140. Alsharari, S. F., Tayel, A. A., & Moussa, S. H., (2018). Soil emendation with nano-fungal chitosan for heavy metals biosorption. *International Journal of Biological Macromolecules, 118*, 2265–2268.

141. Lipatova, I. M., Makarova, L. I., & Yusova, A. A., (2018). Adsorption removal of anionic dyes from aqueous solutions by chitosan nanoparticles deposited on the fibrous carrier. *Chemosphere, 212*, 1155–1162.
142. Jawad, A. H., Mubarak, N. S. A., & Abdulhameed, A. S., (2020). Tunable Schiff's base-cross-linked chitosan composite for the removal of reactive red 120 dye: Adsorption and mechanism study. *International Journal of Biological Macromolecules, 142*, 732–741.
143. Moradi, D. S., Rahmanifar, B., Moradi, A. M., & Azar, P. A., (2014). Removal of permethrin pesticide from water by chitosan–zinc oxide nanoparticles composite as an adsorbent. *Journal of Saudi Chemical Society, 18*(4), 348–355.
144. Kumar, G. P., Sai, R. S., Venkatesh, P. D., Venkat, P., Shree, V., Chithananthan, C., Choudhary, S., et al., (2019). An Update on Overview of Cellulose, Its Structure and Applications, in *Celulose*, Publisher IntechOpen, Chapter 4, DOI:10.5772/intechopen.75244, ISBN 978-1-83968-057-1.
145. Sun, X., Yang, L., Li, Q., Zhao, J., Li, X., Wang, X., & Liu, H., (2014). Amino-functionalized magnetic cellulose nanocomposite as adsorbent for removal of Cr(VI): Synthesis and adsorption studies. *Chemical Engineering Journal, 241*, 175–183.
146. Hokkanen, S., Repo, E., Suopajärvi, T., Liimatainen, H., Niinimaa, J., & Sillanpää, M., (2014). Adsorption of Ni(II), Cu(II) and Cd(II) from aqueous solutions by amino modified nanostructured microfibrillated cellulose. *Cellulose, 21*(3), 1471–1487.
147. El Nemr, A., El-Assal, A. A. M., El Sikaily, A., Mahmoud, M. E., Amira, M. F., & Ragab, S., (2021). New magnetic cellulose nanobiocomposites for Cu(II), Cd(II) and Pb(II) ions removal: Kinetics, thermodynamics and analytical evaluation. *Nanotechnology for Environmental Engineering, 6*(3), 42.
148. Anirudhan, T. S., Deepa, J. R., & Binusreejayan, (2015). Synthesis and characterization of multi-carboxyl-functionalized nanocellulose/nanobentonite composite for the adsorption of uranium(VI) from aqueous solutions: Kinetic and equilibrium profiles. *Chemical Engineering Journal, 273*, 390–400.
149. Jin, L., Li, W., Xu, Q., & Sun, Q., (2015). Amino-functionalized nanocrystalline cellulose as an adsorbent for anionic dyes. *Cellulose, 22*(4), 2443–2456.
150. Mohammed, N., Grishkewich, N., Waeijen, H. A., Berry, R. M., & Tam, K. C., (2016). Continuous flow adsorption of methylene blue by cellulose nanocrystal-alginate hydrogel beads in fixed bed columns. *Carbohydrate Polymers, 136*, 1194–1202.
151. Azmi, A., Lau, K. S., Chin, S. X., Khiew, P. S., Zakaria, S., & Chia, C. H., (2021). Zinc oxide-filled polyvinyl alcohol–cellulose nanofibril aerogel nanocomposites for catalytic decomposition of an organic dye in aqueous solution. *Cellulose, 28*(4), 2241–2253.
152. Bashir, K., & Aggarwal, M., (2019). Physicochemical, structural and functional properties of native and irradiated starch: A review. *J Food Sci Technol, 56*(2), 513–523.
153. Okuo, J., Emina, A., Omorogbe, S., & Anegbe, B., (2018). Synthesis, characterization and application of starch stabilized zerovalent iron nanoparticles in the remediation of Pb-acid battery soil. *Environmental Nanotechnology, Monitoring & Management, 9*, 12–17.
154. Wang, Y., Fang, Z., Kang, Y., & Tsang, E. P., (2014). Immobilization and phytotoxicity of chromium in contaminated soil remediated by CMC-stabilized nZVI. *Journal of Hazardous Materials, 275*, 230–237.
155. Ahamad, T., Naushad, M., Mousa, R. H., & Alshehri, S. M., (2020). Fabrication of starch-salicylaldehyde based polymer nanocomposite (PNC) for the removal of pollutants

from contaminated water. *International Journal of Biological Macromolecules, 165*, 2731–2738.
156. Gad, E. S., Owda, M., & Mousa, R. Y., (2020). A novel starch nanoparticle citrate based adsorbent for removing of crystal violet dye from aqueous solution. *Egyptian Journal of Chemistry, 63*(6), 2075–2095.
157. Chen, Q., Gao, X., Zhao, Y., Liu, Z., Xie, G., Liang, C., & Wang, J., (2021). Starch nanocrystals grafted with epichlorohydrin dimethylamine for methyl blue dye removal. *Starch-Stärke, n/a* (n/a), 2100200.
158. Kim, H., & Lim, S. I., (2021). Adsorption of antibiotics on serum albumin nanoparticle. *Clean Technology [Internet], 27*(1), 55–60. Available from: https://doi.org/10.7464/KSCT.2021.27.1.55.

CHAPTER 3

Algae-Mediated Freshwater Phytoremediation

BRAULIO EDGAR HERRERA-CABRERA,[1] JORGE MONTIEL-MONTOYA,[2] RAFAEL SALGADO-GARCIGLIA,[3] CÉSAR REYES REYES,[4] LUIS GERMÁN LÓPEZ-VALDEZ,[5] LUZ MARÍA BASURTO GONZÁLEZ,[6] GONZALO GUILLERMO LUCHO-CONSTANTINO,[7] FABIOLA ZARAGOZA MARTÍNEZ,[8] and HEBERT JAIR BARRALES-CUREÑO[6]

[1]*Colegio de Postgraduados, Campus Puebla, Puebla, México*

[2]*Centro Interdisciplinario de Investigación para el Desarrollo Integral Regional, Instituto Politécnico Nacional, Guasave, Sinaloa, México*

[3]*Universidad Michoacana de San Nicolás de Hidalgo, Morelia, Michoacán, México*

[4]*Universidad Intercultural del Estado de Puebla, Huehuetla, Puebla, México*

[5]*Universidad Autónoma Chapingo, Texcoco, Edo. de México*

[6]*Instituto Tecnológico de Estudios Superiores de Zamora, Carrera en Ingeniería en Innovación Agrícola Sustentable, Zamora, Michoacán, México*

[7]*Instituto Tecnológico Superior de Jesús Carranza, Veracruz, México*

[8]*Centro de Investigación y Estudios Avanzados, CINVESTAV*

Phytoremediation and Biofortification: Strategies for Sustainable Environmental and Health Management. Nand Kumar Singh, Shadma Afzal, & Tariq Aftab (Eds.)

ABSTRACT

Algae are a highly diversified group of photosynthetic organisms, varying in size and shape, and occur in almost all marine and freshwater habitats. The removal of heavy metals by conventional methods is costly and inefficient when the concentration of metals is not very high (10 to 50 ppm), so the use of biological systems for the removal of heavy metals from dilute solutions has the potential to do so at lower cost. Algae are considered to have great potential for use in the removal of nutrients (e.g., trace metals) due to their high capacity to accumulate metals, in addition to the fact that they are sensitive to different test materials, their nutritional requirements are known, they have a high growth rate that allows to know in a few days the density and the effect caused by the toxic agent and their handling is relatively simple in the laboratory. Algae are very sensitive to metal toxicity and some species are used as biological biosensors to detect potential toxic effects of heavy metals. Toxic effects in microalgae can be caused by several mechanisms: (i) blocking of functional groups of biologically important molecules such as enzymes in the essential nutrient and ion transport system; (ii) displacement or substitution of essential metal ions from biomolecules and cellular functional units. The above may result in modification and inactivation of enzymes or disruption of cell membrane integrity. Therefore, the objectives of this chapter deal with the main algae used in the phytoremediation of heavy metals, as well as the main mechanisms of molecular interaction and biosorption.

3.1 INTRODUCTION

In freshwaters, the term algae refer to microscopic organisms, in principle unicellular, some of which form colonies and reach dimensions visible to the naked eye. Some of which form colonies and reach dimensions visible to the naked eye as tiny green particles [1]. Generally, these organisms are finely dispersed throughout the water, and at high densities, can produce considerable turbidity. Cyanobacteria are organisms that possess characteristics of bacteria and a little of algae. They resemble algae in size and unlike other bacteria, contain blue-green or green pigments and therefore photosynthesize [2].

Therefore, they are also referred to as 'algae,' they are also referred to as 'blue-green algae.' Unlike most other algae, many species of cyanobacteria species of cyanobacteria can accumulate in surface foams, usually referred to as 'blooms.' called 'blooms,' with an extremely high cell density. Biosorption is defined as a technology for the removal of metals and metalloids from effluents by biological materials [3]. Large amounts of metals can accumulate in biological materials through a wide variety of metabolism-dependent or metabolism-independent processes. Biosorption utilizes living or dead biomasses such as bacteria, fungi, yeasts, algae, as well as cellular products such as polysaccharides. The mechanisms of sequestration of heavy metals from solution by biomaterials are mainly due to physical adsorption, complexation, and ion exchange [4].

During the biosorption process, interactions occur between the metal and the functional groups present in the cell wall of the biomaterials. Biosorption is one of the most promising technologies, not only because of its low cost, but also because it is a fast process that can treat large volumes of water with low metal concentrations efficiently [5]. It is a fast process that allows treating large volumes of water with low metal concentrations in an efficient way. In addition, the possibility of using dead biomass or products derived from its metabolism overcomes toxicity problems and even allows the regeneration and reuse of the biomaterial through several adsorption/desorption cycles. Microalgae are used in the elimination of toxic substances such as heavy metals, herbicides, and pesticides, among others, due to their capacity to accumulate significant concentrations of toxic compounds without affecting their biological activity [6]. Another benefit of microalgae is their tendency to release oxygen through photosynthesis, degrading organic compounds and generating an accumulation of biomass, which can be reused in biofuel production systems and/or in poultry, aquaculture or agricultural processes [7]. Therefore, the objectives of this chapter deal with the main algae used in the phytoremediation of heavy metals, as well as the main mechanisms of molecular interaction and biosorption.

3.2 BIOSORPTION

The history of research on the interactions between metals and biomass dates back to the 1960s. In the environment it was found that when soluble chemical compounds are present in aqueous environments, they make contact and interact with biological materials in two ways, they bind to the cell wall in a process called biosorption or they can accumulate inside the cell in a process called bioaccumulation [8].

Both biosorption and bioaccumulation participate in the matter cycle and occur all the time in nature, humans can interpret them and find an industrial application under controlled operating conditions. Biosorption is defined as a sorbate concentration process and the prefix "bio" refers to the fact that the sorbent is of biological origin, therefore the adsorption surface has a specific chemical-biological composition that will depend on the biological material used. The process is simple to operate and is very similar to conventional adsorption or ion exchange columns. It is highly effective in the treatment of dilute effluents [9, 10]. The only difference is that the material used is biological, metabolically inactive, selective, and regenerable [11]. While the bioaccumulation process involves a first stage which is biosorption and then followed by other states that have to do with the transport of pollutants through an active transport system that involves the consumption of energy inside the cell with its consequent increase in concentration in the cells. Biosorption has certain advantages over bioaccumulation processes [8] (Table 3.1).

TABLE 3.1 Comparison between Biosorption and Bioaccumulation Processes

Biosorption	Bioaccumulation
Passive process	Active process
Lifeless biomass	Biomass with life
Metals bind to the surface of the cell wall.	Metals bind to the surface of the cell wall and also accumulate inside the cell.
Adsorption	Absorption
Reversible process	Partially reversible process
No nutrients required	Nutrient requirements
A one-step process	Two-step process
Fast	Slow
Not controlled by metabolism.	Controlled by metabolism.
Unaffected by the toxic effect of pollutants.	It is affected by the toxic effect of pollutants.
No cell growth	Involves cell growth
Reaches intermediate equilibrium concentrations of the contaminants.	Reaches very low equilibrium concentrations of the pollutants.
Possibility of recovery and reuse of pollutants through a desorption process.	Metals cannot be recovered
The biomass can be regenerated and used in various adsorption-desorption cycles.	Biomass cannot be recovered

Source: Ref. [8].

The use of a metabolically active organism (bioaccumulation) can be affected by high concentrations of the contaminant, interrupting the process by their death. This can be overcome by using dead biomass (biosorption), which is flexible to different environmental conditions and contaminant concentrations.

3.2.1 METAL RETENTION MECHANISMS

A set of mechanisms show the sequestration or retention of metals in different areas of the biosorbent. The mechanisms of metal accumulation are physical adsorption, chelation or complexation, ion exchange, precipitation. In the case of physical adsorption, this is related to phenomena related to Vander Waals forces, being the adsorption reversible and efficient. In the case of chelation, it consists of the binding of the metal to the active centers of the cell walls by means of chemical bonds with the formation of complexes. Ion exchange occurs when metal ions are exchanged with ions of the polysaccharides that are part of the biomass. The process is usually reversible and efficient. Finally, precipitation involves the association of a complex forming on the cell wall, which is then hydrolyzed. In the case of biosorption, it can occur by more than one of the above mechanisms [13]. According to Tsezos [14], several stages are proposed in the adsorption mechanism related to the transfer of the metals through the layers bordering the cells: the following cells [14].

- Transport of the metal from within the solution to the boundary layer around the cell wall;
- Transport from the boundary layer to the cell wall;
- Transport of the metal from the cell wall to the active points of the bond;
- Binding phases: complexation, adsorption, and precipitation in the intramembrane.

For the study of the mechanisms mentioned above, it is necessary to have information on the physical and chemical structure of the cell wall of the biomass. The cell wall of the different types of biosorbents (fungi, algae, higher plants, bacteria) differs significantly from each other, in terms of the functional groups present, type, and size of pores, polysaccharide chains, etc. To establish the role of the different functional groups in the biosorption process through ion exchange, adsorption, complexation, etc., the concept of strong and weak acids and bases and the pKa values of the different functional groups present in the biomass are used. More advanced instrumental analysis such as ESEM-EDAX, FTIR, each of which reveals some information as to structural changes and/or functional groups involved during biosorption, can also be performed which may help to elucidate the possible mechanisms involved [15].

3.3 HEAVY METAL PHYTOREMEDIATION

3.3.1 FACTORS AFFECTING THE BIOSORPTION PROCESS

Knowledge and analysis of the factors affecting the efficiency of the biosorption process are important to optimize the operating conditions of the process in order to biosorption process are important for optimizing the operating conditions of the process itself, as well as for the regeneration of the biosorbent itself, as well as for the regeneration of the biosorbent. One of the most important factors is pH. Describing the effect of pH requires an understanding of the chemistry of the solution (hydrolysis, complexation, redox reactions, precipitation) (Figure 3.1) [16].

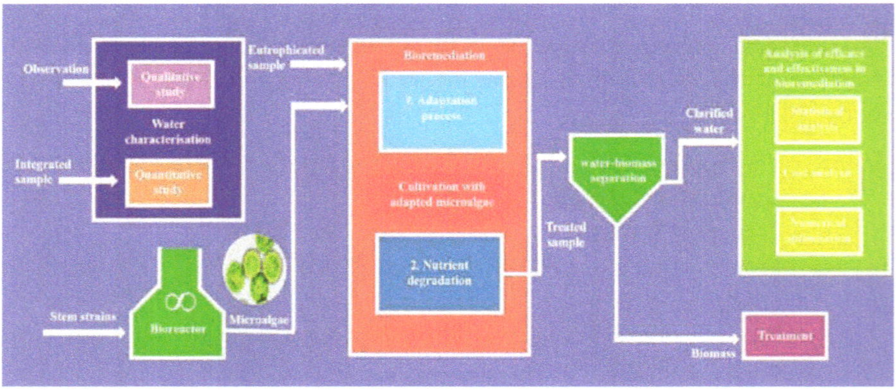

FIGURE 3.1 Stages developed in the process of bioremediation of toxic metals by algae.

Source: Corresponding author.

It determines the speciation and solubility of metal ions and also affects biomass properties by protonation and deprotonation of their active sites [17] which are available to metals in their deprotonated state. The effect of temperature on biosorption depends on the enthalpy change that accompanies the adsorption process. If physical adsorption is involved the $\Delta H < 0$, therefore the adsorption reaction is exothermic and occurs preferentially at low temperatures. If the adsorption is chemical, the $\Delta H > 0$, then the adsorption reaction is endothermic and is favored at high temperatures. In general, the energy for adsorption of heavy and light metals lies in the following range 7–11 kJ.mol^{-1} and 2.1–6 kJ.mol^{-1}, respectively [18]. These ranges are so that the effect of temperature is negligible for the practical purposes of applying a biosorption application of a biosorption process. Finally, two further factors must be taken into account: the ionic strength ionic strength and the presence of other ions. An increase in ionic strength suppresses biosorption, as a result of the increased electrostatic charge, and the presence of other ions in solution can presence of other ions in solution may compete with the metal of interest for binding sites. Some examples of algae with heavy metal accumulation capacity are shown in further subsections.

3.3.2 CADMIUM (CD) ACCUMULATING ALGAE

The global emission of cadmium into the environment comes mostly from anthropogenic sources, being associated with Zn extraction and wastes from plastics, paints, metal alloys and other industrial operations. Cadmium has been recognized as a hazardous environmental pollutant and is accumulated by phytoplankton. In the sea it is found in concentrations of 0.04 μg l^{-1} and 78% of it is in organic form. Meanwhile in wastewater cadmium can be found in concentrations of 0.6 μg l^{-1}. Some species that accumulate cadmium are the algae *Chlamydomonas reinh ardtii*, *Chlorella regularis*, *Scenedesnus bijuga*, *Scenedesmus obliquus*, *Chlamydomonas angulosa* and *Scenedesmus choreloides* [19]. Nagano et al. [20] reported that *Chlorella ellipsoidea* can accumulate cadmium with a concentration coefficient around 166. Also, *Chaetomorpha brychagona* and *Enteromorpha crinita* accumulate high concentrations of zinc from polluted waters [20].

3.3.3 ARSENIC ACCUMULATING ALGAE

Chlorella vulgaris accumulates high levels of arsenic, as well as other metals and it is because it has detoxification mechanisms by which it transforms the metal it is accumulating. This group of algae could be used to arsenic concentrations in natural waters and in industrial wastewater. Other genera of freshwater (*Chlorella* sp. and *Oscillatoria* sp.) and seawater (*Phaeodactyla* sp.) microalgae could be used to control arsenic concentrations in natural and industrial wastewater. *Phaeodactylum* sp. and *Skeletonema* sp., accumulated arsenic from an aqueous solution containing 1–30 mg L^{-1} arsenic, with a concentration factor of 240–2,800 and 710–2,900, respectively [21].

3.3.4 URANIUM ACCUMULATING ALGAE

Sakaguchi et al. [23] demonstrated that some species of microalgae, including *Chlorella regularis*, *Chlamydomonas* sp. and *Scenedesmus* sp. can absorb uranium [23]. Edgington et al. [24] examined uranium accumulation in 20 species of microalgae, finding that those with the highest uranium concentration factor are: *Dictyoca divaricata*, *Halimeda opuntia*, and *Penicillus capitatus* [24]. Greene et al. [25] found that the cells of *Chlorella vulgaris* accumulate significant amounts of uranium(VI) [25].

3.3.5 CROMO ACCUMULATING ALGAE

Chlamydomonas sp. acclimated in high concentrations of chromium, shows a great ability to remove this metal [26]. Also, *Chlorella pyrenoidosa* and Scenedesmus obliquus can accumulate chromium from wastewater (50 μg l^{-1}) with an accumulation factor of 320 and 230, respectively [27].

3.4 CELL WALL COMPONENTS OF DIFFERENT TYPES OF BIOMASSES RELEVANT TO BIOSORPTION PROCESSES

The composition of the cell wall of Chlorophyta algae (green algae) contains a CL fibrillar component and an amorphous matrix amorphous matrix composed of polysaccharides such as, galactose, ribose, xylose, glucuronic acid, which can be sulfonated. Proteins represent 10 to 70% of the dry weight (DW) of the of the cell wall. Provides functional groups such as hydroxyl, carboxyl, carboxyl, sulfhydryl such as hydroxyls, carboxyls, sulfhydryls, amines, imines, amides, imidazoles. The composition of the cell wall of the Rhodophyta (red algae) is composed of CL fibrillar component and the main components of the main components of the amorphous matrix are sulfonated polysaccharides such as fucoids, galactases (agar, carrageenans, etc.) constituting up to 70% of the DW of their cell wall. Proteins account for 35 to

50% of the cell wall. The composition of the cell wall of Phaeophyta (brown algae) is composed of a CL fibrillar component and an amorphous alginate matrix alginate matrix, which accounts for 10–40% of the DW. 10–40% of the DW. Proteins are present in are present in 35–50% [22] (Figure 3.2).

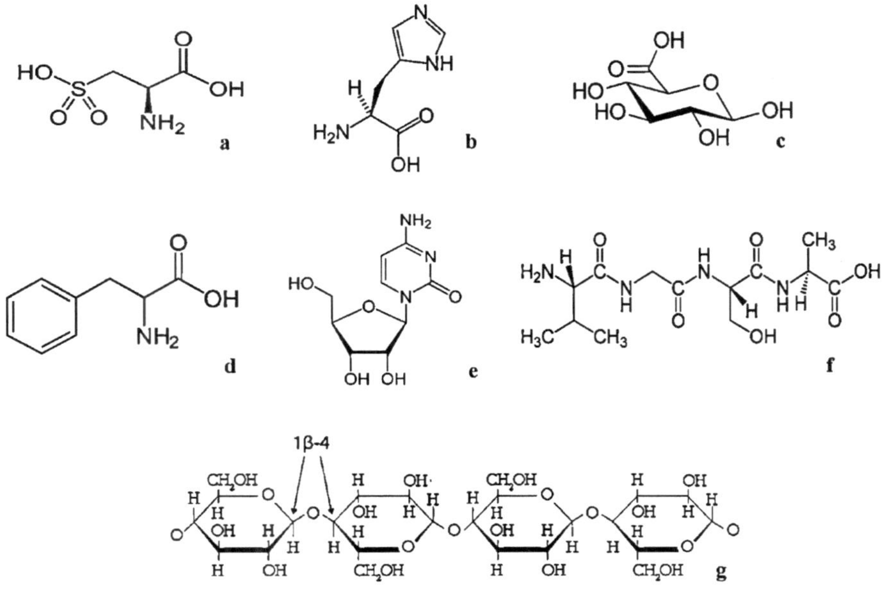

FIGURE 3.2 Functional groups present in the cell wall of algae involved in heavy metal adsorption heavy metal. (a) Sulfate in cysteic acid (169.16 g/mol, $C_3H_7NO_5S$, pKa 1.3); (b) imidazol in histidine (155.16 g/mol, $C_6H_9N_3O_2$, pKa 6–7); (c) carboxylic acid in uronic acid (194.14 g/mol, $C_6H_{10}O_7$, pKa 3–4.4); (d) hydroxyl in phenol-tyrosine (181.19 g/mol, $C_9H_{11}NO_3$, pKa 9.5–10.5); (e) amino in cytidine (243.21 g/mol, $C_9H_{13}N_3O_5$, pKa 4.1); (f) imino in peptide (pKa 13); and (g) phosphate in polysaccharides.

Source: Corresponding author.

Nakajima et al. [28] indicated that several green microalgae can accumulate copper, because they show tolerance to this metallic ion (*Chlorella vulgaris*, *Chlamydomonas angulosa*, and *Scenedesmus bijuga*) [28]. There is a history of using microalgae to remove molybdenum from aqueous systems [29]. Species such as *Scenedesmus chlorelloides*, *Chlamydomonas reinhardtii*, *Chlorella regularis*, *Scenedesmus bijuga*, *Chlamydomonas angulosa*, and *Scenedesmus obliquus* were found to be efficient for molybdenum accumulation. About 15 species belonging to 6 phyla

(Chlorophycea, Cryptophycea, Xanthophycea, Bacillariophycea, Chrysophycea, and Prasinophycea) accumulate aluminum from artificial seawater [30]. The freshwater microalga Chlorella vulgaris has a high affinity for gold, so this microalga can accumulate both gold(I) and gold(III) from aqueous solutions [19].

Microalgae exposed to cadmium and under cadmium stress exhibit a variety of strategies to mitigate the adverse effects of the metal [32]. Precisely, there is previous information on the emergence of several microalgae genera (Stigeoclonium, Chlorella, Chlamydomonas, and Scenedesmus), resistant to exposure to different toxic metals (e.g., Cd, Pb, Cr) [33]. In other studies, the sensitivity of *Dictyosphaerium chlrelloides* the sensitivity of *Dictyosphaerium chlrelloides* to chromium exposure as Cr(VI) was exposure to chromium as Cr(VI), showing massive cell destruction in most of the cellular mass destruction in most of the population, however, some resistant variant strains with viable However, some resistant variant strains with viability to Cr(III)-contaminated environments were found [34]. Among the basic mechanisms induced by exposure to Cd, intracellular chelation through polypeptides such as Cd-metallothioneins (MTs) such as class III metallothioneins (Mt[III]).

3.4.1 MT(III) EXPRESSION

One of the best-defined mechanisms in protists, plants, algae, and some fungi for protists, plants, algae, and some fungi on the process of toxic metal detoxification by toxic metals, is through metal chelation via induction of reduced metal chelation via induction of reduced glutathione (GSH) [35] and in the synthesis of class III metallothioneins (Mt[III]) via phytochelatin synthase (PCS), an enzyme originally discovered and characterized originally discovered and characterized in the plant *Silene cucubalus* by Grill et al. [36] with Cd being the major inducer of this Cd being the major inducer of this system (also activated by Ag, Pb, Cu, Cu, Cd, and by Ag, Pb, Cu, Hg, Au, Zn) [37].

The first studies on the identification of Mt(III) isoforms in microalgae were carried out in 1977 by Stokes et al. [38] on the Chlorophyceae *Scenedesmus acutiformis*. Subsequently, Mt(III) isoforms have been identified in Mt(III) have subsequently been identified in other microalgae species in the genus *Stichococcus* and in diatoms such as *Thalassiosira* [39]. These proteins can bind to a wide variety of metals in the cytosol, and, depending on the organism and, depending on the organism, Mt(III)-metal and GSH-metal complexes are actively transported into the vacuole as first observed in vacuole as was first observed in the microalgae *Dunaliella bioculata* [40]. Like other metallothioneins (e.g., Mt(II)), phytochelatins (PCs) function as apoenzymes with function as apoenzymes with consequent activation by metal ion cofactors with essential metal ion cofactors and are involved in the detoxification of toxic metals [41]. Mt(III) complexes according to the inclusion of sulfide ions are divided into two categories: (i) the low molecular weight (LMW) form where the metal has a preference for binding to thiol groups (-SH); and (ii) the high molecular weight complex (HMW) where inorganic sulfide ions (S inorganic sulfide (S_2^-) ions are incorporated to form nanoparticles (biomineralization) [42].

Thus, the formation of Mt(III) can take place mainly by three processes: (i) Mt(III) are synthesized in the cytosol where they synthesized in the cytosol where they sequester Cd(II), the Cd-Mt(III) complexes are then Cd-Mt(III) complexes are subsequently transported to the chloroplast and mitochondria forming the mitochondria forming LMW complexes; (ii) Mt(III) are synthesized within the vacuole where they bind to Cd which is transported as a free ion to form LMW complexes is transported as a free ion to form HMW complexes; and (iii) both processes co-exist and Mt(III) are synthesized in all three cellular compartments three cellular compartments [43]. Previously, it has been shown that Mt(III) synthesis is related to the mechanism of cadherin resistance related to the mechanism of cadmium resistance and an increased in photosynthetic protists such as *Euglena gracilis* [37].

In previous genetic studies, such as the one conducted by Ha et al. [45] it was shown that PC-deficient mutants of cruciferous plants of the genus Arabidopsis, behaved as hypersensitive to Cd. However, it has been confirmed that the synthesis of Mt(III) synthesis is not induced under heat, cold, ultraviolet radiation (UV), anoxia and oxidative stress [46]. In contrast, microorganisms typically undergo a two-stage process for the entry and accumulation of metal ions. This process involves an initial rapid, reversible, and passive adsorption onto the cell surface. Metal ions are adsorbed through electrostatic interactions with functional groups present on the cell wall, such as OH^-, SH^-, COO^-, $PO_4{}^{3-}$, NO_3-RNH_2^-, RS^-, and RO^- [47]. Followed by a much slower, irreversible process, which involves the active transport of metal cations across the cell membrane in the cytoplasm the cell membrane into the cytoplasm, with subsequent binding to intracellular biomolecules [48] (Table 3.2).

TABLE 3.2 Examples of Results Obtained in the Adsorption of Certain Heavy Metals with Different Types of Biomasses

Organism	Biomass	Metal	Amount Adsorbed (mg.g–1)	References
Algae	*Rhizopus nigricans*	Pb	166	[49]
–	*Fucus vesiculosus*	Pb	270–371	[50]
–	*Ascophyllum nodosum*	Cd	215	[51]
–	*Spirogyra* sp.	Pb	140	[52]
–	*Apanothece halophutica*	Zn	133	[53]
–	*Sargassum* sp.	Zn	118	[54]

3.5 FRESHWATER ALGAE

Chlorella vulgaris is a spherical, unicellular, eukaryotic Chlorophyta, with chlorophyll a and b has chlorophyll a and b. It can be found in marine and freshwater environments, due to its cell wall being its cell wall is composed of a complex mixture of sugars, glucosamine, proteins, and uric acid. This microalga is capable of incorporating large amounts of metals amounts of metals ($Cr^{2+,3+,6+}$, $Fe^{2+,3+}$, $Cu^{1+,2+}$, Zn^{2+}, $Pb^{2+,4+}$, and $Hg^{1+,2+}$) by means of absorption and accumulation. Scenedesmus obliquus, like Chlorella vulgaris, are freshwater microalgae that form colonies of up to 16 cells. The cells are generally eukaryotic cylindrical and have at each end a spine up to 200 μm long. *Synechococcus* sp., unlike the previous two, are prokaryotic cells. As a cyanobacterium being a cyanobacterium, it has metabolic and ecophysiological versatility, which allows it to adapt to extreme conditions of light, temperature, salinity, and nutrients [55].

3.5.1 ADSORPTION CENTERS OF MARINE ALGAE

The effectiveness of algae in selectively adsorbing heavy metal cations is due to the presence of heavy metal cations is due to the presence of high electron density functional groups high electron density functional groups: alginates and fucoidans. The alginates are mainly responsible for biosorption [56].

Alginate belongs to the polysaccharide family of family of linear polysaccharides composed of 1,4-β-D mannuronic acid (M) and α-L guluronic acid (G) units arranged in an irregular manner, with the sequences MM, GM, MG, and GG sequences coexist. This particular geometry confers the property of melting in the presence of Calcium(II) ions. The gel formation of alginates is favored by the dimerization of alginate chains in solution, forming the so-called "egg box" [57] where divalent cations are "trapped" within the dimeric structure of the alginate. It should also be noted that the most accessible active centers for divalent cations are the accessible for divalent cations are found in the sequence, GG, which has the sequence, GG, which has a concave geometry and allows the formation of chelate formation [58].

The second important functional groups in algae are the fucoidans, which are polysaccharides with pKa-esterified with esterified sulfate groups with pKa of around 1 to 2.5. Their adsorption capacity is relatively low, their effectiveness is manifested at low pH, through deprotonation of their acidic groups. Thus, while many adsorbents, such as chitosan (CS), chitin, and fungi, have almost zero adsorption capacities at low pH, their effectiveness is manifested at low pH by deprotonation of their acidic groups almost zero adsorption capacity at pH less than 3 [59, 60] seaweeds still maintain some adsorptive capacity for heavy metals under these conditions, even at values close to heavy metals under these conditions, even at values close to 2 [61, 62].

Fucanoids could be a powerful tool against extreme (strongly acidic) wastewater conditions as they adsorb toxic metals under these conditions, but unfortunately, the alginate polysaccharide alginate polysaccharide chains (glycosidic linkages) are unstable and hydrolysis is a limiting factor for their extensive use, deteriorating their mechanical properties by increasing their solubility in water as a result of their fractionation. Modifications of seaweeds in Biosorption From an economic point of view, it is important to know the modifications to which seaweeds modifications to which marine algae are subjected in order to enhance their adsorption capacity or thermal-mechanical properties as well as their selectivity. The main objective of chemical modification of seaweeds is to facilitate the contact of heavy metal ions with the functional groups responsible for adsorption, and to create new biopolymer chains or to cross-link them [22, 64].

Among the main modifications, the following stand out major modifications include:

1. **$CaCl_2$:** It causes cross-linking of poly-alginate chains [65]. Formaldehyde and glutaraldehyde: also cause cross-linking of poly-alginate chains between adjacent functional groups, mainly hydroxyl groups.
2. **Formaldehyde and Glutaraldehyde:** Also cause cross-linking of poly-alginate chains between adjacent functional groups, mainly the

hydroxyl groups [66]. NaOH: generally, this treatment decreases the protonation of adsorbent functional groups.

3. **NaOH:** This treatment generally decreases the protonation of adsorbent functional groups. It substitutes sodium ions in the functional groups, increasing the electrostatic attraction to heavy metal cations, facilitating the heavy metal cations, facilitating ion exchange [67]. HCl: Acid washing leads to a reduction of the sodium ionization.
4. **HCl:** Acid washing leads to the displacement of light metals by protons in the light metals by protons on the adsorbent surface.

In addition, acid treatment can dissolve damaged outer polysaccharides from the outer cell wall, creating new and additional adsorption centers [68]. However, the modification of the biomass of the adsorbent can also lead to a reduction of the adsorption capacity of the adsorbent. However, biomass modification does not always improve adsorption capacity. Sometimes, heavy metal removal capacity decreases due to the solubilization of different forms of heavy metals. Solubilization of the different forms of the alginate, as has been shown [69] in the presence of NaOH.

From the results, it has been shown that the chemical modification is not cost-effective and does not result in the removal of heavy metals lead to noticeable increases in the adsorption capacity of the algae. It should be noted that such an investment in chemical modification could only be beneficial when it is desired to improve the physical stability of the biomass, especially in continuous biomass, especially in continuous processes (columns).

3.6 INSTRUMENTAL TOOLS AND TECHNIQUES USED IN BIOSORPTION STUDIES OF METALS

Biosorption studies have been carried out using numerous analytical techniques under different modalities or modifications, including equilibrium study, mechanism elucidation, adsorbent characterization, adsorption kinetics, etc.; the main analytical techniques include atomic absorption spectroscopy (AAS), ion selective electrode (ISE), Fourier transform infrared spectroscopy (FTIR), scanning electron microscopy (SEM), electron titration electron microscopy (SEM), potentiometric, and conductometric titrations, transmission electron microscopy (TEM), surface determination by Brunauer-Emmet-Teller (BET) and X-ray photoelectron spectroscopy.

Fourest and Volesky found that cadmium ions and alginate carboxyl groups alginate carboxyl groups form a bidentate complex using FTIR [70].

Ramos et al. also determined the functional groups involved in the adsorption of cadmium ions on Lentinus edodes by FTIR through changes in the vibrational modes and bathochromic displacement of cadmium ions vibrational modes and bathochromic displacement of amides and methylenes [71].

Doshi et al. SEM, that copper ions adhere to the surface of the adsorbent, forming adsorbent surface forming adsorption nuclei and do not rule out the penetration of ions into the adsorbent. Furthermore, for the adsorption of chromium for the adsorption of hexavalent chromium [63].

The characterization of *Sargassum* sp. was conducted using scanning electron microscopy (SEM), revealing a surface with microtubules. The presence of these microtubules could be attributed to the deposition of calcium and other crystalline salts. This observation provides supporting evidence for the hypothesis of a potential ion exchange occurring between divalent heavy metals and calcium ions, both on the surface of the adsorbent and within its structure following the intraparticle diffusion of pollutants through its pores. Ionization constants of marine algae have been determined by potentiometric and conductometric titrations [31, 44], this information is of utmost importance because, complementing the FTIR results, it allows for the identification of acidic functional groups that may be involved in the committed to the adsorption process. It is necessary to is necessary to know the pH at which the biomass is deprotonated, because it is under these conditions where adsorption is maximized (maximum electrostatic interactions) [62].

The results obtained show that functional groups results show that the functional groups of marine algae ionize algae ionize at pH 4 and above, a fact that is confirmed by the lack of maximum adsorption values for marine algae in the literature at pH below 4 do not exist in the literature. Finally, aware of the fact that biosorption is a physical-chemical adsorption process chemical adsorption process, the surface area of the adsorbent also plays an important surface area of the adsorbent also plays an important role in its efficiency in its efficiency for the removal of heavy metals. The BET technique, by means of nitrogen adsorption, makes it possible to determine the exact the exact accessible surface area of an adsorbent and to make comparisons with other available biomaterials can be made.

3.6.1 CELLULAR MECHANISM OF THE BIOSORPTION OF METAL IONS IN MICROALGAE

Among the molecular mechanisms that determine the toxicity of heavy metals include:

- Displacement of essential metal ions from biomolecules and blocking of their functional groups (Figure 3.3);
- Modification of the active conformation of biomolecules, especially enzymes and polynucleotides;
- Disruption of the integrity of biomolecules; and
- Modification of other biologically active agents [12].

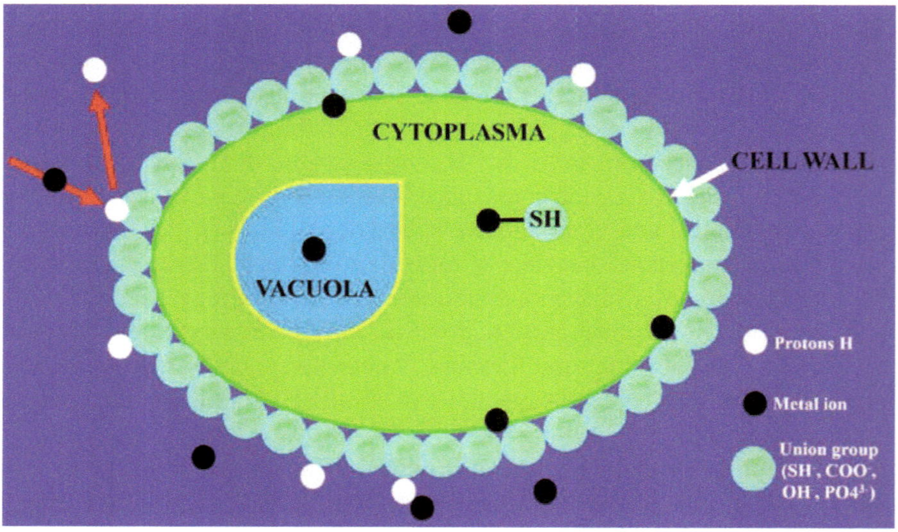

FIGURE 3.3 Cellular mechanism and different interacting functional groups in the biosorption of metal ions in microalgae.

Source: Corresponding author.

3.7 CONCLUSION

Phytoremediation allows improving the quality of freshwater in a natural way, which means avoiding the use of chemical products, in an effective, safe, and profitable way. However, the choice of algae is key to the preservation or recovery of different ecosystems, so it is important to take care of this aspect. Unlike other decontamination systems, which consist of using chemicals or moving the problem to another location, phytoremediation represents an environmentally friendly solution. Specifically, algae can clean the water of heavy metals at low cost and in a comprehensive manner. Notwithstanding the fact that there are really expensive projects. Algae-assisted phytoremediation is a particularly attractive bioalternative due to their high photosynthetic capacity, allowing them to convert solar energy into biomass, which incorporates nutrients such as nitrogen and phosphorus that cause eutrophication. Processes using algae and cyanobacteria are mainly focused on the removal of nutrients, radioactive elements from effluents and mainly heavy metals. Considering the potential of microalgae in phycoremediation, plus the multiple uses of microalgae biomass, it is suggested that their cultivation is economically and environmentally attractive.

KEYWORDS

- **biosorption**
- **cell wall**
- **freshwater algae**
- **heavy metal phytoremediation**
- **heavy metals**
- **metal retention mechanisms**
- **phytoremediation**

REFERENCES

1. Timsit, Y., Lescot, M., Valiadi, M., & Not, F., (2021). Bioluminescence and photoreception in unicellular organisms: Light-signaling in a bio-communication perspective. *International Journal of Molecular Sciences, 22*(21), 11311. https://doi.org/10.3390/ijms222111311.
2. Okechukwu, I., Okogwu, O., & Ugwumba, A., (2009). Cyanobacteria abundance and its relationship to water quality in the mid-cross river floodplain, Nigeria. *Revista de Biología Tropical, 57*(1, 2), 33–43. http://www.scielo.sa.cr/scielo.php?script=sci_arttext&pid=S0034-77442009000100004&lng=en&tlng=en (accessed on 22 May 2023).
3. Elgarahy, M. A., Elwakeel, K. Z., Mohammad, S. H., & Elshoubaky, G. A., (2021). A critical review of biosorption of dyes, heavy metals and metalloids from wastewater as an efficient and green process. *Cleaner Engineering and Technology.*

4. Mosa, K. A., Saadoun, I., Kumar, K., Helmy, M., & Dhankher, O. P., (2016). Potential biotechnological strategies for the cleanup of heavy metals and metalloids. *Frontiers in Plant Science, 7*, 303. doi: 10.3389/fpls.2016.00303.
5. Bilal, M., Rasheed, T., Sosa-Hernández, J. E., Raza, A., Nabeel, F., & Iqbal, H., (2018). Biosorption: An interplay between marine algae and potentially toxic elements: A review. *Marine Drugs, 16*(2), https://doi.org/10.3390/md16020065.
6. *Saravanan, A., SenthilKumar, P., Hemavathy, R. V., Jeevanantham, S., Harikumar, P., Priyanka, G., & Devakirubai, D. R. A., (2022). A comprehensive review on sources, analysis and toxicity of environmental pollutants and its removal methods from water environment.* Science of The Total Environment, *812*.
7. Mehrabadi, A., Craggs, R., & Farid, M. M., (2015). Wastewater treatment high-rate algal ponds (WWT HRAP) for low-cost biofuel production. *Bioresources Technology, 184*, 202–214. doi: 10.1016/j.biortech.2014.11.004.
8. Chojnacka, K., (2010). Biosorption and bioaccumulation-the prospects for practical applications. *Environ. Int., 36*, 299–307.
9. Orhan, Y., Hrenovic, J., & Büyükgüngör, H., (2006). Biosorption of heavy metals from wastewater by biosolids. *Eng. Life Sci., 6*, 399–402.
10. Norton, L., Baskaran, K., & McKenzie, T., (2004). Biosorption of zinc from aqueous solutions using biosolids. *Environ. Res., 8*, 629–635.
11. Gadd, G. M., (2009). Biosorption: Critical review of scientific rationale, environmental importance and significance for pollution treatment. *J. Chem. Technol., 84*, 13–28.
12. Ochiai, E. I., (1987). *General Principles of Biochemistry of the Elements* (p. 648). Plenum Press, New York.
13. Ho, Y. S., Ng, J. C. Y., & Mackay, G., (2001). Removal of lead (II) from effluents by sorption on peat using second order kinetics. *Sep. Purif. Technol., 36*, 241–261.
14. Tsezos, M., (2001). Biosorption of metals. The experience accumulated and the outlook for technology development. *Hydrometallurgy, 59*, 241–243.
15. Michalak, I., Mironiuk, M., & Marycz, K., (2018). A comprehensive analysis of biosorption of metal ions by macroalgae using ICP-OES, SEM-EDX and FTIR techniques. *PLoS One, 13*(10), e0205590. https://doi.org/10.1371/journal.pone.0205590.
16. Ofomaja, A. E., & Ho, Y. S., (2007). Effect of pH on cadmium biosorption by coconut copra meal. *Journal of Hazardous Materials, 139*, 356–362.
17. Xu, H., Liu, Y., & Tay, J. H., (2006). Effect of pH on nickel biosorption by aerobic granular sludge. *Bioresource Technology, 97*, 359–363.
18. Volesky, B., Weber, J., & Park, J. M., (2003). Continuous-flow metal biosorption in regenerable sargassum column. *Water Research, 37*(2), 297–306.
19. Maeda, S., & Sakaguchi, T., (1990). Accumulation and detoxification of toxic metal elements by algae. En: Akatsuka, I., (ed.). *Introduction to Applied Phycology* (pp. 137–149). SBS Academic Publishing. The Netherlands.
20. Nagano, T., Hattori, S., Nagai, T., Ukishim, A. Y., Unno, C., & Honm, A. T., (1977). Studies on accumulation of heavy metals by chlorella. II. Uptake of cadmium by *Chlorella ellipsoidea* (C-27) and its distribution in chlorella cells. *J. Hyg. Chem., 23*, 1–6.
21. Lunde, G., (1973). The synthesis of fat and water soluble arseno organic com pounds in marine and limnetic algae. *Acta Chem Scand., 27*, 1586–1594.
22. Davis, T. A., Volesky, B., & Mucci, A., (2003). A review of the biochemistry of heavy metal biosorption by brown algae. *Water Research, 37*, 4311–4330.

23. Sakaguchi, T., Horikoshi, T., & Nakajim, A. A., (1978). Uptake of uranium from sea water by microalgae. *J. Ferment. Technol., 56*, 561–565.
24. Edgington, D. N., Gordon, S. A., & Thomm, E. S., (1970). The concentration of radium, thorium, and uranium by tropical marine algae. *Limnol. Oceanogr., 15*, 945–955.
25. Green, B., Henzl, M. T., & Darnall, D. W., (1986). Elimination of bicarbonate interference in the binding of U (VI) in waters to freeze-dried *Chlorella vulgaris. Biotechnol. Bioeng., 28*, 764–767.
26. Ryder, R. A., (1979). Chromium tannery wastewater effects on algae oxidation pond and physical-chemical processes. *Process Industrial Waste Conference, 33*, 706–737.
27. Becker, E. W., (1983). Limitations of heavy metal rem oval from wastewater by means of algae. *Water Res., 17*, 459–466.
28. Nakajima, A., Horikoshi, T., & Sakaguchi, T., (1981). Studies on the ac cumulation of heavy metal elements in biological systems. Selective accumulation of heavy metal ions by *Chlorella regularis. Eur. J. Appl. Microbiol. Biotech., 16*, 88–91.
29. Sakaguchi, T., Nakajim, A., & Horikoshi, T., (1981). Studies on the accumulation of heavy metals elements in biological system. Accumulation of molybdenum by green microalgae. *Eur. J. Appl. Microbiol. Biotech., 12*, 84–89.
30. Madena, S., Nakashim, S., Takeshita, T., & Higashi, S., (1985). Bioaccumulation of arsenic by freshwater algae and the application to the removal of inorganic arsenic from an aqueous phase. Part 2. by *Chlorella vulgaris* isolated from arsenic-polluted environment. *Science, 20*, 153–161.
31. Murphy, V., Hughes, H., & Mcloughlin, P., (2007). Cu(II) binding by dried biomass of red, green and brown macroalgae. *Water Res., 41*, 731−740.
32. Zouari, M., Ben, A. C., Elloumi, N., & Ben, R. B., (2016). Impact of proline application on cadmium accumulation, mineral nutrition and enzymatic antioxidant defense system of *Olea europaea* L. cv chemlali exposed to cadmium stress. *Ecotoxicology and Environmental Safety., 128*, 195–205.
33. Lavoie, I., Campeau, S., Zugic-Drakulic, N., Winter, J. G., & Fortin, C., (2014). Using diatoms to monitor stream biological integrity in Eastern Canada: An overview of 10 years of index development and ongoing challenges. *Science of the Total Environment, 475*, 187–200.
34. Pereira, M., Bartolomé, M. C., & Sánchez-Fortún, S., (2014). Photosynthetic activity and protein overexpression found in Cr(III)-tolerant cells of the green algae *Dictyosphaerium chlorelloides. Chemosphere, 108*, 274–280. doi: 10.1016/j.chemosphere.2014.01.052.
35. Pantoja, M. L., Purchase, D., Jones, H., Raab, A., Urgast, D., Feldmann, J., & Garelick, H., (2016). The mechanisms of detoxification of As(III), dimethylarsinic acid (DMA) and As(V) in the microalga *Chlorella vulgaris. Aquat. Toxicol., 175*, 56–72. doi: 10.1016/j.aquatox.2016.02.020.
36. Grill, E., Löffler, S., Winnacker, E. L., & Zenk, M. H., (1989). Phytochelatins, the heavy-metal-binding peptides of plants, are synthesized from glutathione by a specific γ-glutamylcysteine dipeptidyl transpeptidase (phytochelatin synthase). *Proc. Natl. Acad. Sci., 86*(18), 6838–6842.
37. Mendoza-Cózatl, D., Devars, S., Loza-Tavera, H., & Moreno-Sánchez, R., (2002). Cadmium accumulation in the chloroplast *of Euglena gracilis. Physiol. Plant, 115*(2), 276–283. doi: 10.1034/j.1399-3054.2002.1150214.x.

38. Stokes, P. M., Maler, T., & Riordan, J. R., (1977). A low molecular weight copper binding protein in a copper tolerant strain of *Scenedesmus acutiformis* [Algae]. *Trace Subst. Environ. Health Conf., 3*(1–4), 667–678. doi: 10.1080/01904168109362869.
39. Pawlik-Skowrońska, B., Pirszel, J., Kalinowska, R., & Skowroński, T., (2004). Arsenic availability, toxicity and direct role of GSH and phytochelatins in As detoxification in the green alga *Stichococcus bacillaris*. *Aquat. Toxicol., 70*(3), 201–212. doi: 10.1016/j.aquatox.2004.09.003.
40. Clemens, S., (2001). Molecular mechanisms of plant metal tolerance and homeostasis. *Planta, 212*(4), 475–486. doi: 10.1007/s004250000458.
41. Thumann, J., Grill, E., Winnacker, E. L., & Zenk, M. H., (1991). Reactivation of metal-requiring apoenzymes by phytochelatin-metal complexes. *FEBS Lett., 284*(1), 66–69. doi: 10.1016/0014-5793(91)80763-S.
42. Perales-Vela, H. V., Peña-Castro, J. M., & Cañizares-Villanueva, R. O., (2006). Heavy metal detoxification in eukaryotic microalgae. *Chemosphere, 64*(1), 1–10. doi: 10.1016/j.chemosphere.2005.11.024.
43. Mendoza-Cózatl, D., Loza-Tavera, H., Hernández-Navarro, A., & Moreno-Sánchez, R., (2005). Sulfur assimilation and glutathione metabolism under cadmium stress in yeast, protists and plants. *FEMS Microbiol. Rev., 29*(4), 653–671. doi: 10.1016/j.femsre.2004.09.004.
44. Navarro, A. E., Ramos, K., Agapito, R., & Cuizano, N. A., (2006). Biosorption of heavy metals by marine algae: Possible solution to contamination at low concentrations. *Rev. Latin America. Rec. Natural, 2*, 45–52.
45. Ha, S. B., Smith, A. P., Howden, R., Dietrich, W. M., Bugg, S., O'Connell, M. J., Goldsbrough, P. B., & Cobbett, C. S., (1999). Phytochelatin synthase genes from *Arabidopsis* and the yeast *Schizosaccharomyces* pombe. *Plant Cell, 11*(6), 1153–1163. doi: 10.1105/tpc.11.6.1153.
46. Steffens, J. C., (1990). The heavy metal-binding peptides of plants. *Annu. Rev. Plant Physiol. Plant Mol. Biol., 41*(1), 553–575. doi: 10.1146/annurev.pp.41.060190.003005.
47. Monteiro, C. M., Castro, P. M. L., & Malcata, F. X., (2012). Metal uptake by microalgae: Underlying mechanisms and practical applications. *Biotechnol. Prog., 28*(2), 299–311. doi: 10.1002/btpr.1504.
48. Zeraatkar, A. K., Ahmadzadeh, H., Talebi, A. F., Moheimani, N. R., & McHenry, M. P., (2016). Potential use of algae for heavy metal bioremediation, a critical review. *J. Environ. Manage, 181*, 817–831. doi: 10.1016/j.jenvman.2016.06.059.
49. Fourest, E., & Roux, J. C., (1992). Heavy metals biosorption by fungal mycelial by-products: Mechanism and influence of pH. *Applied Microbiology Biotechnology, 37*, 399–403.
50. Holan, Z. R., & Volesky, B., (1995). Accumulation of cadmium, lead, and nickel by fungal and wood biosorbents. *Appl. Biochem. Biotechnol., 53*, 133–146.
51. Holland, J. L., Johnston, J. A., Johnston, J. A., & Asama, N. F., (1993). The vocational identity scale: A diagnostic and treatment tool. *J. Career Assess, 1*, 1–12. doi: 10.1177/106907279300100102.
52. Gupta, V. K., & Rastogi, A., (2008). Equilibrium and kinetic modelling of cadmium (II) biosorption by nonliving algal biomass *Oedogonium* sp. from aqueous phase. *J. Hazard. Mater.,* ***153*(1, 2)**, 759–766. doi: 10.1016/j.jhazmat.2007.09.021.

53. Incharoensakdi, A., & Kitjaharn, P., (2002). Zinc biosorption from aqueous solution by a halotolerant cyanobacterium *Aphanothece halophytica*. *Curr Microbiol., 45*(4), 261–264. doi: 10.1007/s00284-002-3747-0. PMID: 12192523.
54. Valdman, E., & Leite, S. G. F., (2000). Biosorption of Cd, Zn and Cu by Sargassum sp. waste biomass. *Bioprocess and Biosystems Engineering, 22*(2), 171–173. 10.1007/PL00009109.
55. Graham, L. E., & Wilcox, L. E., (2000). *Algae* (p. 420). Prentice Hall International, London.
56. Davis, T., Volesky, B., & Mucci, A., (2003). A review on biochemistry of heavy metal biosorption by brown algae. *Water Res., 37*, 4311–4330.
57. Morris, E., Rees, D., & Thom, D., (1978). Chioptical and stoichiometric evidence of a specific, primary dimerization process in alginate gelation. *Carbohydrate Res., 66*, 145–154.
58. Emmerichs, N., Wingender, J., Flemming, H., & Mayer, C., (2004). Interaction between alginates and manganese cations: Identification of preferred cation binding sites. *I Int. J. Biol. Macromol., 34*, 73–79.
59. Silva, J., Flores, J., Rodríguez, A., Ly, M., & Maldonado, H., (2005). Adsorption of chromium onto cross-linked chitosan. *Separ. & Purif. Technol., 44*, 31–36.
60. Murphy, V., Hughes, H., & McLoughlin, P., (2007). Comparative study of chromium biosorption by red, green and brown seaweed biomass. *Water Res., 41*, 731–740.
61. Navarro, A. E., Blanco, D., Llanos, B., Flores, J., & Maldonado, H., (2004). Bioremoción de Cadmio (II) por desechos de algas marinas. Optimización del equilibrio y propuesta de mecanismo. *Rev. Soc. Quím, Perú, 70*, 147–157.
62. Navarro, A. E., Ramos, K., Campos, K., & Maldonado, H., (2006). Elucidación del efecto del pH en la adsorción de metales pesados mediante biopolímeros naturales: Cationes divalentes y superficies activas. *Rev. Iberoamer. Polím., 7*, 113–126.
63. Yang, Y., & Chen, J., (2008). Biosorption of hexavalent chromium onto raw and chemically modified *Sargassum* sp. *Bioresour. Technol., 99*, 297–307.
64. Yan, G., Viraraghavan, T., & Water, S. A., (2000). *Effect of Pretreatment on the Bioadsorption of Heavy Metals on Mucor Rouxii, 26*, 119–123.
65. Matheickal, J., Yu, Q., & Woodburn, G., (1999). Cellulosic substrates for removal of pollutants from aqueous systems: A review. *Metals Biores. Technol., 69*, 223–229.
66. Kaewsarn, P., & Yu, Q., (2001). Cadmium(II) removal from aqueous solutions by pre-treated biomass of marine alga *Padina* sp. *Environ. Pollut., 112*, 209–213.
67. Yan, G., & Viraraghavan, T., (2003). Removal of copper from aqueous solutions using biosolids. *Water Res., 37*, 4486–4496.
68. Huang, C. H., & Huang, C. P., (1996). Application of *Aspergillus oryzae* and *Rhizopus oryzae* for Cu(II) removal. *Water Res., 30*, 1985–1990.
69. Figuera, M., Volesky, B., Ciminelli, V., & Roddick, F., (2000). Biosorption of metals in brown seaweed biomass. *Water Res., 34*, 196–204.
70. Fourest, E., & Volesky, B., (1996). Biosorción para la eliminación de metales pesados en aguas de desecho. *Environ. Sci. Technol., 30*, 277–282.
71. Ramos, K. P., Llanos, B. P., Maldonado, H. J., & Navarro, A. E., (2007). Evidencias del mecanismo de adsorción del cadmio divalente en Lentinus edodes. *An. Quím., 103*(2), 36–40.

CHAPTER 4

Algae as Agents for the Reclamation of Wastewater: A Green Surrogate for Decontamination

A. M. SHACKIRA,[1] KRICELLE MOSQUERA DEAMICI,[2] HOUDA TAIMOURYA,[3] EDAPPAYIL JANEESHMA,[4] and MOHAMMAD SARRAF[5]

[1]*Department of Botany, Sir Syed College, Kannur University, Kerala, India*

[2]*Faculty of Pharmacy, Federal University of Bahia, Salvador, Brazil*

[3]*Department of Production, Protection and Plant Biotechnology, Agronomy and Veterinary Institute Hassan II, Rabat, Morocco*

[4]*Department of Botany, MES KEVEEYAM College, Valanchery, Kerala, India*

[5]*Department of Horticultural Science, Faculty of Agriculture, Shahid Chamran University of Ahvaz, Ahvaz, Iran*

ABSTRACT

Freshwater contamination and related health issues are a growing concern of the human population due to the extensive reduction and scarcity of freshwater. One of the major reasons for the scarcity is the contamination of water bodies and it is essential to reclaiming the contaminated water resources. Algal-mediated remediation, algal-phytoremediation or phycoremdiation is a good strategy to decontaminate the wastewater due to the duel benefits of this process which is with the decontamination this algal biomass able to utilize for biofuel production. This work focuses on different algal candidates with metal tolerance potential and the mechanisms underlying pycoremediation. This comprehensive study could help to identify effective algae by developing high tolerance in them.

Phytoremediation and Biofortification: Strategies for Sustainable Environmental and Health Management.
Nand Kumar Singh, Shadma Afzal, & Tariq Aftab (Eds.)

4.1 INTRODUCTION

The rapid urbanization and industrialization are the main cause of the ecological imbalance resulting in a massive pollution of the soil, air, and water across the world. Koul et al. [1] reported that there may be a significant rise in the level of global waste production annually to 3.40 billion tons from the existing 2.01 billion tons, on current levels by 2050. Most of the different types of wastes may have an utter destination in the aquatic stream as water bodies may serve as a sink making the water unfit for consumption. In this scenario, there is an urgent need to focus on the effective recycling processes and operations of wastewater which must be sustainable, cost-effective, and operating efficiently with minimal infrastructures [1, 2]. Generally, the conventional wastewater treatment consists of one or a combination of physical, chemical, and biological methods. Out of this, physical and chemical methods are costly as well as alter the structure and quality of water. Moreover, based on the types of pollutants the treatment strategy may be cataloged as primary, secondary, and tertiary which includes various purification schemes like sedimentation, filtration, adsorption, precipitation, etc. Compared to these strategies, the biological method is the most promising and efficient one as it offers a sustainable method for wastewater reclamation [3].

The most promising tactic for enhancing the efficacy of wastewater reclamation is the employment of native aquatic members for the purification process. In this aspect, algae serve as an ideal candidate as it naturally acts as a net carbon sink and can reduce the carbon footprint thereby mitigating global warming to a certain extent [1, 2, 4]. Naturally, algae are exceptionally talented for flourishing in adverse environmental conditions because of their adaptation to a large array of abiotic and biotic factors and are hence making it a superior agent of decontamination. Algae are rich in biodiversity and comprise about 8 lakhs species on average [5] and can be grouped as microalgae and macroalgae. Another striking feature is that besides exploiting them as agents of remediation, they can be also employed for biofuels generation in an eco-friendly manner [6, 7].

4.2 FRESHWATER POLLUTION-SOURCES AND SORT OF POLLUTANTS

Water bodies act as the basis of life on earth as every organism needs water in one or the other way. However, it is terrible to note that every source of water is heavily infested with an array of diverse pollutants which may range from non-degradable plastic to hazardous chemical wastes making the water bodies a poison pool. The main sources and kinds of pollutants in the freshwater showed in Figure 4.1. Also, we can have categorized them as in subsections.

4.2.1 SEWAGE WASTE

Sewage is organic waste and accounts for a quarter of the serious water incidents all over the World. The main sources of sewage wastes include household and municipal wastes. Sewage sludge is the byproduct of wastewater treatment and the main strategies for the disposal of sludge include recycling for agricultural and landscaping purposes [8]. Another advanced application is the recovery of various industrially important organic components from the sludge.

4.2.2 AGRICULTURAL WASTE

Agricultural processes like unrestrained usage of manure, use of pesticides and fertilizers can cause severe pollution in the water bodies. The employment of toxic chemicals in modern fertilizers and pesticides to evade pests and invasive species are a source of potential chemicals in the soil. Once they reach the water bodies from the soil, it enters and contaminates the groundwater which may then enter the food chain [1].

4.2.3 OIL SPILLAGE

Every year there may be several pollution incidents about oil and fuels which may alter the water quality and make it unfit for specific purposes. Oil spillage can ultimately negatively affect biodiversity and it is a very life-threatening contaminant to the aquatic flora and fauna. The main feature associated with oil pollution is diminishing the supply of oxygen which makes the water unfit for the existence of living beings [9].

4.2.4 RADIOACTIVE SUBSTANCES

Radioactive substances are nowadays widely used in several medical and industrial appliances like X-ray machines, watches, clocks, television, etc., in addition to nuclear power plants. But the problem is with the disposal of these hazardous substances once it is used, and in many instances, lack of proper management may result in serious pollution problems [1].

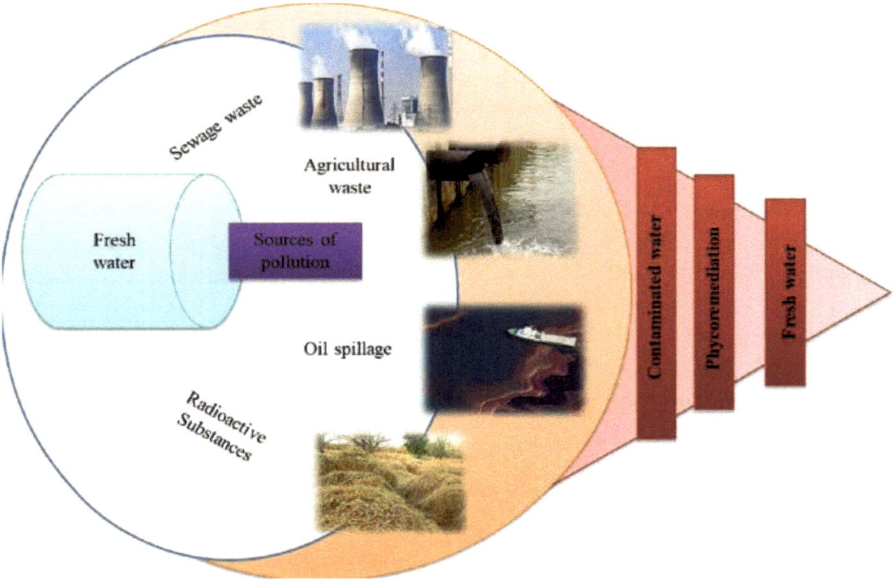

FIGURE 4.1 Sources of freshwater contamination.

4.3 ALGAE AS POTENTIAL AGENTS OF PHYCOREMEDIATION

Algae correspond to the group of micro- and macro-algae, photosynthetic organisms capable of contributing to the greenhouse gas (GHS) mitigation through atmospheric carbon dioxide (CO_2) fixation, with potential for use in wastewater treatment and producing rich biomass, used for increasing food quality and nutritional value. Microalgae are microorganisms unicellular that can convert solar energy into chemical energy through CO_2 fixation, more efficiently than higher plants [9]. The microalgae produce a large variety of biocompounds, like lipids, proteins, carbohydrates that can be used in many areas.

Microalgae cultivation requires many nutrients to grow. Otherwise for commercial purposes, they are cultivated in synthetic media (i.e.: Zarrouk, BG 11 media, among others). These cultural media present a lot of compounds that confer a high cost in the cultivation. Usually, wastewater contains nitrogen and phosphorus in high concentrations [10, 11], which are nutrients required by the microalgae for structure development. In this way, microalgae and macroalgae are good candidates to treat wastewaters. In fact, they act by degrading the contaminants present in order to obtain nutrients for their growth [12], by producing biomass with added-value compounds, including biopolymers and lipids/carbohydrates which facilitate the use of this biomass as an alternative for bioenergy production, in a biorefinery approach. The idea of exploiting algae for wastewater treatment was first proposed by Oswald and Gotaas [13] approximately 40 years before.

There are three different processes for the treatment of wastewater, which are classified as: physical, chemical, and biological. The use of micro- and macroalgae are considered as the most eco-friendly biological method, besides being the less expensive one [14]. In addition, there is no secondary pollution, since the biomass produced is recycled as a nutrient and reused [15, 16].

Otherwise, some wastewater's composition presents barriers to achieving an optimal biological process. We cite as an example: high organic loads, presence of antimicrobial or biostatic compounds (phenol and lipids), low pH, and unbalanced composition of nutrients [17]. In this way, the study of the use of micro- and macroalgae needs to be evaluated, considering the nutrient supply and the composition of the wastewater used.

Phycoremediation is defined as the use specifically of these organisms to make the wastewater treatment. John [18] is the one who created this term and is used afterwards to define the use of algae (micro- or macroalgae) for the removal or biotransformation of pollutants (organic and inorganic substances, nitrogen, phosphorus, and xenobiotics) from wastewater and simultaneously produce biomass [15, 19–21].

Phycoremediation has many applications. It is principally used to remove nutrients from municipal wastewater and effluents. We can cite many acting mods: using algae-based sorbents to remove nutrient and xenobiotic compounds; using algae-based biosensors to detect toxic compounds and the transformation and degradation of xenobiotics [22].

4.3.1 MICROALGAE AND MACRO-MEDIATED PHYCOREMEDIATION

Literally, the use of algae for phycoremediation started to get more insight since the 1950s, when Oswald & Gotaas [13] presented the first studies of using microalgae for the tertiary treatment of municipal wastewater. Nowadays, phycoremediation is a widespread concept, since there are many studies in the literature involving microalgae and macroalgae mediated phycoremediation.

It has been reported that the most widely used algae for phycoremediation are *Chlorella* sp., *Botryococcus* sp., *Phormidium* sp., *Scenedesmus* sp., and *Spirulina* sp. [21, 23]; they are demonstrating the efficacy of this biological method especially in domestic wastewater treatment. Olguin et al. [20] have been extensively evaluating the *Spirulina* microalga in phycoremediation. Since this microalga demonstrates better resistance to physiological stresses and a good nutrient removal efficiency. According to Olguin et al. [20], Spirulina presents high tolerance to ammonia-nitrogen, besides presenting the characteristic of auto-flocculate, making easier the harvesting, besides it adapts well to different culture conditions, as temperatures range of 29–34°C and luminosity intensity ranging from 476–1.784 μmol photon m^{-2} s^{-1}.

It has been proven that Spirulina has a high capacity to eliminate other contaminants from wastewater, like heavy metals. This ability is called heavy metal adsorption and it is due to the fact that spirulina's cell surfaces contain a high-affinity metal bindings [24]. According to Chojnacka et al. [25] the maximum biosorption capacity provided by Spirulina is reported for the metals: Cr, Cd, and Cu, moreover this capacity is improving with high light intensity.

Ajayan et al. [26] demonstrated that microalga *Scenedesmus* sp. isolated from a local habitat presented a good phycoremediation of tannery wastewater. During 12 days of cultivation, the microalga was capable of reducing the pollution load of heavy metals like Cr, Zn, Cu, and Pb as well as nutrients (NO_3 and PO_4). *Scenedesmus obliquus* was also cultivated by Pan et al. [27] to biodegrade the antibacterial/antifungal agent climbazole (CBZ) from an aqueous medium. The results revealed that this microalga was able to remove more than 88% of CBZ after 12 days of cultivation. On the other hand, *Monoraphidium braunii* can tolerate high concentrations of Bisphenol A (BPA) in the aqueous medium. According to Gatullo et al. [28], *M. braunii* was exposed to 2, 4, and 10 mg L^{-1} and reduced the initial concentration in 39, 48, and 35%, respectively.

Macroalga also has been used extensively to indicate the pollution in marine environments through heavy metals. Species of green algae *Cladophora* and/or *Enteromorpha* have been studied for this purpose [29].

For this purpose, many enterprises in the world use wastewater sources for microalgal biomass production. For example, Algae Systems and Sundine Enterprises, Inc. (United States), I-PHYC; Australia—Algae Enterprises (United Kingdom), Aquonos Energy Ltd. (Israel).

4.4 MECHANISM BEHIND PHYCOREMEDIATION

The basic phycoremediation mechanism is inherent to the algal metabolism, therefore, the removal process will be variable with microalga strain and the different characteristics of the effluents [30]. Each strain presents a characteristic behavior, and it reflects on the metabolism used in the phycoremediation.

According to latest studies, microalgal biofilms and algal turf scrubbers (ATS), or suspension culture system that uses high-rate algal ponds or naturally sedimentation ponds represent different categories of phycoremediation that are based on the attached system [31]. Even if the microalgal biofilms and ATS method is limited for large scale operation, it represents many advantages as: low cost, high versatility and adaptability, easier and cheaper to harvest algal biomass. Additionally, the suspension system presents as advantages the high nutrient removal and simplicity but is difficult and expensive to harvest algal biomass and the effluent quality is poor and inconsistent [30]. Figure 4.2 illustrated the mechanism of phycoremediation.

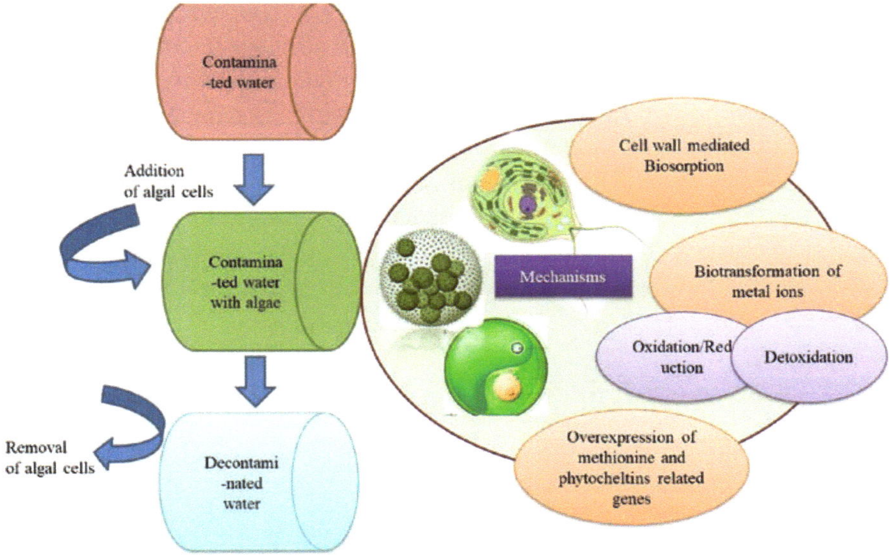

FIGURE 4.2 Insight into the mechanisms involved in phycoremediation.

4.4.1 PHYTOREMEDIATION OF HEAVY METALS BY ALGAE

The main steps for the mechanisms of Heavy metal removal are sedimentation, flocculation, absorption, anion, and cations exchange, complexation, precipitation, reduction/oxidation, microbiological activity, and uptake. Generally, the microalgae remove heavy metals directly from polluted water via two main mechanisms: biosorption or a metabolism dependent uptake into their cells [32].

Macroalgae can accumulate metals in their tissues, which has the use of these organisms as biomonitors of metal availability in marine systems [33, 34].

Examples of algae with the ability to accumulate heavy metals are algae from the phylum *Chlorophyta* and *Cyanophyta.* They have the characteristic to be absorbents and accumulators of Boron and Arsenic, accumulating these elements into their cells, reducing the presence of both pollutants in the waters [35]. The brown algae can be considered as a high accumulator of heavy metals. Otherwise Phaeophytas present a high affinity with heavy metals because they have a high level of sulfated polysaccharides and alginates in the cell walls [36].

Mehta & Gaur [37] suggested the link between metal uptake, metal toxicity and proline accumulation in *Chlorella Vulgaris* microalgae since the results showed the proline accumulation was higher in *Chlorella* cells treated with Cr and Cu, besides it showed a protective effect on metal toxicity by the deterrence of lipid peroxidation [38, 39].

According to Gonzalez-Dkila et al. [40], the phycoremediation can be realized through different mechanisms, depending on different factors involved, such as the heavy metal element, its speciation, the microalgae strain, and if the operational condition is realized with living or non-living systems.

4.4.1.1 PHYCOREMEDIATION BY LIVING MICROALGAE

The heavy metals phycoremediation by living microalgae consists of both extracellular and intracellular strategies [39]. Generally, the extracellular strategy is based on the bio-removal of heavy metal by biosorption whether in the cell wall or in the extracellular polysaccharides substances (EPS) in fact these substances are principally formed in response to stress conditions. The mechanism of biosorption, in the case of the cell wall, is independent of the microalga metabolism. Actually, absorption in the cell wall refers to a physicochemical property of its surface, which links to heavy metal ions from the solution. On the other hand, the mechanism is different for the EPS case [41]. Given that the microalgae cells can change the properties of these biopolymers and also regulate EPS synthesis. The heavy metals biosorption into the EPS is then considered as a microalga metabolism-dependent process [41, 42].

Additionally, some macromolecules (lipids, proteins, and carbohydrates) contained in the cell wall represent an important factor in heavy metal biosorption; which is due to the different negatively charged functional groups, such as amino, hydroxyl, carboxyl, sulfate, phenol presented in the surface of these macromolecules [43]. Then these macromolecules permit the link of ions from the environment around, making the last layer of the cell wall the first involved in the removal of heavy metals [3, 44–46].

4.4.1.2 BIOTRANSFORMATION AND MITIGATION OF HEAVY METALS

For the detoxification pathways, preferring enzymatic and biochemical transformations of toxic heavy metal, biotransformation referred to heavy metals can be applied [47]. The enzymatic transformation is based on an oxidation-reduction reaction where the toxic element is transformed into one less harmful element. The objective is to mitigate the toxic impact of the heavy metal by being changed from an oxidation state to another one in an inorganic complex form. Lee et al. [48]; and Yen et al. [49], reported that via an enzymatic reaction catalyzed if using the enzyme chromate reductase, Chlorella Vulgaris strains can convert Cr(VI) to Cr(III). While *Chlorella fusca*, *Selenastrum minutum*, and *Galdiera sulfuraria* present the ability to catalyze the biotransformation of Hg_2þ in elemental Hg0 and meta-cinnabar (β-HgS) through mercuric reductase [50].

The heavy metal mitigation by microalgae is through biochemical mechanisms. As an example, the mitigation of the toxicity of inorganic arsenic can be the result of several other detoxification mechanisms [49, 51–54]. As reported in the literature, many microalgae species can reduce As(V) to As(III). *Chlorella salina* after 72 h of exposure to As(V), converted 32% of total intracellular concentration to As(III) [52].

According to Checkroun et al. [38], the genetic engineering study can be an alternative to improve the possibilities of accumulation of heavy metals by microalgae. These studies can open the way of developing transgenic species overexpressing PCs and metallothioneins (MTs) able to form complexes with heavy metals, aiming to maximize the phycoremediation and accelerate the removal of toxic elements from marine systems.

4.5 FUTURE PROSPECT

Employment of a green alternative method for the sustainable reclamation of our polluted ecosystem has turned into a demanding need of our society as the conventional techniques are costly and also pose threat to the flora and fauna of the site affected. Phycoremediation in this aspect has a leading role in stabilizing the balance of the ecosystem because of its eco-friendly nature and cost-effectiveness. Characterization of the existing native pollutant tolerant varieties, as well as the manipulation of the sensitive varieties through transgenic approaches, are two crucial aspects of the technique. Moreover, there is an additional advantage for the algal agents being exploited for the technique as it can be served as an excellent candidate for bioenergy crops.

KEYWORDS

- **algal culture**
- **algal-phytoremediation**
- **biotransformation**
- **environmental safety**
- **fresh water pollution**
- **oil spillage**
- **phycoremediation**
- **radioactive substances**
- **wastewater**

REFERENCES

1. Koul, B., Sharma, K., & Shah, M. P., (2022). Phycoremediation: A sustainable alternative in wastewater treatment (WWT) regime. *Environmental Technology & Innovation, 25*, 102040.
2. Ding, G. T., Yasin, N. H. M., Takriff, M. S., Kamarudin, K. F., Salihon, J., Yaakob, Z., & Hakimi, N. I. N. M., (2020). Phycoremediation of palm oil mill effluent (POME) and CO_2 fixation by locally isolated microalgae: *Chlorella sorokiniana* UKM2, *Coelastrella* sp. UKM4 and *Chlorella pyrenoidosa* UKM7. *Journal of Water Process Engineering, 35*, 101202.
3. Ihtisham, M., Noori, A., Yadav, S., Sarraf, M., Kumari, P., Brestic, M., Imran, M., et al., (2021). Silver nanoparticle's toxicological effects and phytoremediation. *Nanomaterials, 11*, 2164.

4. Arbib, Z., Ruiz, J., Álvarez-Díaz, P., Garrido-Perez, C., & Perales, J. A., (2014). Capability of different microalgae species for phytoremediation processes: Wastewater tertiary treatment, CO_2 bio-fixation and low cost biofuels production. *Water Research, 49*, 465–474.
5. Starckx, S., (2012). *A Place in the Sun-Algae is the Crop of the Future, According to Researchers in Geel.* Flanders Today.
6. Hussain, F., Shah, S. Z., Ahmad, H., Abubshait, S. A., Abubshait, H. A., Laref, A., Manikandan, A., et al., (2021). Microalgae an ecofriendly and sustainable wastewater treatment option: Biomass application in biofuel and bio-fertilizer production. A review. *Renewable and Sustainable Energy Reviews, 137*, 110603.
7. Poonia, A. K., Kajla, S., Koul, B., & Panwar, J. S., (2022). Algae: The high potential resource for biofuel production. In: *An Integration of Phycoremediation Processes in Wastewater Treatment* (pp. 155–176). Elsevier.
8. Kacprzak, M., Neczaj, E., Fijałkowski, K., Grobelak, A., Grosser, A., Worwag, M., Rorat, A., et al., (2017). Sewage sludge disposal strategies for sustainable development. *Environmental Research, 156*, 39–46.
9. Sathasivam, R., Radhakrishnan, R., Hashem, A., & Abd_Allah, E. F., (2019). Microalgae metabolites: A rich source for food and medicine. *Saudi Journal of Biological Sciences, 26*, 709–722.
10. Abdel-Raouf, N., Al-Homaidan, A., & Ibraheem, I., (2012). Microalgae and wastewater treatment. *Saudi Journal of Biological Sciences, 19*, 257–275.
11. De La Noüe, J., Laliberté, G., & Proulx, D., (1992). Algae and waste water. *Journal of Applied Phycology, 4*, 247–254.
12. Kensa, V. M., (2011). Bioremediation-an overview. *Journal of Industrial Pollution Control, 27*, 161–168.
13. Oswald, W. J., & Gotaas, H. B., (1957). Photosynthesis in sewage treatment. *Transactions of the American Society of Civil Engineers, 122*, 73–97.
14. Mantzavinos, D., & Kalogerakis, N., (2005). Treatment of olive mill effluents: Part I. Organic matter degradation by chemical and biological processes—An overview. *Environment International, 31*, 289–295.
15. Mulbry, W., Kondrad, S., Pizarro, C., & Kebede-Westhead, E., (2008). Treatment of dairy manure effluent using freshwater algae: Algal productivity and recovery of manure nutrients using pilot-scale algal turf scrubbers. *Bioresource Technology, 99*, 8137–8142.
16. Pizarro, C., Mulbry, W., Blersch, D., & Kangas, P., (2006). An economic assessment of algal turf scrubber technology for treatment of dairy manure effluent. *Ecological Engineering, 26*, 321–327.
17. Morillo, J., Antizar-Ladislao, B., Monteoliva-Sánchez, M., Ramos-Cormenzana, A., & Russell, N., (2009). Bioremediation and biovalorization of olive-mill wastes. *Applied Microbiology and Biotechnology, 82*, 25–39.
18. John, J. (2000). A self-sustainable remediation system for acidic mine voids. In: *Proceedings of 4th International Conference of Diffuse Pollution* (pp. 506–511).
19. Olguín, E. J., (2003). Phycoremediation: Key issues for cost-effective nutrient removal processes. *Biotechnology Advances, 22*, 81–91.
20. Olguín, E. J., Sánchez, G., & Mercado, G., (2004). Cleaner production and environmentally sound biotechnology for the prevention of upstream nutrient pollution in the Mexican coast of the Gulf of Mexico. *Ocean & Coastal Management, 47*, 641–670.

21. Moreno-Garrido, I., (2008). Microalgae immobilization: Current techniques and uses. *Bioresource Technology, 99*, 3949–3964.
22. Lavoie, A., & De La Noüe, J., (1985). Hyperconcentrated cultures of *Scenedesmus* obliquus: A new approach for wastewater biological tertiary treatment? *Water Research, 19*, 1437–1442.
23. Danquah, M. K., Gladman, B., Moheimani, N., & Forde, G. M., (2009). Microalgal growth characteristics and subsequent influence on dewatering efficiency. *Chemical Engineering Journal, 151*, 73–78.
24. Naz, M., Ghani, M. I., Sarraf, M., Liu, M., & Fan, X., (2022). Ecotoxicity of nickel and its possible remediation. In: *Phytoremediation* (pp. 297–322). Elsevier.
25. Chojnacka, K., Chojnacki, A., & Gorecka, H., (2005). Biosorption of Cr^{3+}, Cd^{2+} and Cu^{2+} ions by blue–green algae *Spirulina* sp.: Kinetics, equilibrium and the mechanism of the process. *Chemosphere, 59*, 75–84.
26. Ajayan, K. V., Selvaraju, M., Unnikannan, P., & Sruthi, P., (2015). Phycoremediation of tannery wastewater using microalgae *Scenedesmus* species. *International Journal of Phytoremediation, 17*, 907–916.
27. Pan, C. G., Peng, F. J., & Ying, G. G., (2018). Removal, biotransformation and toxicity variations of climbazole by freshwater algae *Scenedesmus* obliquus. *Environmental Pollution, 240*, 534–540.
28. Gattullo, C. E., Bährs, H., Steinberg, C. E., & Loffredo, E., (2012). Removal of bisphenol A by the freshwater green alga *Monoraphidium braunii* and the role of natural organic matter. *Science of the Total Environment, 416*, 501–506.
29. Al-Homaidan, A. A., Al-Ghanayem, A. A., & Alkhalifa, A. H., (2011). Green algae as bioindicators of heavy metal pollution in Wadi Hanifah Stream, Riyadh, Saudi Arabia. *International Journal of Water Resources and Arid Environments, 1*, 10–15.
30. Pacheco, D., Rocha, A. C., Pereira, L., & Verdelhos, T., (2020). Microalgae water bioremediation: Trends and hot topics. *Applied Sciences, 10*, 1886.
31. Magdouli, S., Brar, S. K., & Blais, J. F., (2016). Co-culture for lipid production: Advances and challenges. *Biomass and Bioenergy, 92*, 20–30.
32. Matagi, S., Swai, D., & Mugabe, R., (1998). A review of heavy metal removal mechanisms in wetlands. *Afr. J. Trop. Hydrobiol. Fish, 8*, 23–35.
33. Gosavi, K., Sammut, J., Gifford, S., & Jankowski, J., (2004). Macroalgal biomonitors of trace metal contamination in acid sulfate soil aquaculture ponds. *Science of the Total Environment, 324*, 25–39.
34. Rainbow, P. S., (1995). Biomonitoring of heavy metal availability in the marine environment. *Marine Pollution Bulletin, 31*, 183–192.
35. Baker, A. J., (1981). Accumulators and excluders-strategies in the response of plants to heavy metals. *Journal of Plant Nutrition, 3*, 643–654.
36. Davis, T. A., Volesky, B., & Mucci, A., (2003). A review of the biochemistry of heavy metal biosorption by brown algae. *Water Research, 37*, 4311–4330.
37. Mehta, S., & Gaur, J., (1999). Heavy-metal-induced proline accumulation and its role in ameliorating metal toxicity in *Chlorella vulgaris*. *New Phytologist, 143*, 253–259.
38. Chekroun, K. B., & Baghour, M., (2013). The role of algae in phytoremediation of heavy metals: A review. *J. Mater. Environ. Sci., 4*, 873–880.
39. Sarraf, M., Vishwakarma, K., Kumar, V., Arif, N., Das, S., Johnson, R., Janeeshma, E., et al., (2022). Metal/metalloid-based nanomaterials for plant abiotic stress tolerance: An overview of the mechanisms. *Plants, 11*, 316.

40. González-Dávila, M., (1995). The role of phytoplankton cells on the control of heavy metal concentration in seawater. *Marine Chemistry, 48*, 215–236.
41. Naveed, S., Li, C., Lu, X., Chen, S., Yin, B., Zhang, C., & Ge, Y., (2019). Microalgal extracellular polymeric substances and their interactions with metal (loid) s: A review. *Critical Reviews in Environmental Science and Technology, 49*, 1769–1802.
42. Ubando, A. T., Africa, A. D. M., Maniquiz-Redillas, M. C., Culaba, A. B., Chen, W. H., & Chang, J. S., (2021). Microalgal biosorption of heavy metals: A comprehensive bibliometric review. *Journal of Hazardous Materials, 402*, 123431.
43. Javanbakht, V., Alavi, S. A., & Zilouei, H., (2014). Mechanisms of heavy metal removal using microorganisms as biosorbent. *Water Science and Technology, 69*, 1775–1787.
44. Leong, Y. K., & Chang, J. S., (2020). Bioremediation of heavy metals using microalgae: Recent advances and mechanisms. *Bioresource Technology, 303*, 122886.
45. Saavedra, R., Muñoz, R., Taboada, M. E., Vega, M., & Bolado, S., (2018). Comparative uptake study of arsenic, boron, copper, manganese and zinc from water by different green microalgae. *Bioresource Technology, 263*, 49–57.
46. Singh, D. V., Bhat, R. A., Upadhyay, A. K., Singh, R., & Singh, D., (2021). Microalgae in aquatic environs: A sustainable approach for remediation of heavy metals and emerging contaminants. *Environmental Technology & Innovation, 21*, 101340.
47. Danouche, M., El Ghachtouli, N., & El Arroussi, H., (2021). Phycoremediation mechanisms of heavy metals using living green microalgae: Physicochemical and molecular approaches for enhancing selectivity and removal capacity. *Heliyon, 7*, e07609.
48. Lee, L., Hsu, C. Y., & Yen, H. W., (2017). The effects of hydraulic retention time (HRT) on chromium (VI) reduction using autotrophic cultivation of *Chlorella vulgaris*. *Bioprocess and Biosystems Engineering, 40*, 1725–1731.
49. Yen, H. W., Chen, P. W., Hsu, C. Y., & Lee, L., (2017). The use of autotrophic *Chlorella vulgaris* in chromium (VI) reduction under different reduction conditions. *Journal of the Taiwan Institute of Chemical Engineers, 74*, 1–6.
50. Kelly, D. J., Budd, K., & Lefebvre, D. D., (2007). Biotransformation of mercury in pH-stat cultures of eukaryotic freshwater algae. *Archives of Microbiology, 187*, 45–53.
51. Hasegawa, H., Papry, R. I., Ikeda, E., Omori, Y., Mashio, A. S., Maki, T., & Rahman, M. A., (2019). Freshwater phytoplankton: Biotransformation of inorganic arsenic to methylarsenic and organoarsenic. *Scientific Reports, 9*, 1–13.
52. Karadjova, I. B., Slaveykova, V. I., & Tsalev, D. L., (2008). The biouptake and toxicity of arsenic species on the green microalga *Chlorella salina* in seawater. *Aquatic Toxicology, 87*, 264–271.
53. Levy, J. L., Stauber, J. L., Adams, M. S., Maher, W. A., Kirby, J. K., & Jolley, D. F., (2005). Toxicity, biotransformation, and mode of action of arsenic in two freshwater microalgae (*Chlorella* sp. and *Monoraphidium arcuatum*). *Environmental Toxicology and Chemistry: An International Journal, 24*, 2630–2639.
54. Wang, Y., Wang, S., Xu, P., Liu, C., Liu, M., Wang, Y., Wang, C., et al., (2015). Review of arsenic speciation, toxicity and metabolism in microalgae. *Reviews in Environmental Science and Bio/Technology, 14*, 427–451.

CHAPTER 5

Plant-Assisted Bioremediation of Heavy Metals in Soil and Water

ASANKA TENNAKOON,[1] W. L. N. WASANA,[2] B. K. A. BELLANTHUDAWA,[3] SALINDA SANDAMAL,[4,5] and DISNA RATNASEKERA[6]

[1]*Department of Biosystems Technology, Faculty of Technology, Eastern University, Sri Lanka*

[2]*Department of Agronomy, Kansas State University, Manhattan, KS, USA*

[3]*Department of Agricultural Engineering, Faculty of Agriculture, University of Ruhuna, Sri Lanka*

[4]*State Key Laboratory of Systematic and Evolutionary Botany, Institute of Botany, Chinese Academy of Sciences, Beijing, China*

[5]*University of Chinese Academy of Sciences, Beijing, China*

[6]*Department of Agricultural Biology, Faculty of Agriculture, University of Ruhuna, Sri Lanka*

ABSTRACT

Environmental pollution is escalating at an alarming rate due to human activities such as rapid urbanization, technological innovation, improper farming practices, and industrialization. Among all concerns, heavy metal contamination has been identified as one of the most critical environmental risks. Heavy metals contamination in the environment is persistent due to their toxicity, and it causes even risks to human lives. Even though some metals are essential components for plants and animals, high quantities of metals disrupt metabolic activities in organisms. Conventional chemical and physical treatments, typically expensive and ineffectual when metal

Phytoremediation and Biofortification: Strategies for Sustainable Environmental and Health Management.
Nand Kumar Singh, Shadma Afzal, & Tariq Aftab (Eds.)

concentrations are low, produce considerable amounts of toxic sludge. As a remedy, bioremediation is an eco-friendly technique that can be employed for cleaning up such polluted areas from heavy metal contamination. It uses bacteria and plants to eliminate pollutants through biological mechanisms to restore the environment. The basic concepts of bioremediation entail modifying pH, redox processes, and adsorption of contaminants from polluted environments. Phytoremediation applies plants and rhizospheric microbes to clear up organic and metal pollutants. This approach is advantageous as it is affordable, sustainable, and effective in restoring polluted areas, especially heavy metals. The effectiveness of phytoremediation is influenced by the amount of metal pollutant content in the soil and the ability of plants to take up metals from the soil actively. The removal of heavy metals from contaminated soils by plants involves several processes. Phytoextraction employs the technique of hyperaccumulation to remove metal contaminants from the soil and transport them into the plants' above-ground components through the roots. Rhizofiltration (the use of plant roots), blastofiltration (the use of seedlings), and caulofiltration (the use of excised plant shoots) are the three types of phytofiltration. It is the use of plant roots or seedlings from aqueous wastes to clean up polluted areas. Rhizofiltration is the process of removing contaminants from groundwater by filtering them through plant roots. Rhizofiltration uses the plant's rhizospheric accumulation mechanism. Plants that can absorb and resist high quantities of hazardous metals, such as hyperaccumulators, are ideal for rhizofiltration. Phytostimulation stands for stimulating microbial activity to break down organic pollutants by exudates from plant roots. Soil contaminants are absorbed and stabilized by plant roots in the rhizosphere, inhibiting their spread into the environment. Plants remove soil pollutants via phytovolatilization, converting them into vapor and releasing them into the atmosphere. The metabolic potentials of plants in conjunction with rhizosphere bacteria convert pollutants into volatile forms during phytovolatilization. Phytodegradation is another process of breaking down pollutants by plant enzymes. Rhizospheric microorganisms could also facilitate the breakdown of organic contaminants in soil via the rhizodegradation process. Furthermore, understanding plant pathways for metal detoxification is critical for developing genetically altered plants with phytoremediation potential. Therefore, this chapter explores the environmental consequences of heavy metal pollution and how phytoremediation can effectively repair the damage. Moreover, it also discusses how these organisms remove heavy metals from the environment or how do they perform that function while the detoxification of heavy metals by plants is also taken into consideration.

5.1 INTRODUCTION

Over the past few decades, rapid industrialization has led to various human health and environmental issues, while environmental contamination of organic and inorganic wastes is one of the major concerns [1]. Heavy metals are a group of well-known soil and water contaminants among many other environmentally important inorganic wastes. Heavy metals pose a much more significant threat than other pollutants due to their persistence in nature due to the bioaccumulation phenomenon [2–4]. For example, Cadmium (Cd) has 18 years of biological half-life [5], whereas the soil retention time of Lead (Pb) is around 150–5,000 years [6]. As a result of bioaccumulation, heavy metals accumulate in the tissues of living organisms. Human activities and natural processes release heavy metals into the environment. They could be detrimental to the organisms at the higher trophic levels due to biomagnification. Heavy metals in soils may adversely affect soil microbes and reduce soil microbial range and quantity [1, 4]. Even though excessive levels of heavy metals are life-threatening to humans and other living organisms, they have an essential role in biological systems. For instance, transport, homeostasis, and binding into important cell constituents involve heavy metals [7–9]. Heavy metals found in biological systems have been classified as essential and non-essential based on their roles. Essential heavy metals include Fe, Cu, Zn, Ni, and Mn, while non-essential heavy metals include Pb, Cd, Hg, and Cr, which are highly poisonous once they enter biological systems [10–14].

5.2 SOURCES OF HEAVY METALS AND THEIR ENVIRONMENTAL CONSEQUENCES

In general, metals released through natural processes cause minor environmental damage compared to those released from anthropogenic activities [9]. Apart from anthropogenic activities such as mining, disposal of industrial effluents and solid waste, sewage sludge disposal, military operations, and usage of agrochemicals, natural contamination occurs because of weathering of rock minerals, volcanic activities, forest fire, and erosion, etc. [1, 15–19]. Heavy metals which are released into the environment as a result of these various reasons end up increasing the extent of contaminated agricultural lands and lowering consumable water percentage, causing threats to living organisms, including humans [19–21] (Table 5.1).

TABLE 5.1 Sources of Different Heavy Metals

Heavy Metal	Sources	References
Hg	Forest fires, volcanic activities, emissions of coal, peat, wood-burning industries.	[22]
Pb	Municipal sewage, burning of leaded gasoline, Pb enriched industrial waste, mining, and smelting.	[2, 23, 24]
Cd	Metal smelting and refining, sewage sludge, phosphorous fertilizer application.	[25, 26]
As	Wood preservatives, petroleum refining, mining, and smelting, coal power plants, herbicides.	[27–29]
Cu	Mining, smelting, and refining, bio-solids, electroplating.	[30]
Zn	Mining, smelting, and refining, bio-solids, electroplating.	[29, 30]

Metallic contamination impacts food quality, ground, surface, and subsurface water quality, other terrestrial and aquatic ecosystems, and human health [7, 31, 32]. One of the significant human health concerns associated with heavy metal contamination is the high possibility to enter and accumulate in the food chain. Even the minute quantities of some heavy metals, including Pb, As, Cd, and Hg, are highly toxic to living organisms, whereas all the heavy metal concentrations beyond threshold levels may cause severe health impacts [18, 33, 34]. Causing oxidative stress by free radical formation [35], replacing essential metals in protein binding sites, and disrupting cell functions [36, 37] are a few problems associated with heavy metals in biological systems. Other than these general consequences, toxic levels of heavy metals could also cause specific human health problems given in Table 5.2.

TABLE 5.2 Specific Health Issues Caused by Toxic Levels of Heavy Metals

Heavy Metal	Health Issues	References
Hg	Depression, drowsiness, fatigue, hair loss, insomnia, memory loss, autoimmune diseases, brain, kidney, and lung damages, etc.	[38–40]
Pb	Excessive exposure may result in impaired development, reduced intelligence, short term memory loss, increased risk for cardiovascular diseases like problems in children.	[17, 41–43]
Cd	Renal failure, chronic anemia, carcinogenic, mutagenic, and teratogenic.	[41, 44]
As	Interferes with oxidative phosphorylation and ATP synthesis.	[18, 45]
Cu	Stomach and intestinal irritation, liver cirrhosis, chronic anemia, bran, and kidney damage.	[17, 41]
Zn	Fatigue and dizziness.	[46]

5.3 CONVENTIONAL METHODS OF HEAVY METAL REMEDIATION IN SOIL AND WATER

Concentrations of heavy metals in the environment increase due to rapid urbanization and industrialization [47]. Even though some heavy metals are essential in minute quantities for the optimum growth of plants and humans, releasing heavy metals to the soil and water continuously in large amounts has made both soils and water unhealthy [48]. Hence, removing heavy metals from contaminated soils and waters is essential. Nevertheless, the removal of metallic contaminants from the environment is a complicated concern because of their non-biodegradable nature. Several physical, chemical, and biological methods for remediation of heavy metal contaminated soils and waters have been researched and practiced in recent decades. When selecting the best remediation method, criteria such as site characteristics, contaminant type, and concentration to be removed, and the final use of remediated medium should be considered [21, 49]. Some common in situ and ex-situ soil heavy metal remediation techniques include soil flushing, solidification, electrophoresis, pneumatic fracturing, soil washing and excavation, soil leaching, and enzymatic degradation [1, 17, 50–52]. Precipitation, reverse osmosis, and ion exchanges are the common physicochemical methods for remediation of heavy metal contaminated water [53]. However, many of these conventional physicochemical methods are ineffective at heavy metal concentrations below 100 mg/L [54]. With this said, biological methods which consume plants and microorganisms in remediation might be a sustainable alternative [55–57].

5.3.1 PHYSICAL REMEDIATION

Physical remediation methods include soil containment, isolation, soil replacement, and thermal treatment [58, 59]. Soil replacement involves mixing large quantities of uncontaminated soil with contaminated soil or covering contaminated soil with uncontaminated soil [60]. This technique is effective for severely contaminated small land areas, yet it is labor and cost intensive. This is not considered as a direct soil remediation method but only a physical measure used to prevent the migration of contaminants into groundwater. Encapsulation, surface capping, and landfilling are some techniques used in soil replacement. Surface capping and encapsulation are applied as an In-situ treatment while landfilling is an Ex-situ remediation method for highly contaminated soils. The use of barrier walls made from cement, steel, bentonite, and grout to prevent further dispersion of contaminants is practiced in physical remediation, making the techniques much expensive [61, 62].

Electrical resistive heating, steam-based heating, conductive heating, and radiofrequency heating are the main thermal treatment methods where contaminant's volatility is used as the principle of the heating soil surface and remediating soil. Researchers have found that thermal treatments are very effective in decontaminating Hg contaminated soils, unfortunately, this may alter soil properties significantly [63, 64].

5.3.2 CHEMICAL REMEDIATION

Chemical remediation uses chemical reagents, reactions, and soil remediation principles. This category includes soil washing, soil flushing, electrokinetic, vitrification, and solidification [62, 64].

Water [65] or some other suitable washing agents namely surfactants [66], chelating agents [67], organic acids [68], saponin [69] are used to remove contaminants in the soil in soil washing and soil flushing techniques. Ethylenediaminetetraacetic acid (EDTA) is the most effective chelating agent in the remediation of heavy metal contaminated soils using chemicals [70]. Appropriate recycling technologies, high removal efficiency, and low biodegradability are some of the major advantages of using EDTA [71].

Solidification is another chemical process cause immobilizing or physically encapsulating contaminants by mixing contaminated soils with reagents or materials [62]. Different types of waste resources, including calcined oyster shells [72] and cockle shells [73], eggshells [74], waste mussel shells [75], have been identified as possible materials to be used in solidification techniques. In addition to heavy metal encapsulation, they contribute to cost-effectiveness, environmental friendliness, and soil quality improvement [62].

Vitrification, also called molten glass, is a technology that applies glass-forming precursors with thermal energy to encapsulate heavy metal contaminants in soils. Contaminated soils are mixed with glass-forming precursors. Heating temperature is the most crucial factor in this process. The mixture is liquidized by heating up to a temperature around 1,400°C–2,000°C. The liquid is allowed to cool down to obtain amorphous homogenous glass in this technique [58]. Chemical bonding and encapsulation are the two main processes that immobilize heavy metals by interactions with glass matrix

[76]. Further, the addition of appropriate additives may increase the efficiency of the vitrification process [77].

Electrokinetic is a comparatively modern soil heavy metal remediating technology that applied a direct current to the contaminated soil using different mechanisms, including electroosmosis, electrophoresis, electromigration, and electrolysis [61]. Past literature has claimed that chelating agents could also be applied to increase the removal efficiency of heavy metals like Cu, Cd, Cr, As, Pb, and Zn, etc. [62, 78].

5.4 BIOREMEDIATION AND ITS' ADVANTAGES

Bioremediation is the scientific approach of using living organisms such as plants and microbes to decrease the release of heavy metals into the environment and minimize their effects on ecosystems. It uses natural biological processes to eliminate the contaminants and could be applied in land and water remediation [79]. Employing plants in bioremediation is termed phytoremediation [80, 81]. Using bacteria, fungi, and enzymes in decontamination are generally termed microbial remediation [81, 82]. Bioremediation is more practical, environmentally friendly, efficient, and affordable than conventional physicochemical methods. Solar-driven nature is another advantage that makes bioremediation more interesting [18, 83]. Another advantage of bioremediation techniques is their ability to be performed on-site. Consequently, the potential for broader exposures during transportation and cleanup activities is reduced, as is the non-invasive nature and ability to combine with physicochemical techniques [81, 84]. Applying phytoremediation techniques in soil remediation minimizes soil erosion caused by wind and water and does not produce secondary wastes [85]. On top of that, bioremediation products are metal-rich recyclable biomass [7]. Given the concerns, one of the significant problems related to bioremediation is the difficulty of predicting the time duration required by the process as it depends on many environmental factors [86].

The bioavailability of heavy metals is an important factor determining the efficiency of phytoremediation techniques. Only a fraction of heavy metals is bioavailable, and the other portion is not bioavailable [87]. Heavy metals in soil are classified into three types on their bioavailability. They are readily bioavailable, moderately bioavailable, and least bioavailable [88]. Cu, Cd, Ni, As, and Zn are readily bioavailable heavy metals. In addition to that, Fe, Mn, and Co are moderately bioavailable, and U, Pb, and Cr are the least bioavailable heavy metals in soil [89]. Significant fractions of heavy metals are insoluble due to either strong binding to soil particles or precipitation [90]. Heavy metal phytoremediation could be accomplished through natural plant uptake of heavy metals, or it could be induced or chelate assisted. Naturally, plants have developed specific mechanisms to solubilize heavy metals in soil. Plant roots secrete phytosiderophores into the rhizosphere to increase metal solubility in soil [29]. These compounds can lower rhizosphere soil pH, which could increase heavy metal concentration in soil by enhancing metal dissolution, and this process is used in natural phytoextraction [91]. Chelating agents such as ammonium sulfate, citric acid, EDTA, cyclohexane-1,2-diamine tetraacetic acid (CDTA), ethylene glycol-bis-(beta-amino-ethyl ether) N,N,N,'N'-tetraacetic acid (EGTA), NTA, and elemental sulfur are applied to the soil in induced phytoremediation techniques to increase the bioavailability of heavy metals in soil [29, 66, 92–94].

Because synthetic chelating agents are non-biodegradable, they may cause groundwater contamination and secondary pollution issues. High concentrations can be toxic to plants [95, 96]. However, the use of natural chelating agents such as citric acid would not result in such secondary issues [97]. Other important factors deciding the rate and effectiveness of bioremediation include nutrients, nature of pollutants, soil structure, temperature, moisture content, pH, and microbial diversity [81]. Carbon to nitrogen (C:N) ratio 10:1 and carbon to phosphorous ratio (C:P) 30:1 are the optimum nutrient requirements for bioremediation [98]. Microbiological bioremediation works best in the pH range of 5.5 to 8. Well-structured soil can help the bioremediation process drive more efficiently. Furthermore, temperatures ranging from 15 to 45°C are ideal for bioremediation techniques [84, 99].

5.5 TECHNIQUES OF BIOREMEDIATION

Bioremediation techniques can be divided into two categories depending on where they are performed: In-situ and Ex-situ. Plant-assisted bioremediation is subdivided into several sections based on the nature of the process. Ex-situ bioremediation is less effective than in-situ bioremediation. Because of its non-destructive nature, in-situ bioremediation has gotten more attention. In-situ phytoremediation may enhance soil fertility by adding organic matter (OM) other than soil remediation [100]. The concept of bioremediation solely depends on the ability of plants and microbes to absorb pollutants from the environment and detoxify those compounds by using different mechanisms.

The concept of phytoremediation was first proposed by Chaney in 1983, and it instantly gained popularity due to its sustainable nature [101]. Phytoextraction, phytofiltration, phytostabilization, phytovolatilization, phytodegradation, and phytodesalination are the basic techniques of phytoremediation [18, 102].

5.5.1 PHYTOEXTRACTION/PHYTOACCUMULATION

Phytoextraction is a phytoremediation technique in which plant roots take up contaminants in the soil and water, translocate, and store in harvestable aerial plant parts like shoots and leaves. The harvested biomass is disposed of afterwards [1, 42, 103, 104]. In phytoextraction, seeds or seedlings of metal accumulating plants are planted in contaminated soils using already established agronomy practices, and once plants absorb contaminants along with water and other plant nutrients and stored, harvestable above-ground plant parts are harvested and destroyed after a satisfactory plant growth and metal extraction. This method ensures a permanent removal of heavy metal contaminants such as Pb, V, Cd, Cr, Ni, and Cu. Nevertheless, phytoaccumulation is only applicable in low to moderately metal contaminated soils as plants don't sustain in heavy metal contaminated soils. Translocation of heavy metals to shoots is very crucial and essential in decontaminating heavy metal contaminated soil or water using phytoextraction as considering roots as the harvestable biomass is not practical [18, 105]. To increase metal solubility in soil solution and to ensure sufficient plant uptake of heavy metals, chelating agents such as Ethylenediamine-N,N'-disuccinic acid (EDDS) are added to the soil [89, 106]. The plant species to be used in phytoextraction successfully should have the characteristics of ability to withstand in highly contaminated soils, rapid growth rate and a dense root system, high metal accumulation capability and ability to store uptake heavy metals in above ground harvestable plant parts [1, 107].

Eichhornia crassipes, *Pistia stratiotes*, *Myrio-phylum aquaticum*, *Mentha aquatic*, *Ludwigina palustris*, *Centella asiatica* are some widely researched aquatic plants for their potential to be used in phytoextraction [108–111]. Phytoaccumulation could also be applied in radionuclide waste such as U, Pl, and Cs decontamination and petroleum hydrocarbon accumulation due to its' low initial investment and environmental feasibility [112–114] (Table 5.3).

TABLE 5.3 Potential Plant Species to be Used in Different Phytoaccumulation

Potential Plant Species	Contaminant	References
Pteris vittata L.	As	[115]
Gypophila sphaerocephala Fenzel	B	[116]
Oryza sativa L.	Cd	[117]
Brassica juncea L.	Cr	[118]
Berkheya coddii Roessler	Co	[119]
Chenopodium album L.	Pb	[120]
Phytolacca americana L.	Mn	[121]
Marrubium vulgare L	Hg	[122]
Brassica rapa L.	Se	[123]

5.5.2 PHYTOFILTRATION

Phytofiltration uses plants to remove contaminants from wastewater or contaminated surface waters [124]. Phytofltration is divided into three categories as rhizofiltration, blastofiltration, and caulofiltration depending on the plant part which perform the task [18]. The principle of phytofiltration is preventing deep penetration of contaminants in the soil by plant absorption or adsorption [18, 125]. The use of plant roots in the contaminant removal process is known as rhizofiltration whereas, blastofiltration is the process that uses seedlings to remove pollutants from wastewater and surface waters. Caulofiltration uses plant shoots to decontaminate surface waters and wastewater. This remediation technique can either be practiced in-situ to decontaminate heavy metals in surface water bodies or ex-situ in constructed systems to treat wastewater by using introduced plants. Studies have found that Cu, Ni, Pb, Cd, and Cr like heavy metals and radionucliids including U, Cs, and Sr can be remediated effectively by phytofiltration [126]. Plant species to be ideally used in phytofiltration should have the characteristics of production of a significant amount of root mass, tolerant to high levels of heavy metal and ability to accumulate large quantities of heavy metals without producing much secondary waste, easy handling, and low maintenance, etc. Considering the production of a large quantity of root biomass with larger surface area terrestrial plants were found to be ideal to be employed in phytofiltration [1, 127] (Table 5.4).

TABLE 5.4 Potential Plant Species to be Used in Different Phytofiltration

Potential Plant Species	Contaminant	References
Lemna minor L.	Cd, Cu, Ni, Se	[128–130]
Brassica juncea L.	Cr, Zn	[126, 131]
Hemidesmus indicus L.	Pb	[132]
Pistia stratiotes L.	Hg	[133]
Chenopodium amaranticolor H. J. Coste & Reyn	U	[134]

5.5.3 PHYTOSTABILIZATION/PHYTOIMMOBILIZATION

Phytostabilization or Phytoimmobilization reduce toxicity and mobility of metal contaminants in soil using certain plants thus, reducing the bioavailability of contaminants by preventing their entry into the food chain or groundwater [134–136]. Phytoimmobilization occurs in the root system or rhizosphere of certain plant species because of root absorption through root hairs or root adsorption of contaminants or precipitation within the rhizosphere [7, 137]. Plant immobilization of heavy metals in soil could be a result of precipitation, complexation, sorption in roots or metal valence reduction in the rhizosphere [17, 18, 138]. Plants convert highly toxic metals into relatively less toxic and less mobile states by using special redox enzymes by metal valence reduction and minimizing possible damage to the environment [139]. Plant species to be used in phytostabilization should have a special ability to grow in contaminated sites with root systems growing into contaminated zones which will alter chemical, physical, and biological conditions of the soil and permit desirable toxicity reduction processes. *Agrostis tenius* and *Festuca rubra* are two widely utilized cultivars in phytostabilization of Pb-, Cu-, and Zn- contaminated soils [1, 140]. According to the literature, deep-rooted plants could be effectively used in the phytostabilization of toxic metals including Cr^{6+} and Cr^{3+} and the remediation technique is very effective in fine-textured soils with high OM content [141, 142]. Nevertheless, the incorporation of soil amendments such as OM, bio-solids, natural or artificial clay minerals, phosphate fertilizers, etc., could enhance soil aggregates and improve the phytostabilization process [1]. Phytostabilization is only a management technique to limit the mobility of toxic substances in the soil as it does not remove contaminants permanently from soil or water and is essentially not a long-term solution for soil and water remediation [143]. Yet, phytostabilization could effectively be employed in the rehabilitation of highly metal-contaminated sites where natural plant growth is not possible.

The major advantage of this technique is reducing the risk of groundwater contamination by limiting contaminants mobilization results in wind, rain, and leaching [7, 144] (Table 5.5).

TABLE 5.5 Potential Plant Species to be Used in Different Phytostabilization

Potential Plant Species	Contaminant	References
Lemna minor L.	Cd, Se, and Cu	[128]
Myriophylhum aquaticum, *Ludwigina palustris*, *Mentha aquatic*	Fe, Zn, Cu, and Hg	[113]
Solanum tuberosum L.	As, Ag, Cr, and Sb	[145]
Amaranthus spinosus L.	Cu, Pb, and Cd	[146]
Solenum nigrum L., *Spinacia oleracea* L.	Pb, Cu, Cd, and Cr	[147]
Populus cathayana, *P. prezwaskii*, *P. yunnanensis*	Cd	[148]
Mentha aquatica L., *Mentha sylvestris* L., *Riccinus communis* L., *Spirogyra irregularis* Nageli	Ni	[108, 149, 150]
Lemna gibba L., *Scirpus mucronatus*	Zn	[151, 152]

5.5.4 PHYTOVOLATILIZATION

Phytovolatilization is the process that uptakes soil pollutants by plants along with water and other plant nutrients during plant growth and volatilizing to the atmosphere by converting them into less toxic volatilized compounds [42, 154]. For example, Hg, Se, and As compounds can be volatilized in hundreds of times fewer toxic forms such as Dimethyl selenide and Mercuric oxide to the atmosphere through phytovolatilization. Some of these compounds are still toxic to living organisms and plants thus, phytovolatilization is not recommended to practice in highly populated areas [7, 155–157]. According to literature, *Typha latifolia* L., *Chara canescens*, *Brassica juncea* are employed in plant assisted volatilization of toxic metals including Cd, Cu, Zn, Se, and Hg, etc. [138, 158]. This is the most controversial among phytoremediation techniques because it cannot completely remove pollutants from the system but only transfer the contaminants from soil to air which still can deposit on another site [42]. However, this bioremediation technique has the advantages of less soil erosion, minimum site disturbances and no production of contaminated biomass [1, 159] (Table 5.6).

TABLE 5.6 Potential Plant Species to be Used in Different Phytovolatilization

Potential Plant Species	Contaminant	References
Typha latifolia L.	Zn, Mn, Co, Cd, Cr, Ni, and As	[160]
Sorghum bicolor L.	Cd and Zn	[161]
Alternanthera philoxeroides, *Artemisia princeps*, *Bidens fron-dosa*, *Bidens pilosa*, *Cynodon dactylon*, *Digitaria sanguinalis*, *Erigeron canadensis*, and *Setaria plicata*	Cd, Mn, Pb, and Zn	[162]
Phragmites australis	Pb/Zn and Cu mine tailings	[163]
Silphium perfoliatum, *Quercus ilex*, *Acanthus ilicifolius* L., *Lupinus uncinatus* Schldl	Cd	[164–167]
Thysanolaena maxima, *Athyrium wardii*	Pb	[168, 169]
Lolium perenne and *Trifolium repens*	Cd, Zn, Pb	[170]

5.5.5 PHYTODEGRADATION/PHYTOTRANSFORMATION

Phytodegradation is the degradation of organic contaminants by plants either using enzymes or their metabolic activities. It depends on the species of rhizospheric microorganisms [171, 172]. Green plants are known as the "green liver of the biosphere" considering the phytodegradation process in which plants accumulate organic xenobiotics and detoxify through their metabolic processes. Plants secrete certain types of enzymes to the rhizosphere during their metabolic activities and those enzymes break down organic contaminants in the rhizosphere into nontoxic forms. Some phytotransformed products with simpler forms are up taken by plants for their rapid growth. Oxygenase, dehalogenase, nitroreductase, nitrilase, and peroxidase (POD) are several enzyme groups contributing to phytodegradation [1]. This bioremediation technique is only applicable for organic pollutants removal including synthetic agrochemicals and cannot be applied in remediation of heavy metal contaminated soils and water as metals are non-biodegradable [18, 173]. Khandare & Govindwar [174] stated that class 4 plants have superior phytodegradation abilities over class 3 plant species. *Eichchornia crassipes*, *Azolla caroliniana*, *Myriophyllum aquaticum*, *Typha angustifolia* L., and *Cyperus papyrus* L. are some of the eminent plant species used in phytodegradation whereas, *Brassica juncea* and *Liliodendron tulipifera* are two genetically modified plants with superior phytodegradation qualities [175–178]. Contaminant removal efficiency depends on phytochemical properties of the plant and contaminant concentration and uptake efficiency [179]. *Eichchornia crassipes* is very economically efficient in the bioremediation of industrial wastewater [180] (Table 5.7).

TABLE 5.7 Potential Plant Species to Be Used in Different Phytodegradation

Potential Plant Species	Contaminant	References
Canna glauca L., *Colocasia esculenta* L. Schott, *Cyperus papyrus* L., *Typha angustifolia* L., *Arundo donax* L., *Pteris vittata*	As	[157, 178, 181]
Azolla caroliniana	Hg	[176]
Liriodendron tulipifera	methyl-Hg	[182]
Brassica juncea, *Lupinus* sp.	Mercury-contaminated mine tailings	[153, 183]

5.5.6 PHYTODESALINATION

Among the methods, phytodesalination is the technique that removes excess salts from salt affected soils using halophytic plants which are well adapted to saline conditions [184, 185]. It is an emerging branch of phytoremediation which is applicable for converting salt affected non-arable lands supportive for normal plant growth [18, 186]. Studies reveal *S. maritima* and *S. portulacastrum* can accumulate NaCl from highly salt affected soils and convert lands suitable for normal cultivation after several consecutive cultivations [187, 188].

5.6 APPLICATIONS OF BIOREMEDIATION IN CONSTRUCTED WETLANDS

Constructed wetlands have been designed by scientists to simulate the function of natural wetlands with the purpose of wastewater treatment and to trap contaminants in contaminated sites. Constructed wetlands are complex ecosystems consisting of plants, microbes, and sediments which as a biogeochemical filter which slow down flow speed of large volumes of contaminated water and removes heavy metals and hydrocarbons like contaminants [143, 189–192]. This technology has been identified as a low cost, environmentally friendly treatment method for the removal of contaminants in municipal wastewater and effluent from oil refineries and electricity generating plants [193–195].

Mostly, aquatic plants are recommended and employed in constructed wetlands over terrestrial plants due to their faster growth rates, higher ability to pollutants uptake due to direct contact with contaminated water and production of more biomass like characteristics [196]. Several types of aquatic plants could be identified which are used in wetland purposes as emergent plants, floating-leaved plants, freely floating plants and submerged plants [197]. Most of the biomass of emergent plants locate above the water surface while roots are located in shallow water and these types of plants produce less biomass compared to submerged plants [198]. Only the leaves of the floating-leaved plants are floating on the water surface while all the other plant parts including roots and stem are located around 0.5 m–3.0 m below the water surface [199]. All the parts of submerged plants are located below the water surface, and they produce more biomass for contaminants absorption [200]. The whole plant is engaged in contaminant absorption in submerged plants while only roots of floating plants actively engage in heavy metal uptake [201]. *Eichhornia crassipes, Pistia stratiotes, Azolla* are some commonly used aquatic plants in constructed wetlands while *Salix* spp. and *Populus* spp. are used along the edges of constructed wetlands [199].

5.7 LIMITATIONS OF BIOREMEDIATION

As with any soil remediation method bioremediation also has its limitations. One of the major concerns related to phytoremediation is its inability to remediate contaminants present in deep soils and aquifers. Usually, plants can only clean up shallow soils due to their short root system and trees could perform slightly higher remediation up to around 15 feet in soil compared to plants. Occasionally, plants and trees could clean shallow groundwater, despite that phytoremediation is not an effective method in decontaminating deep aquifers [42, 202]. Also, the efficiency of phytoremediation processes is another issue as it could require several growing seasons to remediate soil or water using these bioremediation techniques [7]. Soil and water pollution becoming air pollution issues while applying phytovolatilization and incomplete removal of air contaminants are another two limitations of bioremediation [157, 203]. The possibility to contaminate the food chain by plant absorbed toxics and producing non-edible plant materials are another human health and agricultural productivity related issue of phytoremediation techniques [204, 205].

5.8 FUTURE TRENDS

Phytoremediation is a modern branch of research and most of the research have been conducted either on a laboratory scale or in greenhouses. Except that the actual results may vary from those situations as actual fields because it is a place where all the environmental factors simultaneously affect the results [18, 206]. Temperature variations, precipitation, moisture, soil structure, soil type, soil pH, nutrients, uniformity of contaminant distribution, and plant pathogens are some key factors affecting the results of phytoremediation techniques in actual field conditions [143, 207, 208]. Therefore, conducting different phytoremediation techniques in actual field conditions is a must to determine suitable plant species for uptake of different heavy metals and to determine the exact applicability and feasibility of those techniques [18].

Future research exposure towards the biochemical processing of heavy metals may provide a sharp platform to understand the phytotoxicity constraints. Also, the aspect of genetic engineering will enable plants to tolerate higher concentrations of metals. With this said, it is justified that biochemistry and biotechnology play a major role in future research in the field of phytoremediation. The development of 'superbug plants' for employed in phytoremediation is another ongoing research area that should be paid more attention to [209]. Genetic composition varies among different plant species even among cultivars of the same species. Thus, metal uptake, accumulation, and translocation like mechanisms vary with the species as well as with the cultivars. Therefore, identification of desirable traits for phytoremediation in natural hyperaccumulators which produce high biomass, selection of those traits either by conventional breeding or hybridization techniques and successful incorporation of those genes to suitable plant species could develop superbug plants [210, 211]. Therefore, identification of different desirable traits and their relevant genes and genetic coding is essential for the improvement of plant species that could serve the purpose best. Nonetheless, sufficient attention should be paid to gene expressions, interactions, and transgenic pathways in order to avoid unintended negative effects to the biosphere from those genetically modified species in future [212].

Conventional disposal approaches are not applicable for produced biomass from phytoremediation due to the possible presence of volatile metals. Thus, methods available for the recovery of heavy metals accumulated in plant biomass resulting from bioremediation are still limited. Therefore, the development of new effective methods for heavy metal recovery from plant materials should pay more attention [29]. Besides, use of genetic manipulation technique of chloroplast genome for some plant species as a substitute for transgene escape through pollens [213], transgenic based study in phytoremediation to overcome combined contamination occurring in polluted areas using multigene approach [214], selection of suitable species with substances to that may interrupt herbivores from feeding and enhanced tolerance capacity [133], suggesting cropping systems with target pollutant and site-based phytoremediation, and developing management plans and policy instruments and tools to control the heavy metal contamination and decontamination at regional farm or lands to the national level [214] can be practiced to combat this environmental concern.

5.9 CONCLUSIONS AND RECOMMENDATIONS

Ever growing diverse industrial applications in the world have fueled the heavy metal contamination of soil and water causing massive environmental pollution. At present, it challenges the research scientists, engineers, and environmentalists to employ the best sustainable pollution control and prevention methods to overcome this global crisis which is heavy metal contamination. Among the current treatment technologies, conventional chemical and physical treatment methods support abate adverse environmental and health effects driven by heavy metal contamination of soil and water, however, this approach costs the expenses and highly advanced technology to remediate the soil and water to minimize the heavy metal pollution. As a sustainable solution, many scientists have come up with the fact that bioremediation methods (application of biological components such as microbes and plant species to remove heavy metal contamination) can serve as a proxy to enhance the efficiency of removal of heavy metals and meanwhile alleviate the drawbacks of other conventional methods. In the context of phytoremediation, the mechanisms involve a change of pH, redox activity or adsorption activity of contaminant from the environment. This heavy metal contaminant removal process is heavily dependent on the nature and fate of the heavy metal, heavy metal transportation process through cells, and tolerance and sensitivity. Further, connectivity and interaction of the rhizosphere of the plant with existing environmental variables also directly affect the remediation process. The phytofiltration that employs plant roots, seeds, and exposed plant shoots to remove pollutants from water has been applied in numerous cases in phytoremediation activities. In addition, phytostimulation, phytodegradation, phytostabilization, phytodesalination, phytovolatilization, and phytoextraction are other major methods of phytoremediation techniques that are available worldwide. However, all these techniques entail their pros and cons, yet it should be carefully concerned in determining the best suitable method to remove heavy metals cases based. The modifications and redirection of the phytoremediation strategy have been considered to enhance its potential efficiency. Hence, the further development of suggested technological strategies can be used to fulfill upcoming research gaps. Moreover, there exists a potent to occupy the synergy between physicochemical technologies combined with phytoremediation techniques to boost effective decontamination and control of heavy metals from soil and water sources. Ultimately, this chapter signifies the importance of site or event specific and technology-specific phytoremediation approaches whereas all the stakeholder involvement should be well addressed.

KEYWORDS

- **bioremediation**
- **heavy metals**
- **metal detoxification**
- **metal pollution**
- **physical remediation**
- **phytofiltration**
- **phytoremediation**

REFERENCES

1. Jabeen, R., Ahmad, A., & Iqbal, M., (2009). Phytoremediation of heavy metals: Physiological and molecular mechanisms. *The Botanical Review, 75*(4), 339–364.
2. Gisbert, C., Ros, R., De Haro, A., Walker, D. J., Bernal, M. P., Serrano, R., & Navarro-Aviñó, J., (2003). A plant genetically modified that accumulates Pb is especially promising for phytoremediation. *Biochemical and Biophysical Research Communications, 303*(2), 440–445.
3. Halim, M., Conte, P., & Piccolo, A., (2003). Potential availability of heavy metals to phytoextraction from contaminated soils induced by exogenous humic substances. *Chemosphere, 52*(1), 265–275.

4. Khan, S., Hesham, A. E. L., Qiao, M., Rehman, S., & He, J. Z., (2010). Effects of Cd and Pb on soil microbial community structure and activities. *Environmental Science and Pollution Research, 17*(2), 288–296.
5. Förstner, U., (1995). Land contamination by metals: Global scope and magnitude of problem. In: Allen, H. E., et al., (ed.) *Metal Speciation and Contamination of Soil* (pp. 1–33). Lewis Publ., Boca Raton, FL.
6. Kumar, P. N., Dushenkov, V., Motto, H., & Raskin, I., (1995). Phytoextraction: The use of plants to remove heavy metals from soils. *Environmental Science & Technology, 29*(5), 1232–1238.
7. Naushad, M., & ALOthman, Z. A., (2015). Separation of toxic Pb^{2+} metal from aqueous solution using strongly acidic cation-exchange resin: Analytical applications for the removal of metal ions from pharmaceutical formulation. *Desalination and Water Treatment, 53*(8), 2158–2166.
8. Naushad, M., Abdullah Al, O. Z., Rabiul, A. M., Alfadul, S. M., & Ahamad, T., (2016). Adsorption of rose Bengal dye from aqueous solution by amberlite Ira-938 resin: Kinetics, isotherms, and thermodynamic studies. *Desalination and Water Treatment, 57*(29), 13527–13533.
9. Muthusaravanan, S., Sivarajasekar, N., Vivek, J. S., Paramasivan, T., Naushad, M., Prakashmaran, J., & Al-Duaij, O. K., (2018). Phytoremediation of heavy metals: Mechanisms, methods and enhancements. *Environmental Chemistry Letters, 16*(4), 1339–1359.
10. Garty, J., (2001). Biomonitoring atmospheric heavy metals with lichens: Theory and application. *Critical Reviews in Plant Sciences, 20*(4), 309–371.
11. Göhre, V., & Paszkowski, U., (2006). Contribution of the arbuscular mycorrhizal symbiosis to heavy metal phytoremediation. *Planta, 223*(6), 1115–1122.
12. Peng, K. J., Luo, C. L., Chen, Y. H., Wang, G. P., Li, X. D., & Shen, Z. G., (2009). Cadmium and other metal uptake by *Lobelia chinensis* and *Solanum nigrum* from contaminated soils. *Bulletin of Environmental Contamination and Toxicology, 83*(2), 260–264.
13. Dabonne, S., Koffi, B., Kouadio, E., Koffi, A., Due, E., & Kouame, L., (2010). Traditional utensils: Potential sources of poisoning by heavy metals. *British Journal of Pharmacology and Toxicology, 1*(2), 90–92.
14. Sharma, G., Pathania, D., & Naushad, M., (2015). Preparation, characterization, and ion exchange behavior of nanocomposite polyaniline zirconium (IV) selenotungstophosphate for the separation of toxic metal ions. *Ionics, 21*(4), 1045–1055.
15. Khan, M. A., Ahmad, I., & Rahman, I. U., (2007). Effect of environmental pollution on heavy metals content of *Withania somnifera*. *Journal of the Chinese Chemical Society, 54*(2), 339–343.
16. Sabiha, J., Mehmood, T., Chaudhry, M. M., Tufail, M., & Irfan, N., (2009). Heavy metal pollution from phosphate rock used for the production of fertilizer in Pakistan. *Microchem. J, 91*, 94–99.
17. Wuana, R. A., & Okieimen, F. E., (2011). Heavy metals in contaminated soils: A review of sources, chemistry, risks and best available strategies for remediation. *International Scholarly Research Notices, 2011*.
18. Ali, H., Khan, E., & Sajad, M. A., (2013). Phytoremediation of heavy metals-concepts and applications. *Chemosphere, 91*(7), 869–881.

19. Sivarajasekar, N., Mohanraj, N., Sivamani, S., Moorthy, I. G., Kothandan, R., & Muthusaravanan, S., (2017). Comparative modeling of fluoride biosorption onto waste *Gossypium hirsutum* seed microwave-bichar using response surface methodology and artificial neural networks. In: *2017 International Conference on Intelligent Computing, Instrumentation and Control Technologies (ICICICT)* (pp. 1631–1635). IEEE.
20. Zubair, M., Shakir, M., Ali, Q., Rani, N., Fatima, N., Farooq, S., Shafiq, S., et al., (2016). Rhizobacteria and phytoremediation of heavy metals. *Environmental Technology Reviews, 5*(1), 112–119.
21. Nazir, M. I., Idrees, I., Idrees, P., Ahmad, S., Ali, Q., & Malik, A., (2020). Potential of water hyacinth (*Eichhornia crassipes* L.) for phytoremediation of heavy metals from wastewater. *Biological and Clinical Sciences Research Journal, 2020*(1).
22. Lindqvist, O., (1991). Mercury in the Swedish environment. *Water Air Soil Bull., 55*(1), 23–32.
23. Seaward, M. R. D., & Richardson, D. H. S., (1990). Atmospheric sources of metal pollution and effects on vegetation. In: Shaw, A. J., (ed.), *Heavy Metal Tolerance in Plants: Evolutionary Aspects* (pp.75–92). CRC Press, Florida.
24. Flora, G., Gupta, D., & Tiwari, A., (2012). Toxicity of lead: A review with recent updates. *Interdisciplinary Toxicology, 5*(2), 47–58.
25. Alloway, B. J., (1995). *Heavy Metals in Soils*. Blackie Academic & Professional, London.
26. Kabata-Pendias, A., (2001). *Trace Elements in Soils and Plants* (3rd edn.) CRC Press, Boca Raton, Florida.
27. Walsh, P. R., Duce, R. A., & Finishing, J. I., (1979). Consideration of the enrichment, sources, and flux of arsenic in the troposphere. *J. Geophysical. Res., 84*, 1719–1726.
28. Nriagu, J. O., (1994). *Arsenic in the Environment, 1. Cycling and Characterization*. John Wiley and Sons, New York.
29. Lone, M. I., He, Z. L., Stoffella, P. J., & Yang, X. E., (2008). Phytoremediation of heavy metal polluted soils and water: Progresses and perspectives. *Journal of Zhejiang University Science B, 9*(3), 210–220.
30. Liu, X. M., Wu, Q. T., & Banks, M. K., (2005). Effect of simultaneous establishment of *Sedum alfridii* and Zea mays on heavy metal accumulation in plants. *Int. J. Phytoremediation, 7*(1), 43–53. doi: 10.1080/16226510590915800.
31. Ayoub, K., Van, H. E. D., Cassir, M., & Bermond, A., (2010). Application of advanced oxidation processes for TNT removal: A review. *J Hazard Mater, 178*, 10–28. https://doi.org/10.1016/j.jhazm at.2010.02.042.
32. Sivarajasekar, N., & Baskar, R., (2015). Biosorption of basic violet 10 onto activated *Gossypium hirsutum* seeds: Batch and fixed-bed column studies. *Chinese Journal of Chemical Engineering, 23*(10), 1610–1619.
33. Arora, M., Kiran, B., Rani, S., Rani, A., Kaur, B., & Mittal, N., (2008). Heavy metal accumulation in vegetables irrigated with water from different sources. *Food Chemistry, 111*(4), 811–815.
34. Memon, A. R., & Schröder, P., (2009). Implications of metal accumulation mechanisms to phytoremediation. *Environmental Science and Pollution Research, 16*(2), 162–175.
35. Mudipalli, A., (2008). Metals (micro nutrients or toxicants) & Global Health. *Indian Journal of Medical Research, 128*(4), 331–334.

36. Malayeri, B. E., Chehregani, A., Yousefi, N., & Lorestani, B., (2008). Identification of the hyper accumulator plants in copper and iron mine in Iran. *Pakistan Journal of Biological Sciences: PJBS, 11*(3), 490–492.
37. Jaishankar, M., Tseten, T., Anbalagan, N., Mathew, B. B., & Beeregowda, K. N., (2014). Toxicity, mechanism and health effects of some heavy metals. *Interdisciplinary Toxicology, 7*(2), 60–72.
38. Neustadt, J., & Pieczenik, S., (2007). Patient handout: Toxic metal contamination: Mercury. *Integrative Medicine-Innovision Communications, 6*(2), 36–37.
39. Ainza, C., Trevors, J., & Saier, M., (2010). Environmental mercury rising. *Water, Air, and Soil Pollution, 205*(1), 47–48.
40. Gulati, K., Banerjee, B., Lall, S. B., & Ray, A., (2010). Effects of diesel exhaust, heavy metals and pesticides on various organ systems: Possible mechanisms and strategies for prevention and treatment. *Indian Journal of Experimental Biology, 48*, 710–721.
41. Salem, H. M., Eweida, E. A., & Farag, A., (2000). Heavy metals in drinking water and their environmental impact on human health. In: *Int. Conference on the Environ. Hazards Mitigation, Cairo Univ. Egypt* (pp. 542–556).
42. Padmavathiamma, P. K., & Li, L. Y., (2007). Phytoremediation technology: Hyper-accumulation metals in plants. *Water, Air, and Soil Pollution, 184*(1), 105–126.
43. Iqbal, M. P., (2012). Lead pollution-a risk factor for cardiovascular disease in Asian developing countries. *Pakistan Journal of Pharmaceutical Sciences, 25*(1), 289–294.
44. Awofolu, O. R., (2005). A survey of trace metals in vegetation, soil and lower animal along some selected major roads in metropolitan city of Lagos. *Environmental Monitoring and Assessment, 105*(1), 431–447.
45. Tripathi, R. D., Srivastava, S., Mishra, S., Singh, N., Tuli, R., Gupta, D. K., & Maathuis, F. J., (2007). Arsenic hazards: Strategies for tolerance and remediation by plants. *Trends in Biotechnology, 25*(4), 158–165.
46. Hess, R., & Schmid, B., (2002). Zinc supplement overdose can have toxic effects. *J. Paediatr. Haematol. Oncol., 24*, 582–584.
47. Govindasamy, C., Arulpriya, M., Ruban, P., Francisca, L. J., & Ilayaraja, A., (2011). Concentration of heavy metals in seagrasses tissue of the Palk Strait, Bay of Bengal. *Int. J. Environ. Sci., 2*, 145–153.
48. Dixit, R., Malaviya, D., Pandiyan, K., Singh, U. B., Sahu, A., Shukla, R., & Paul, D., (2015). Bioremediation of heavy metals from soil and aquatic environment: An overview of principles and criteria of fundamental processes. *Sustainability, 7*(2), 2189–2212.
49. Mulligan, C. N., Yong, R. N., & Gibbs, B. F., (2001). Remediation technologies for metal-contaminated soils and groundwater: An evaluation. *Eng. Geol., 60*, 193–207. https://doi.org/10.1016/S0013-7952(00)00101-0.
50. Tanaka, T., Yamada, K., Tonosaki, T., et al., (2000). Enzymatic degradation of alkylphenols, bisphenol A, synthetic estrogen and phthalic ester. *Water Sci. Technol., 42*, 89–95.
51. Guangde, L., Zhongwen, Z., Pei, J., Nannan, Z., Li, L., Yufei, Y., & Miao, Y., (2009). Leaching remediation of heavy metal contaminated fluvio-aquatic soil with tea-saponin. *Trans. Chinese Soc. Agri. Eng., 10*, 1, 2.
52. Veselý, T., Tlustoš, P., & Száková, J., (2012). Organic acid enhanced soil risk element (Cd, Pb and Zn) leaching and secondary bioconcentration in water lettuce (*Pistia stratiotes* L.) in the rhizofiltration process. *International Journal of Phytoremediation, 14*(4), 335–349.

53. Akpor, O. B., & Muchie, M., (2010). Remediation of heavy metals in drinking water and wastewater treatment systems: Processes and applications. *International Journal of Physical Sciences, 5*(12), 1807–1817.
54. Ahluwalia, S. S., & Goyal, D., (2007). Microbial and plant derived biomass for removal of heavy metals from wastewater. *Bioresour. Technol., 98*, 2243–2257.
55. Kapoor, A., & Viraraghavan, T., (1995). Fungal biosorption—An alternative treatment option for heavy metal bearing wastewaters: A review. *Bioresource Technology, 53*(3), 195–206.
56. Katiyar, P., Pandey, N., & Sahu, K. K., (2020). Biological approaches of fluoride remediation: Potential for environmental clean-up. *Environmental Science and Pollution Research, 27*(12), 13044–13055.
57. Borah, S. N., Sen, S., Sarma, H., & Pakshirajan, K., (2021). Biological remediation of selenium in soil and water. *Handbook of Assisted and Amendment: Enhanced Sustainable Remediation Technology*, 403–421.
58. Yao, Z. T., Li, J. H., Xie, H. H., & Yu, C. H., (2012). Review on remediation technologies of soil contaminated by heavy metals. *Procedia Environ. Sci., 16*, 722–729. doi: 10.1016/j. proenv.2012.10.099.
59. Khalid, S., Shahid, M., Niazi, N. K., Murtaza, B., Bibi, I., & Dumat, C., (2017). A comparison of technologies for remediation of heavy metal contaminated soils. *Journal of Geochemical Exploration, 182*, 247–268.
60. Wan, X., Lei, M., & Chen, T., (2020). Review on remediation technologies for arsenic-contaminated soil. *Frontiers of Environmental Science & Engineering, 14*(2), 1–14.
61. Jankaite, A., & Vasarevicius, S., (2005). Remediation technologies for soils contaminated with heavy metals. *J. Environ. Eng. Landscape Manage. 13*, 109–113. doi: 10.1080/16486897.2005.9636854.
62. Li, C., Zhou, K., Qin, W., Tian, C., Qi, M., Yan, X., & Han, W., (2019). A review on heavy metals contamination in soil: Effects, sources, and remediation techniques. *Soil and Sediment Contamination: An International Journal, 28*(4), 380–394.
63. Chang, T. C., & Yen, J. H., (2006). On–site mercury–contaminated soils remediation by using thermal desorption technology. *J. Hazard. Mater., 128*, 208–217. doi: 10.1016/j. jhazmat.2005.07.053.
64. Song, B., Zeng, G. M., Gong, J. L., Liang, J., Xu, P., Liu, Z. F., Zhang, Y., et al., (2017). Evaluation methods for assessing effectiveness of in situ remediation of soil and sediment contaminated with organic pollutants and heavy metals. *Environ. Int., 105*, 43–55. doi: 10.1016/j.envint.2017.05.001.
65. Dermont, G., Bergeron, M., Mercier, G., & Richer–Lafleche, M., (2008). Soil washing for metal removal: A review of physical/chemical technologies and field applications. *J. Hazard. Mater. 152*, 1–31. doi: 10.1016/j.jhazmat.2007.10.043.
66. Sun, Y. B., Sun, G. H., Zhou, Q. X., Xu, Y. M., Wang, L., Liang, X. F., & Qing, X., (2011). Induced-phytoextraction of heavy metals from contaminated soil irrigated by industrial wastewater with Marvel of Peru (*Mirabilis jalapa* L.). *Plant, Soil and Environment, 57*(8), 364–371.
67. Jiang, W., Tao, T., & Liao, Z. M., (2011). Removal of heavy metal from contaminated soil with chelating agents. *Open J. Soil Sci., 1*, 70–76. doi: 10.4236/ojss.2011.12010.
68. Kim, J. O., Lee, Y. W., & Chung, J., (2013). The role of organic acids in the mobilization of heavy metals from soil. *KSCE J. Civil Eng., 17*, 1596–1602. doi: 10.1007/ s12205-013-0323-z.

69. Maity, J. P., Huang, Y. M., Fan, C. W., Chen, C. C., Li, C. Y., Hsu, C. M., Chang, Y. F., et al., (2013). Evaluation of remediation process with soapberry derived saponin for removal of heavy metals from contaminated soils in Hai-Pu, Taiwan. *J. Environ. Sci., 25,* 1180–1185. doi: 10.1016/S1001-0742(12)60162-4.
70. Lestan, D., Luo, C. L., & Li, X. D., (2008). The use of chelating agents in the remediation of metal-contaminated soils: A review. *Environ. Pollut., 153,* 3–13. doi: 10.1016/j.envpol.2007.11.015.
71. Qiao, J. B., Sun, H. M., Luo, X. H., Zhang, W., Mathews, S., & Yin, X. Q., (2017). EDTA–Assisted leaching of Pb and Cd from contaminated soil. *Chemosphere, 167,* 422–428. doi: 10.1016/j. chemosphere.2016.10.034.
72. Yong, S. O., Oh, S. E., Ahmad, M., Hyun, S., Kim, K. R., Moon, D. H., & Lee, S. S., (2010). Effects of natural and calcined oyster shells on Cd and Pb immobilization in contaminated soils. *Environ. Earth Sci., 61,* 1301–1308. doi: 10.1007/s12665-010-0674-4.
73. Islam, M. N., Taki, G., Xuan, P. N., Jo, Y. T., Kim, J., & Park, J. H., (2017). Heavy metal stabilization in contaminated soil by treatment with calcined cockle shell. *Environ. Sci. Pollut. Res., 24,* 7177–7183. doi: 10.1007/s11356–016-8330-5.
74. Soares, M. A., Quina, M. J., & Quinta–Ferreira, R. M., (2015). Immobilization of lead and zinc in contaminated soil using compost derived from industrial eggshell. *J. Environ. Manage., 164,* 137–145. doi: 10.1016/j.jenvman.2015.08.042.
75. Otero, M., Cutillasbarreiro, L., Nóvoamuñoz, J. C., Arias–Estévez, M., Fernández–Sanjurjo, M. J., Álvarez–Rodríguez, E., & Núñez–Delgado, A., (2015). Cr (VI) sorption/desorption on untreated and mussel-Shell-Treated soil materials: Fractionation and effects of pH and chromium concentration. *Solid Earth, 6,* 3419–3444. doi: 10.5194/se-6-373-2015.
76. Navarro, A., (2012). Effect of sludge amendment on remediation of metal contaminated soils. *Minerals, 2,* 473–492. doi: 10.3390/min2040473.
77. Guo, G. L., Zhou, Q. X., & Ma, L. Q., (2006). Availability and assessment of fixing additives for the in-situ remediation of heavy metal contaminated soils: A review. *Environ. Monit. Asses., 116,* 513–528. doi: 10.1007/s10661-006-7668-4.
78. Song, Y., Ammami, M. T., Benamar, A., Mezazigh, S., & Wang, H. Q., (2016). Effect of EDTA, EDDS, NTA and citric acid on electrokinetic remediation of As, Cd, Cr, Cu, Ni, Pb and Zn contaminated dredged marine sediment. *Environ. Sci. Pollut. Res., 23,* 10577–10586. doi: 10.1007/ s11356-015-5966-5.
79. Chakraborty, R., Wu, C. H., & Hazen, T. C., (2012). Systems biology approach to bioremediation. *Current Opinion in Biotechnology, 23*(3), 483–490.
80. Conesa, H. M., Evangelou, M. W., Robinson, B. H., & Schulin, R., (2012). A critical view of current state of phytotechnologies to remediate soils: Still a promising tool? *The Scientific World Journal, 2012.*
81. Mani, D., & Kumar, C., (2014). Biotechnological advances in bioremediation of heavy metals contaminated ecosystems: An overview with special reference to phytoremediation. *International Journal of Environmental Science and Technology, 11*(3), 843–872.
82. Jin, Y., Luan, Y., Ning, Y., & Wang, L., (2018). Effects and mechanisms of microbial remediation of heavy metals in soil: A critical review. *Applied Sciences, 8*(8), 1336.
83. Watt, N. R., (2007). Testing amendments for increasing soil availability of radionuclides. In: *Phytoremediation* (pp. 131–137). Humana Press.

84. Vidali, M., (2001). Bioremediation. an overview. *Pure and Applied Chemistry, 73*(7), 1163–1172.
85. Lili, L., & Hui, S., (2007). Advance of research on phytoremediation of petroleum-polluted soil. *Environ. Prot. Chem. Ind., 3*(11).
86. Machackova, J., Wittlingerova, Z., Vlk, K., & Zima, J., (2012). Major factors affecting in situ biodegradation rates of jet-fuel during large-scale biosparging project in sedimentary bedrock. *Journal of Environmental Science and Health, Part A, 47*(8), 1152–1165.
87. Lasat, M. M., (2002). Phytoextraction of toxic metals: A review of biological mechanisms. *J. Environ. Qual., 31*, 109–120.
88. Wang, F., Huang, C., Chen, Z., & Bao, K., (2019). Distribution, ecological risk assessment, and bioavailability of cadmium in soil from Nansha, Pearl River Delta, China. *International Journal of Environmental Research and Public Health, 16*(19), 3637.
89. Prasad, M. N. V., (2003). Phytoremediation of metal-polluted ecosystems: Hype for commercialization. *Russian Journal of Plant Physiology, 50*(5), 686–701.
90. Sheoran, V., Sheoran, A. S., & Poonia, P., (2010). Role of hyperaccumulators in phytoextraction of metals from contaminated mining sites: A review. *Critical Reviews in Environmental Science and Technology, 41*(2), 168–214.
91. Thangavel, P., & Subbhuraam, C. V., (2004). Phytoextraction: Role of hyperaccumulator in metal contaminated soils. *Proceedings of the Indian National Science Academy Part B, 70*, 109–130.
92. Bagga, D. K., & Peterson, S., (2001). Phytoremediation of arsenic-contaminated soil as affected by the chelating agent CDTA and different levels of soil pH. *Remediation, 12*, 77–85.
93. Song, L. S., Wang, S. Q., Xiao, R. P., Spurgeon, H., Lakatta, E. G., & Cheng, H., (2001). β-Adrenergic stimulation synchronizes intracellular Ca^{2+} release during excitation-contraction coupling in cardiac myocytes. *Circulation Research, 88*(8), 794–801.
94. Wenger, K., Kayser, A., Gupta, S. K., Furrer, G., & Schulin, R., (2002). Comparison of NTA and elemental sulfur as potential soil amendments in phytoremediation. *Soil and Sediment Contamination, 11*(5), 655–672.
95. Zhao, H. Y., Lin, L. J., Yan, Q. L., Yang, Y. X., Zhu, X. M., & Shao, J. R., (2011). Effects of EDTA and DTPA on lead and zinc accumulation of ryegrass. *Journal of Environmental Protection, 2*(07), 932–939.
96. Song, X., Hu, X., Ji, P., Li, Y., Chi, G., & Song, Y., (2012). Phytoremediation of cadmium-contaminated farmland soil by the hyperaccumulator *Beta vulgaris* L. var. cicla. *Bulletin of Environmental Contamination and Toxicology, 88*(4), 623–626.
97. Smolińska, B., & Król, K., (2012). Leaching of mercury during phytoextraction assisted by EDTA, KI and citric acid. *Journal of Chemical Technology & Biotechnology, 87*(9), 1360–1365.
98. Atagana, H. I., Haynes, R. J., & Wallis, F. M., (2003). Optimization of soil physical and chemical conditions for the bioremediation of creosote-contaminated soil. *Biodegradation, 14*(4), 297–307.
99. Thapa, B., Kc, A. K., & Ghimire, A., (2012). A review on bioremediation of petroleum hydrocarbon contaminants in soil. *Kathmandu University Journal of Science, Engineering and Technology, 8*(1), 164–170.
100. Mench, M., Schwitzguébel, J. P., Schroeder, P., Bert, V., Gawronski, S., & Gupta, S., (2009). Assessment of successful experiments and limitations of phytotechnologies:

Contaminant uptake, detoxification and sequestration, and consequences for food safety. *Environmental Science and Pollution Research, 16*(7), 876–900.
101. Garbisu, C., & Alkorta, I., (2003). Basic concepts on heavy metal soil bioremediation. *EJMP & EP (European Journal of Mineral Processing and Environmental Protection), 3*(1), 58–66.
102. Alkorta, I., Hernández-Allica, J., Becerril, J. M., Amezaga, I., Albizu, I., & Garbisu, C., (2004). Recent findings on the phytoremediation of soils contaminated with environmentally toxic heavy metals and metalloids such as zinc, cadmium, lead, and arsenic. *Reviews in Environmental Science and Biotechnology, 3*(1), 71–90.
103. Yoon, J., Cao, X., Zhou, Q., & Ma, L. Q., (2006). Accumulation of Pb, Cu, and Zn in native plants growing on a contaminated Florida site. *Science of the Total Environment, 368*(2, 3), 456–464.
104. Rafati, M., Khorasani, N., Moattar, F., Shirvany, A., Moraghebi, F., & Hosseinzadeh, S., (2011). Phytoremediation potential of *Populus alba* and *Morus alba* for cadmium, chromium and nickel absorption from polluted soil. *Int. J. Environ. Res., 5*, 961–970.
105. Tangahu, B. V., Sheikh, A. S. R., Basri, H., Idris, M., Anuar, N., & Mukhlisin, M., (2011). A review on heavy metals (As, Pb, and Hg) uptake by plants through phytoremediation. *International Journal of Chemical Engineering, 2011.*
106. Meers, E., Ruttens, A., Hopgood, M., Lesage, E., & Tack, F. M. G., (2005). Potential of *Brassic rapa, Cannabis sativa, Helianthus annuus* and *Zea mays* for phytoextraction of heavy metals from calcareous dredged sediment derived soils. *Chemosphere, 61*(4), 561–572.
107. Blaylock, M. J., (2000). Phytoextraction of metals. *Phytoremediation of Toxic Metals: Using Plants to Clean up the Environment,* 53–70.
108. Zurayk, R., Sukkariyah, B., Baalbaki, R., & Abi, G. D., (2002). Ni phytoaccumulation in *Mentha aquatica* L. and *Mentha sylvestris* L. *Water, Air, and Soil Pollution, 139*(1), 355–364.
109. Anawar, H. M., Garcia-Sanchez, A., Kul, A. M. T., & Rahman, M. M., (2008). Phytofiltration of water polluted with arsenic and heavy metals. *International Journal of Environment and Pollution, 33*(2, 3), 292–312.
110. Mokhtar, H., Morad, N., & Fizri, F. F. A., (2011). Phytoaccumulation of copper from aqueous solutions using *Eichhornia crassipes* and *Centella asiatica. International Journal of Environmental Science and Development, 2*(3), 205.
111. Harguinteguy, C. A., Schreiber, R., & Pignata, M. L., (2013). *Myriophyllum aquaticum* as a biomonitor of water heavy metal input related to agricultural activities in the Xanaes River (Córdoba, Argentina). *Ecological Indicators, 27*, 8–16.
112. Hossner, L. R., Loeppert, R. H., Newton, R. J., & Szaniszlo, P. J., (1998). Phytoaccumulation of chromium, uranium, and plutonium in plant systems. In: Attrep Jr., M. (Ed.), *Amarillo National Resource Center for Plutonium, 3*, 1–31.
113. Kamal, M., Ghaly, A. E., Mahmoud, N., & Cote, R., (2004). Phytoaccumulation of heavy metals by aquatic plants. *Environment International, 29*(8), 1029–1039.
114. Un Nisa, W., & Rashid, A., (2015). Potential of vetiver (*Vetiveria Zizanioides* L.) grass in removing selected PAHs from diesel contaminated soil. *Pakistan Journal of Botany, 47*(1), 291–296.
115. Gonzaga, M. I. S., Santos, J. A., & Ma, L. Q., (2006). Arsenic chemistry in the rhizosphere of *Pteris vittata* L. and *Nephrolepis exaltata* L. *Environmental Pollution, 143*(2), 254–260.

116. Babaoğlu, M., Gezgin, S., Topal, A., Sade, B., & Dural, H., (2004). *Gypsophila sphaerocephala* Fenzl ex Tchihat.: A boron hyperaccumulator plant species that may phytoremediate soils with toxic B levels. *Turkish Journal of Botany, 28*(3), 273–278.
117. Murakami, M., Ae, N., & Ishikawa, S., (2007). Phytoextraction of cadmium by rice (*Oryza sativa* L.), soybean (*Glycine max* (L.) Merr.), and maize (*Zea mays* L.). *Environmental Pollution, 145*(1), 96–103.
118. Zhang, X. H., Liu, J., Huang, H. T., Chen, J., Zhu, Y. N., & Wang, D. Q., (2007). Chromium accumulation by the hyperaccumulator plant *Leersia hexandra* Swartz. *Chemosphere, 67*(6), 1138–1143.
119. Keeling, S. M., Stewart, R. B., Anderson, C. W. N., & Robinson, B. H., (2003). Nickel and cobalt phytoextraction by the hyperaccumulator *Berkheya coddii*: Implications for polymetallic phytomining and phytoremediation. *International Journal of Phytoremediation, 5*(3), 235–244.
120. Del Río-Celestino, M., Font, R., Moreno-Rojas, R., & De Haro-Bail6n, A., (2006). Uptake of lead and zinc by wild plants growing on contaminated soils. *Industrial Crops and Products, 24*(3), 230–237.
121. Min, Y., Meizhen, T., & Aoyama, I., (2007). Accumulation and uptake of manganese in a hyperaccumulator *Phytolacca americana. Minerals Engineering, 20*(2), 188–190.
122. Moreno-Jiménez, E., Gamarra, R., Carpena-Ruiz, R. O., Millán, R., Peñalosa, J. M. & Esteban, E., (2006). Mercury bioaccumulation and phytotoxicity in two wild plant species of Almadén area. *Chemosphere, 63*(11), 1969–1973.
123. Moreno, D. A., Víllora, G., Soriano, M. T., Castilla, N., & Romero, L., (2005). Sulfur, chromium, and selenium accumulated in Chinese cabbage under direct covers. *Journal of Environmental Management, 74*(1), 89–96.
124. Mukhopadhyay, S., & Maiti, S. K., (2010). Phytoremediation of metal mine waste. *Applied Ecology and Environmental Research, 8*(3), 207–222.
125. Mesjasz-Przybyłowicz, J., Nakonieczny, M., Migula, P., Augustyniak, M., Tarnawska, M., Reimold, W. U., & Głowacka, E., (2004). Uptake of cadmium, lead nickel and zinc from soil and water solutions by the nickel hyperaccumulator *Berkheya coddii. Acta Biologica Cracoviensia Series Botanica, 46*, 75–85.
126. Dushenkov, V., Kumar, P. N., Motto, H., & Raskin, I., (1995). Rhizofiltration: The use of plants to remove heavy metals from aqueous streams. *Environmental Science & Technology, 29*(5), 1239–1245.
127. Dushenkov, D., & Kapulnik, Y., (2000). Phytofiltration of metals. In: Raskin, I., & Ensley, B. D., (eds.). *Phytoremediation of Toxic Metals – Using Plants to Clean up the Environment* (pp. 89–106). Wiley, New York.
128. Zayed, A., Gowthaman, S., & Terry, N., (1998). Phytoaccumulation of trace elements by wetland plants: I. Duckweed. *American Society of Agronomy, Crop Science Society of America, and Soil Science Society of America, 27*(3), 715–721.
129. Axtell, N. R., Sternberg, S. P., & Claussen, K., (2003). Lead and nickel removal using *Microspora* and *Lemna* minor. *Bioresource Technology, 89*(1), 41–48.
130. Hou, W., Chen, X., Song, G., Wang, Q., & Chang, C. C., (2007). Effects of copper and cadmium on heavy metal polluted waterbody restoration by duckweed (*Lemna minor*). *Plant Physiology and Biochemistry, 45*(1), 62–69.
131. Diwan, H., Ahmad, A., & Iqbal, M., (2008). Genotypic variation in the phytoremediation potential of Indian mustard for chromium. *Environmental Management, 41*(5), 734–741.

132. Chandrashekhar, K., Kamala, C. T., Chary, N. S., Sastry, A. R. K., Rao, T. N., & Vairamani, M., (2004). Removal of lead from aqueous solutions using an immobilized biomaterial derived from a plant biomass. *J. Haz. Mat., 108*, 111–117.
133. Skinner, K., Wright, N., & Porter-Goff, E., (2007). Mercury uptake and accumulation by four species of aquatic plants. *Environmental Pollution, 145*(1), 234–237.
134. Eapen, S., & D'souza, S. F., (2005). Prospects of genetic engineering of plants for phytoremediation of toxic metals. *Biotechnology Advances, 23*(2), 97–114.
135. Erakhrumen, A. A., & Agbontalor, A., (2007). Phytoremediation: An environmentally sound technology for pollution prevention, control and remediation in developing countries. *Educational Research and Review, 2*(7), 151–156.
136. Singh, S., (2012). Phytoremediation: A sustainable alternative for environmental challenges. *Int. J. Gr. Herb. Chem., 1*, 133–139.
137. Mendez, M. O., & Maier, R. M., (2008). Phytostabilization of mine tailings in arid and semiarid environments—An emerging remediation technology. *Environmental Health Perspectives, 116*(3), 278–283.
138. Ghosh, M., & Singh, S. P., (2005). A review on phytoremediation of heavy metals and utilization of it's by products. *Asian J. Energy Environ., 6*(4), 18.
139. Wu, G., Kang, H., Zhang, X., Shao, H., Chu, L., & Ruan, C., (2010). A critical review on the bio-removal of hazardous heavy metals from contaminated soils: Issues, progress, eco-environmental concerns and opportunities. *Journal of Hazardous Materials, 174*(1–3), 1–8.
140. Smith, R. A. H., & Bradshaw, A. D., (1979). The use of metal tolerant plant populations for the reclamation of metalliferous wastes. *Journal of Applied Ecology*, 595–612.
141. Berti, W. R., & Cunningham, S. D., (2000). Phytostabilization of metals. In: Raskin, I., & Ensley, B. D., (eds). *Phytoremediation of Toxic Metals: Using Plants to Clean up the Environment* (pp. 71–88). Wiley, New York.
142. James, B. R., (2001). Remediation-by-reduction strategies for chromate-contaminated soils. *Environmental Geochemistry and Health, 23*(3), 175–179.
143. Vangronsveld, J., Herzig, R., Weyens, N., Boulet, J., Adriaensen, K., Ruttens, A., Thewys, T., et al., (2009). Phytoremediation of contaminated soils and groundwater: Lessons from the field. *Environmental Science and Pollution Research, 16*(7), 765–794.
144. Regvar, M., Vogel-Mikuš, K., Kugonič, N., Turk, B., & Batič, F., (2006). Vegetational and mycorrhizal successions at a metal polluted site: Indications for the direction of phytostabilization? *Environmental Pollution, 144*(3), 976–984.
145. Baghour, M., Moreno, D. A., Hernandez, J., Castilla, N., & Romero, L., (2001). Influence of root temperature on phytoaccumulation of As, Ag, Cr, and Sb in potato plants (*Solanum tuberosum* L. var. Spunta). *Journal of Environmental Science and Health, Part A, 36*(7), 1389–1401.
146. Chinmayee, M. D., Mahesh, B., Pradesh, S., Mini, I., & Swapna, T. S., (2012). The assessment of phytoremediation potential of invasive weed *Amaranthus spinosus* L. *Applied Biochemistry and Biotechnology, 167*(6), 1550–1559.
147. Dinesh, M., Kumar, M. V., Neeraj, P., & Shiv, B., (2014). Phytoaccumulation of heavy metals in contaminated soil using Makoy (*Solenum nigrum* L.) and Spinach (*Spinacia oleracea* L.) plant. *Sciences (New York), 2*, 350–354.
148. Chen, L., Han, Y., Jiang, H., Korpelainen, H., & Li, C., (2011). Nitrogen nutrient status induces sexual differences in responses to cadmium in *Populus yunnanensis*. *Journal of Experimental Botany, 62*(14), 5037–5050.

149. Jafari, N., & Ahmady-Asbchin, S., (2011). Ecotoxicological effect of nickel on phytoaccumulation by *Spirogyra irregularis Nageli* (*Chlorophyta*). *International Journal on Algae, 13*(4), 392–404.
150. Adhikari, T., & Kumar, A., (2012). Phytoaccumulation and tolerance of *Riccinus communis* L. to nickel. *International Journal of Phytoremediation, 14*(5), 481–492.
151. Marbaniang, D., & Chaturvedi, S. S., (1994). Phytoaccumulation of zinc by *Scirpus mucronatus* (L.) Palla ex Kerner. *Keanean J. Sci., 1*, 69–75.
152. Khellaf, N., & Zerdaoui, M., (2009). Phytoaccumulation of zinc by the aquatic plant, *Lemna gibba* L. *Bioresource Technology, 100*(23), 6137–6140.
153. Moreno, F. N., Anderson, C. W. N., Stewart, R. B., Robinson, B. H., Nomura, R., & Gomshei, M., (2004). Mercury phytoextraction and phytovolatilization from hg-contaminated artisanal mine sites. *Phytoremediat Mercur. Mine Wastes*, 147–159.
155. Banuelos, G. S., (2000). Phytoextraction of Se from soils irrigated with selenium-laden effluent. *Plant Soil, 224*, 251–258.
156. Deesouza, M. P., Pilon-Smits, E. A. H., & Terry, N., (2000). In: Raskin, I., & Ensley, B. D., (eds.), *The Physiology and Biochemistry of Selenium Volatilization by Plants* (pp. 171–190).
157. Sakakibara, M., Watanabe, A., Inoue, M., Sano, S., & Kaise, T., (2010). Phytoextraction and phytovolatilization of arsenic from As-contaminated soils by *Pteris vittata*. In: *Proceedings of the Annual International Conference on Soils, Sediments, Water and Energy, 12*(1), 26.
158. LeDuc, D. L., & Terry, N., (2005). Phytoremediation of toxic trace elements in soil and water. *Journal of Industrial Microbiology and Biotechnology, 32*(11, 12), 514–520.
159. Rugh, C. L., Bizily, S. P., & Meagher, R. B., (2000). Phytoreduction of environmental mercury pollution. *Phytoremediation of Toxic Metals: Using Plants to Clean-up the Environment* (pp. 151–170). New York, John Wiley & Sons, Inc.
160. Varun, M., D'Souza, R., Pratas, J., & Paul, M. S., (2011). Evaluation of phytostabilization, a green technology to remove heavy metals from industrial sludge using *Typha latifolia* L. *Biotechnology Bioinformatics and Bioengineering, 1*(1), 137–145.
161. Soudek, P., Petrová, Š., & Vanek, T., (2012). Phytostabilization or accumulation of heavy metals by using of energy crop *Sorghum* sp. In: *3rd International Conference on Biology, Environment and Chemistry IPCBEE.* IACSIT Press, Singapore.
162. Yang, S., Liang, S., Yi, L., Xu, B., Cao, J., Guo, Y., & Zhou, Y., (2014). Heavy metal accumulation and phytostabilization potential of dominant plant species growing on manganese mine tailings. *Frontiers of Environmental Science & Engineering, 8*(3), 394–404.
163. Ali, N. A., Bernal, M. P., & Ater, M., (2002). Tolerance and bioaccumulation of copper in *Phragmites australis* and *Zea mays*. *Plant and Soil, 239*(1), 103–111.
164. Domínguez, M. T., Madrid, F., Marañón, T., & Murillo, J. M., (2009). Cadmium availability in soil and retention in oak roots: Potential for phytostabilization. *Chemosphere, 76*(4), 480–486.
165. Ehsan, M., Delgado, K. S., Alarcón, A. V., Chavez, A. A., & De La Cruz, L. N., (2009). Phytostabilization of cadmium contaminated soils by "*Lupinus uncinatus*" Schldl. *Spanish Journal of Agricultural Research*, (2), 390–397.
166. Zhang, X., Xia, H., Li, Z., Zhuang, P., & Gao, B., (2010). Potential of four forage grasses in remediation of Cd and Zn contaminated soils. *Bioresource Technology, 101*(6), 2063–2066.

167. Shackira, A. M., & Puthur, J. T., (2017). Enhanced phytostabilization of cadmium by a halophyte—*Acanthus ilicifolius* L. *International Journal of Phytoremediation, 19*(4), 319–326.
168. Rotkittikhun, P., Chaiyarat, R., Kruatrachue, M., Pokethitiyook, P., & Baker, A. J. M., (2007). Growth and lead accumulation by the grasses *Vetiveria zizanioides* and *Thysanolaena maxima* in lead-contaminated soil amended with pig manure and fertilizer: A glasshouse study. *Chemosphere, 66*(1), 45–53.
169. Zhao, L., Li, T., Zhang, X., Chen, G., Zheng, Z., & Yu, H., (2016). Pb uptake and phytostabilization potential of the mining ecotype of *Athyrium wardii* (Hook.) grown in Pb-contaminated soil. *CLEAN–Soil, Air, Water, 44*(9), 1184–1190.
170. Bidar, G., Garçon, G., Pruvot, C., Dewaele, D., Cazier, F., Douay, F., & Shirali, P., (2007). Behavior of trifolium repens and *Lolium perenne* growing in a heavy metal contaminated field: Plant metal concentration and phytotoxicity. *Environmental Pollution, 147*(3), 546–553.
171. Newman, L. A., & Reynolds, C. M., (2004). Phytodegradation of organic compounds. *Current Opinion in Biotechnology, 15*, 225–230.
172. Vishnoi, S. R., & Srivastava, P. N., (2007). Phytoremediation–green for environmental clean. In: *Proceedings of Taal 2007: The 12th World Lake Conference* (Vol. 1060, p. 1021).
173. Doty, S. L., Shang, T. Q., Wilson, A. M., Tangen, J., Westergreen, A. D., Newman, L. A., & Gordon, M. P., (2000). Enhanced metabolism of halogenated hydrocarbons in transgenic plants containing mammalian cytochrome P450 2E1. *Proceedings of the National Academy of Sciences, 97*(12), 6287–6291.
174. Khandare, R. V., & Govindwar, S. P., (2015). Phytoremediation of textile dyes and effluents: Current scenario and future prospects. *Biotechnology Advances, 33*(8), 1697–1714.
175. Kärenlampi, S., Schat, H., Vangronsveld, J., Verkleij, J. A. C., Van, D. L. D., Mergeay, M., & Tervahauta, A. I., (2000). Genetic engineering in the improvement of plants for phytoremediation of metal polluted soils. *Environmental Pollution, 107*(2), 225–231.
176. Bennicelli, R., Stępniewska, Z., Banach, A., Szajnocha, K., & Ostrowski, J., (2004). The ability of *Azolla caroliniana* to remove heavy metals (Hg (II), Cr (III), Cr (VI)) from municipal waste water. *Chemosphere, 55*(1), 141–146.
177. Rajakaruna, N., Tompkins, K. M., & Pavicevic, P. G., (2006). Phytoremediation: An affordable green technology for the clean-up of metal-contaminated sites in Sri Lanka. *Ceylon Journal of Science, 35*(1), 25.
178. Jomjun, N., Siripen, T., Maliwan, S., Jintapat, N., Prasak, T., Somporn, C., & Petch, P., (2010). Phytoremediation of arsenic in submerged soil by wetland plants. *International Journal of Phytoremediation, 13*(1), 35–46.
179. Zayed, A. M., & Terry, N., (2003). Chromium in the environment: Factors affecting biological remediation. *Plant and Soil, 249*(1), 139–156.
180. Paz-Alberto, A. M., & Sigua, G. C., (2013). Phytoremediation: A green technology to remove environmental pollutants. *American Journal of Climate Change, 2*, 71–86.
181. Mirza, N., Pervez, A., Mahmood, Q., Shah, M. M., & Shafqat, M. N., (2011). Ecological restoration of arsenic contaminated soil by *Arundo donax* L. *Ecological Engineering, 37*(12), 1949–1956.
182. Greipsson, S., (2011). Phytoremediation. *Nature Education Knowledge. 3*(10), 7.

183. Moreno, F. N., Anderson, C. W., Stewart, R. B., & Robinson, B. H., (2004). Phytoremediation of mercury-contaminated mine tailings by induced plant-mercury accumulation. *Environmental Practice, 6*(2), 165–175.
184. Manousaki, E., & Kalogerakis, N., (2011). Halophytes present new opportunities in phytoremediation of heavy metals and saline soils. *Industrial & Engineering Chemistry Research, 50*(2), 656–660.
185. Sakai, Y., Ma, Y., Xu, C., Wu, H., Zhu, W., & Yang, J., (2012). Phytodesalination of a salt-affected soil with four halophytes in China. *Journal of Arid Land Studies, 22*(1), 17–20.
186. Zorrig, W., Rabhi, M., Ferchichi, S., Smaoui, A., & Abdelly, C., (2012). Phytodesalination: A solution for salt-affected soils in arid and semi-arid regions. *Journal of Arid Land Studies, 22*(1), 299–302.
187. Ravindran, K. C., Venkatesan, K., Balakrishnan, V., Chellappan, K. P., & Balasubramanian, T., (2007). Restoration of saline land by halophytes for Indian soils. *Soil Biology and Biochemistry, 39*(10), 2661–2664.
188. Rabhi, M., Ferchichi, S., Jouini, J., Hamrouni, M. H., Koyro, H. W., Ranieri, A., Abdelly, C., & Smaoui, A., (2010). Phytodesalination of a salt-affected soil with the halophyte *Sesuvium portulacastrum* L. to arrange in advance the requirements for the successful growth of a glycophytic crop. *Bioresource Technology, 101*(17), 6822–6828.
189. Xiaoyong, H. L. S. G. Q., (2006). A study on root expansibility of seven constructed wetland plants. *Shanghai Environ Sci, 4,* 11.
190. Lin, Z. Q., Terry, N., Gao, S., Mohamed, S., & Ye, Z. H., (2010). Vegetation changes and partitioning of selenium in 4-year-old constructed wetlands treating agricultural drainage. *International Journal of Phytoremediation, 12*(3), 255–267.
191. Tromp, K., Lima, A. T., Barendregt, A., & Verhoeven, J. T. A., (2012). Retention of heavy metals and poly-aromatic hydrocarbons from road water in a constructed wetland and the effect of deicing. *J. Hazard Mater., 203e204*, 290e298. https://doi.org/10.1016/j.jhazmat.2011.12.024.
192. Langergraber, G., (2013). Are constructed treatment wetlands sustainable sanitation solutions. *Water Sci. Technol., 67*, 2133e2140.
193. Azaizeh, H., Salhani, N., Sebesvari, Z., Shardendu, S., & Emons, H., (2006). Phytoremediation of selenium using subsurface flow constructed wetland. *International Journal of Phytoremediation, 8*(3), 187–198.
194. Olguín, E. J., & Sánchez-Galván, G., (2010). Aquatic phytoremediation: Novel insights in tropical and subtropical regions. *Pure and Applied Chemistry, 82*(1), 27–38.
195. Rai, P. K., (2012). An eco-sustainable green approach for heavy metals management: Two case studies of developing industrial region. *Environmental Monitoring and Assessment, 184*(1), 421–448.
196. Sood, A., Uniyal, P. L., Prasanna, R., & Ahluwalia, A. S., (2012). Phytoremediation potential of aquatic macrophyte, *Azolla. Ambio, 41*(2), 122–137.
197. Saeed, T., & Sun, G., (2012). A review on nitrogen and organics removal mechanisms in subsurface flow constructed wetlands: Dependency on environmental parameters, operating conditions and supporting media. *J. Environ. Manag., 112*, 429e448.
198. Deng, H., Ye, Z. H., & Wong, M. H., (2004). Accumulation of lead, zinc, copper and cadmium by 12 wetland plant species thriving in metal-contaminated sites in China. *Environ. Pollut. 132*, 29–40. https://doi.org/10.1016/j.envpol.2004.03.030.

199. Abdullah, S. R. S., Al-Baldawi, I. A., Almansoory, A. F., Purwanti, I. F., Al-Sbani, N. H., & Sharuddin, S. S. N., (2020). Plant-assisted remediation of hydrocarbons in water and soil: Application, mechanisms, challenges and opportunities. *Chemosphere, 247*, 125932.
200. Chen, G., Liu, X., Brookes, P. C., & Xu, J., (2015). Opportunities for phytoremediation and bioindication of arsenic contaminated water using a submerged aquatic plant: *Vallisneria natans* (lour.) Hara. *Int. J. Phytoremediat., 17*, 249–255. https://doi. org/10. 1080/15226514.2014.883496.
201. Rahman, M. A., & Hasegawa, H., (2011). Aquatic arsenic: Phytoremediation using floating macrophytes. *Chemosphere, 83*(5), 633–646.
202. Pulford, I. D., & Watson, C., (2003). Phytoremediation of heavy metal-contaminated land by trees—A review. *Environment International, 29*(4), 529–540.
203. Garbisu, C., Allica, J. H., Barrutia, O., Alkorta, I., & Becerril, J. M., (2002). Phytoremediation: A technology using green plants to remove contaminants from polluted areas. *Reviews on Environmental Health, 17*(3), 173–188.
204. Mejáre, M., & Bülow, L., (2001). Metal-binding proteins and peptides in bioremediation and phytoremediation of heavy metals. *Trends in Biotechnology, 19*(2), 67–73.
205. Arthur, E., Crews, H., & Morgan, C., (2000). Optimizing plant genetic strategies for minimizing environmental contamination in the food chain: Report on the MAFF funded joint JIC/CSL workshop held at the John Innes Centre, October 21–23, 1998. *International Journal of Phytoremediation, 2*(1), 1–21.
206. Ji, P., Sun, T., Song, Y., Ackland, M. L., & Liu, Y., (2011). Strategies for enhancing the phytoremediation of cadmium-contaminated agricultural soils by *Solanum nigrum* L. *Environmental Pollution, 159*(3), 762–768.
207. Liu, C., Lin, H., Li, B., Dong, Y., & Yin, T., (2020). Responses of microbial communities and metabolic activities in the rhizosphere during phytoremediation of Cd-contaminated soil. *Ecotoxicology and Environmental Safety, 202*, 110958.
208. Bhantana, P., Rana, M. S., Sun, X. C., Moussa, M. G., Saleem, M. H., Syaifudin, M., & Hu, C. X., (2021). Arbuscular mycorrhizal fungi and its major role in plant growth, zinc nutrition, phosphorous regulation and phytoremediation. *Symbiosis, 84*(1), 19–37.
209. Patel, K., Tripathi, I., Chaurasia, M., & Rao, K. S., (2021). Phytoremediation: Status and outlook. Pollutants and water management: Resources, *Strategies and Scarcity,* 67–94.
210. Pollard, A. J., Powell, K. D., Harper, F. A., & Smith, J. A. C., (2002). The genetic basis of metal hyperaccumulation in plants. *Critical Reviews in Plant Sciences, 21*(6), 539–566.
211. Ojuederie, O. B., & Babalola, O. O., (2017). Microbial and plant-assisted bioremediation of heavy metal polluted environments: A review. *International Journal of Environmental Research and Public Health, 14*(12), 1504.
212. Saraswat, S., & Rai, J. P. N., (2011). Complexation and detoxification of Zn and Cd in metal accumulating plants. *Reviews in Environmental Science and Bio/Technology, 10*(4), 327–339.
213. Moreno-Jiménez, E., Gamarra, R., Carpena-Ruiz, R. O., Millán, R., Peñalosa, J. M., & Esteban, E., (2006). Mercury bioaccumulation and phytotoxicity in two wild plant species of Almadén area. *Chemosphere, 63*(11), 1969–1973.
214. Yadav, K. K., Gupta, N., Kumar, A., Reece, L. M., Singh, N., Rezania, S., & Khan, S. A., (2018). Mechanistic understanding and holistic approach of phytoremediation: A review on application and future prospects. *Ecological Engineering, 120*, 274–298..

CHAPTER 6

Mangroves as an Effective Tool of Phytoremediation and Its Implications on Agricultural Land in Estuarine Zones

DEBARATI CHAKRABORTY[1,2]

[1]*International Crop Research Institute for Semi-Arid Tropics, India*

[2]*ASDC Naturophilia Pvt. Ltd., India*

ABSTRACT

Phytoremediation, a treatment technique for getting rid of toxic trace metal pollution is being explored for a long time owing to its low-cost compared to physicochemical methods. Mangrove estuaries are often high metal pollution due to the high affinity of sedimentary organic matter (OM) for trace metals and capacity of mangroves to efficiently trap suspended material from the water and sediments. The functionality of mangroves to store metallic pollutants have been proven to be an effective natural purification process for toxic wastewater treatment. The Hooghly estuary supporting world's largest continuous mangrove forest, the Sundarbans have also reported significantly high trace metal pollution. Particularly Al, As, Cd, Cr, Cu, Co, Fe, Hg, Pb, Mn, Ni, V, and Zn have shown higher shale values when compared with the global standard. Regular disposal of domestic sewage, semi treated industrial effluents, surface, and agricultural run-off are the potential sources of contamination in the zone. Studies have proven some of the dominant mangroves of Sundarbans namely *Avicennia officinalis*, *Sonneratia apetala* and *Excoecaria agallocha* have high toxic metal phytoremediation ability. This is of particular significance as Indian Sundarbans is home to over 4.5 million people and trace metals are non-biodegradable,

Phytoremediation and Biofortification: Strategies for Sustainable Environmental and Health Management. Nand Kumar Singh, Shadma Afzal, & Tariq Aftab (Eds.)

shows bioaccumulation tendency in flora and fauna and toxic on exposure above threshold limits. Agriculture is the main occupation among the inhabitants and hence a balance between mangrove population and metal toxic waste free agricultural land is of baffling importance. Lastly, the post-harvest management of phytoremediation derived waste have received scarce focus to date. Hence strategies for removal of mangrove plant biomass with concentrated metal wastes are discussed as a way forward.

6.1 INTRODUCTION

The ever-increasing demands of modern civilization have been taking its noxious toll on environment through multifarious agents like organic and gaseous pollutants, organometallic compounds, radioactive isotopes and nanoparticles. One amongst them are toxic metal pollutants which are commonly called heavy metals, although the term actually refers to metallic elements and metalloids with a density higher than 5 g cm^{-3} and capable to form sulfides [1]. However, it is incorrect as some heavy metals, e.g., gold (Au), bismuth (Bi) is generally not poisonous whereas metals with lighter density, e.g., selenium (Se), arsenic (As), aluminum (Al) are poisonous [2, 3]. Thus, trace metal is an appropriate term, as it denotes the concentration of these metals in environment which is ~0.1% or ppb range to less than 10 ppm in environment or 100 mg kg^{-1} in dry matter of biological samples [4]. Being nonbiodegradable and having extreme harmful effects on humans, there are rising public health concerns about effects of trace metal pollution in aquatic and terrestrial environments. The history of trace metal emissions particularly lead (Pb), cadmium (Cd), mercury (Hg), zinc (Zn), tin (Sn) through anthropogenic activity dates back to the time of Roman Empire [5]. The condition is further exacerbated since Industrial Revolution of 1760 [5]. Apart from natural sources of emissions, e.g., mineral degradation, volcanoes, evaporation from soil and water, forest fires; the epidemic of toxic trace metals pollution is highly influenced by plethora of anthropogenic sources, e.g., industrial waste inputs from mining, tannery, paint, metallurgy, leather manufacturing, electroplating, chemical pesticides and fertilizers, coal combustion residues, landfills using solid wastes, waste incineration, irrigation by untreated wastewater, mine ore waste, sewage sludge, spillage of petroleum distillates [5].

However, the toxicity of trace metals depends on their bioavailability. This is governed by physical factors, e.g., temperature, phase association, adsorption, and sequestration and chemical factors influencing metal speciation and oxidation states. Upon entering into body though ingestion or inhalation, these metals tend to bioaccumulate into systems causing several biological complications. Trace metals can interact with DNA and nuclear proteins causing site-specific damage. These damages can be either "direct damage" by imposing conformational changes to the biomolecules or "indirect" damage caused by the generation of free radicals which can lead to DNA mutations and thereby carcinogenicity, altered sulfydryl homeostasis, membrane damage and lipid peroxidation. Inhibition of DNA repair mechanisms by cadmium (Cd), nickel (Ni), arsenic (As); carcinogenicity imposed by nickel (Ni), chromium (Cr) and cadmium; oxidative damage to DNA by Cr, Ni, Cd are few examples of several types of potential damages imposed upon trace metal exposure above threshold limit [2].

6.2 TOXIC TRACE POLLUTION IN MANGAL SEDIMENTS

Estuarine sediments are complex systems playing important roles in fate of trace metal biogeochemical cycle. Industrial wastes and wastewater are often discharged to the adjacent waterbodies through which these toxic trace metals reach nearby estuaries and coastal areas (Figure 6.1). Fine-grained sediment particles can carry trace metals due to their greater specific surface area. This results in the high accumulation of toxic metals in the sediments of urban coastal areas via particle scavenging and settling [6–9]. Mangrove vegetation colonizing the tropical and subtropical intertidal estuarine ecosystem act as trace metal repositories and can play a substantial role in restoration of water quality [10, 11]. Mangal sediments are anoxic due to frequently waterlogged conditions, saline, reduced, rich in sulfides and organic matter (OM) [12–16]. These along with tangled roots of mangroves aids to efficiently entrap suspended matter with high metal pollutants increasing the TTM levels in the mangal sediments [17, 18]. Thus, mangroves act as a filter to reduce trace metal pollution load reaching to adjacent oceans through rivers [14, 19, 20]. Mangroves can also successfully restore water quality by enhancing suspended solid deposition and can stabilize and accumulate trace metals from river, tidal water and storm water run-off [21].

Owing to the high residence time, i.e., average time required for the particles to escape the estuary, of metalliferous pollutants in estuarine sediments probabilities of high toxicological effects occur [24, 25]. These metals are potent abiotic stressor imposing several hazardous effects on plants by interrupting their major physiological and metabolic processes [26–28]. Despite this, mangrove vegetations show high tolerance towards trace metal pollutants [29, 30].

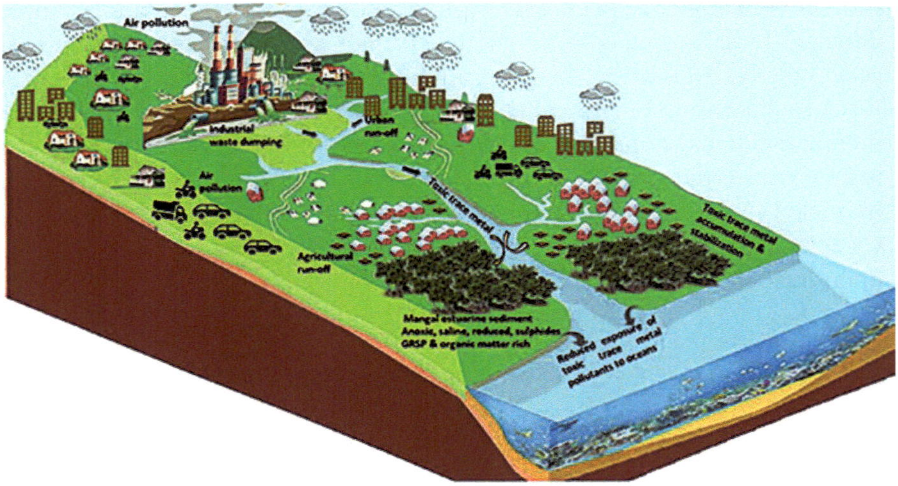

FIGURE 6.1 Toxic trace metal accumulation pathways in mangal sediments and its fate.

Further a recalcitrant glue-like, water insoluble glycoprotein called glomalin-related soil protein (GRSP) generated by arbuscular mycorrhizal fungi (AMF) in mangal sediments can acts as bioflocculant, i.e., aggregate-binding agent. GRSP can bind and sequester high amounts of trace metal [22, 23].

6.3 PHYTOREMEDIATION OF TOXIC TRACE METALS

Trace metals in soil generally exists in non-bioavailable form. Its bioavailability is enhanced through root exudates released by plants which can alter pH of rhizosphere increasing trace metal solubility [31]. This bioavailable metal is consequentially absorbed at the root surface and translocate into the root cells either through passive diffusion, i.e., apoplastic pathway or active transport, against concentration and electrochemical potential gradients across the cell membrane, i.e., symplastic pathway which is more common. The latter is an energy-dependent process involving metal ion carriers or complexing agents [32]. Various process involved in uptake and translocation of metals in plants after entering into plants are xylem loading, root-to-shoot transport, cellular compartmentation, and sequestration. Post entrance into root cells, metal ions forms complexes with different chelators and are immobilized in the extracellular apoplastic cellular walls or intracellular symplastic compartments, e.g., vacuoles [33]. Phytoremediation, a low-cost, eco-friendly TTM remediation technique uses the natural autotrophic abilities of vegetations to uptake, accumulate, store, or degrade organic and inorganic substances utilizing solar energy. It involves the utilization of plants or plant products to restore or stabilize soil and water quality of contaminated sites [34]. This technique has gained substantial attention from researchers worldwide [35]. Phytoremediation also reduces soil erosion and metal leaching through heavy metal stabilization thereby diminishing contaminant spreading threats and also improves soil fertility through increasing the amount of various OM in the soil [36–39]. The two defense strategies plants adopt to confront the trace metal toxicity are metal avoidance and tolerance. These aids the plants to keep the concentrations of toxic trace metals in cells under the toxicity threshold levels [40]. The different types of phytoremediation strategies are phytoextraction employing plants to extract and remove TTMs from water or ground and store in biomass; phytostabilization utilizing tolerant plant species to decrease bioavailability of TTMs in soil by immobilizing metals in or near the rhizosphere through the release of organic acids, polysaccharides; phytovolatilization utilizing plants to absorb TTMs from soil and releasing it into the atmosphere via transpiration as less toxic volatile compounds, phytofiltration uses hydroponically cultured plants to absorb or adsorb TTM ions from polluted surface water and aqueous waste; phytodegradation and rhizodegradation employing plants for breaking down of organic pollutants [35, 41–43]. The first four strategies of these are most widely utilized.

6.4 POTENTIAL OF MANGROVES FOR PHYTOREMEDIATION

Native plants are often regarded to be superior to non-native plants for phytoremediation because of their capacity to better survive, grow, and reproduce on being exposed to environmental stress [44]. Mangroves and mangrove associates which are natural residents of estuarine ecosystem have high antioxidant potential aiding their metal tolerance ability [45–47]. The mangal metal tolerance capacity may be also owing to exclusion or regulated metal uptake at root level and its limited translocation to the shoot [29]. Accumulation of metals like Cu, Zn, Pb, Fe, Mn, and Cd in root tissue instead of foliage are reported in mangroves such as *Avicennia* sp., *Rhizophora* sp.

and *Kandelia* sp., but the concentration were lower than the adjacent sediments [48]. The potential of mangrove for trace metal phytoremediation has been explored for multiple mangrove species, e.g., *Rhizophora apiculata, R. stylosa, R. mucronata, Sonneratia alba, S. caseolaris, Avicennia marina, A. alba, A. lanata* [49–51]. Two categories of phytoremediation explored for use of mangroves are phytoextraction and phytostabilization [52, 53]. The efficiency of particular mangrove species for specific type of metal removal is determined based on Bioconcentration (BCF) and Translocation (TF) Factors [54, 55]. BCF is calculated by the capability of mangrove species for the removal of metal compounds from the surrounding environment and represents the amount of extractable metal. It represents progressive metal accumulation in vegetation [56]. Whereas TF value measures the ability of the mangroves to be translocate the absorbed metals from plant roots to the different aerial parts of the plants including stem, leaves, and fruits [57, 58]. Both of these ratios determine the phytoextraction potential of the plant species. The following equations are used to calculate BCF and TF [59]:

$$BCF_{\text{Leaf}} = Metal\ Concentration_{\text{Leaf}}/Metal\ Concentration_{\text{Sediment}}$$

$$BCF_{\text{stem}} = Metal\ Concentration_{\text{Stem}}/Metal\ Concentration_{\text{Sediment}}$$

$$BCF_{\text{root}} = Metal\ Concentration_{\text{Root}}/Metal\ Concentration_{\text{Sediment}}$$

$$TF_{\text{Leaf}} = Metal\ Concentration_{\text{Leaf}}/Metal\ Concentration_{\text{Root}}$$

$$TF_{\text{Stem}} = Metal\ Concentration_{\text{Stem}}/Metal\ Concentration_{\text{Root}}$$

Where the metal concentrations are expressed in mg kg^{-1}. Plants are categorized into three groups based on their BCF values, e.g., accumulators have BCF > 1, indicator have BCF = 1, and excluders are plant species having BCF < 1 [60]. Mangroves having BCF and TF values >1 can be used as trace metal bioaccumulators. They can be successfully employed for phytoremediation through phytoextraction [53, 61–63]. Plants which have BCF >1 and TF <1 can be employed as phytostabilizers and those having BCF and TF both >1 can be used as phytoextractors [64].

6.5 TOXIC TRACE POLLUTION SCENARIO IN INDIAN SUNDARBANS

The Indian Sundarbans located on the southern fringe of the state of West Bengal, India extends from 21° 30′ N to 22° 30′ N latitude and 87° 25′ E to

89° 10′ E, represents a major part of world's largest contiguous mangrove tract. The region is geographically delineated by the Dampier and Hodges line in the north, the Bay of Bengal to the south, Bangladesh border on its east, and the Hooghly River on its west. It covers an area of about 9,630 sq.km, of which the mangrove forest flourishes in 4,200 sq.km [65]. The Indian Sundarbans spread over 19 administrative blocks is located at the apex of Bay of Bengal. It provides a unique location for deciphering the impacts of pollution on mangrove ecosystems as trace metal pollutants from nearby industrial and anthropogenic sources are continuously contaminating the surrounding aquatic environment. Each day almost 1,125 million liters of wastewater is dispensed through the Hooghly estuary [65]. Of this 1,112 MLD of sewage water is produced by Kolkata Municipal Corporation (KMC) region, which moves through the underground sewers to the terminal pumping stations and is thereafter discharged into open channels enters the East Kolkata Wetland complex via dry weather flow (DWF) channels [66]. It finally meets with Kulti-Bidhyadhari river at Ghusighata (30 km away from the city border) which again meets with river Raimangal in the Indian Sundarbans. These along with pollutant from Haldia Petrochemicals and port complex; industrial disposal and sewage generated from urban settlements in Kolkata and North 24 Paraganas; chain of factories and industries with unmonitored pollutant disposal system situated on the western bank of the Hooghly estuary; presence of paper, fertilizer, plastic, shellac, textile, jute, tire, chemical, and pharmaceutical, leather, and cycle rim industries in the lower part of the Hooghly estuary contribute towards the continuous TTM pollution load in Hooghly estuary which flowing through the Sundarbans drains into the Bay of Bengal [67–71]. Regular use of antifouling paints in boats, trawlers, ferries, and extensive use of chemical fertilizers and pesticides for agricultural needs, coal combustion, dredging, sewage, emission of smelters, fossil fuel burning, disposal of rechargeable batteries, phosphate fertilizers and pesticides are other sources of TTM particularly Cu, Ni, and Hg in Indian Sundarbans [72–74].

Occurrence of 34 true mangroves and 62 associated species of flora and fauna are reported in the Sundarbans [65]. Some of the dominant mangroves of Indian Sundarbans are *A. alba*, *A. officinalis*, *A. marina*, *Excoecaria agallocha*, *S. apetala*, *S. caseolaris*, *R. apiculata*, *Bruguiera gymnorrhiza*, *Aegiceras corniculatum*, *Aegialitis rotundifolia* [65]. There exist several literatures on accumulation of TTMs in Indian Sundarbans and areas surrounding it, its effect on mangrove vegetations and potential of mangroves for phytoremediation [67, 71, 75–77]. However, along with numerous mangroves, and mangrove associates, flora, and fauna the Indian

Sundarbans is also home to 4.5 million people [78]. The major occupation of the inhabitants of the region is agriculture along with landless laborers working in agricultural lands, aquaculture farms locally known as bheries; coastal fishing on full time and part time basis and procurement non-timber forest products like honey; prawn, crab, shrimps [79]. Effect of the TTM pollution load on the agriculture, aquaculture, public health of the people of Indian Sundarbans and how the capability of mangroves for metal removal from contaminated sites can be successfully implemented for remediating the conditions of locals are topics much less explored. Drinking water, rice and vegetables, poultry, and livestock are reported as potential sources of As contamination in North 24 Parganas of West Bengal [80, 81]. Increased Cu, Fe, and Zn in monsoon and of Cd, Ni, and Pb in pre-monsoon season is reported as major TTMs in surface water of the region with CD and Pb being the main players [82]. Also, high concentrations of Pb (27.79–47.07 mg/kg) and Cd (0.8–3.3 mg/kg) in places of Indian Sundarbans dependent on motorboats running on fossil fuels for transport and rechargeable batteries for coping with the electricity shortage [75]. This high TTM load directly affects tiger prawn seed (locally called meen) collectors who are reported to be suffering from orthopedic disorders, infections on skin and in lower body parts exposed to contaminated water [82]. Bhattacharya et al. [83] reported high carcinogenic risk for dermal absorption of Cr, Pb, and Cd with two sites namely Canning and Dhamakhali having huge TTM load through domestic sewage effluents coming from the megacity of Kolkata.

The potential of mangrove species for extracting TTM depends on several factors including their metal specific BCF and TF, metal availability in the adjacent sediment, sediment pH and organic carbon content. The last two factors control TTM availability and mobility [84, 85]. Several research articles exist on phytoremediation ability of mangroves for reclamation of TTM contaminated sites. The different metal accumulating ability of different mangrove species as reported in different studies are enlisted in Table 6.1.

6.6 WAY FORWARD

A plethora of research article on TTM phytoremediation ability of mangroves exists yet there is no real effort to use the obtained knowledge in reality in Indian Sundarbans region. Thus, the way forward should be a collaboration between research and application of the obtained knowledge. For this foremostly a detailed geo-referenced map showing accurately identified

TABLE 6.1 Bioconcentration Factors of Different Mangrove Species Having High Potential for Phytoremediation of Toxic Trace Metals

Mangrove	Trace Metal	BCF in Plant Parts	References
A. marina	Cu	9.08 L, 4.55 S, 5.23 R	[74]
	Cd	2.36 L, 0.9 S, 0.0 R	
	Cr	6.43 L, 5.91 S, 8.04 R	
	Zn	4.71 L, 5.07 S, 4.08 R	
	Ni	9.08 L, 1.03 S, 2.81 R	
E. agallocha	Cd	15.5 L	[86]
	Zn	1.61 L	
	Cu	1.38 L	
	Mg	2.02 L	
S. apetala	Mn	2.34 L, 1.44 B, 1.04 P	[82]
	Co	1.50 B, 1.08 P	
A. officinalis	Mn	2.14 L, 2.20 B, 1.39 P	
	Co	1.15 P	
R. apiculata	Co	1.61 B, 1.64 R	
E. agallocha	Co	2.43 B, 3.57 R	
Xylocarpus mecongenesis	Co	1.43 L, 1.09 B	
R. apiculata	Mn	3.52	[53]
	Zn	1.88	
	Cu	1.33	
Acanthus ilicifolius	Zn, Cu	> 1 L, R*	[87]

where; 'L' is leaf; 'B' is bark; and 'P' is pneumatophore; 'S' is shoot; 'R' is root; '' exact value not mentioned in study. The mangroves represented here are available in Indian Sundarbans too and thus their TTM phytoextraction or phytostabilization potential can be successfully implemented in the region to save degraded agricultural land.*

different toxic metal polluted zones of Indian Sundarbans along with the metal concentrations should be developed and be available in public domain. Also, mangrove nurseries may be developed with suitable mangrove species having phytoextraction and phytostabilization potential with locals will also generate income possibilities. Further, the potential role of nanotechnology for phytoremediation cannot be neglected. While transplanting mangroves on polluted soils use of additives like nano-zero valent ions can be explored further to check its efficiency as an aided phytostabilization strategy [88]. This approach is fruitful to prevent deeper penetration of TTMs in soil profile. However, its production cost is high, and research is still ongoing in this field. Hence natural options for aided phytostabilization, e.g., biochar and compost can be also explored [89]. The production of biochar and compost can be done by training local people where soil restoration work is being done which will provide another way of revenue earning. Last and most importantly the disposal of plant biomass used for phytoremediation must be keep in mind. Approaches like high temperature pyrolysis can be considered for making biochar with greater oxidation resistance, reduced metal leachability and carbon stability [90].

ACKNOWLEDGMENT

The author is immensely grateful to the farming families of Indian Sundarbans who have shared their insights, experiences, and different kind of problems they face in their daily life and to all the researchers who have worked on the topic. The author is also thankful to architect Abhijit Sarbadevabhatla for kindly developing the figure.

KEYWORDS

- **agricultural soil**
- **Indian Sundarbans**
- **mangroves**
- **phytoextraction**
- **phytoremediation**
- **phytostabilization**
- **toxic trace metals**

REFERENCES

1. Adriano, D. C., (2001). *Trace Elements in Terrestrial Environments: Biogeochemistry, Bioavailability, and Risks of Metals*. New York: Springer.
2. Briffa, J., Sinagra, E., & Blundell, R., (2020). Heavy metal pollution in the environment and their toxicological effects on humans. *Heliyon, 6*(9), e04691.
3. Mohan, R., (2010). Green bismuth. *Nature Chemistry, 2*, 336.
4. Kabata-Pendia, A., (2001). *Trace Elements in Soils and Plants.* Boca Raton, FL: CRC Press.

5. Nriagu, J. O., (1996). A history of global metal pollution. *Science, 272*(5259), 223.
6. Bopp, R. F., & Simpson, H. J., (1989). Contamination of the Hudson River: The sediment record. In: *Contaminated Marine Sediments Assessment and Remediation* (pp. 401–416).
7. Valette-Silver, N. J., (1993). The use of sediment cores to reconstruct historical trends in contamination of estuarine and coastal sediments. *Estuaries, 16*, 577–588.
8. Bothner, M., Buchholtz, T. B. M., & Manheim, F., (1998). Metal concentrations in surface sediments of Boston Harbor—Changes with time. *Marine Environmental Research, 45*, 127–155.
9. Santos, I. R., Silva-Filho, E. V., Schaefer, C. E., Albuquerque-Filho, M. R., & Campos, L. S., (2005). Heavy metal contamination in coastal sediments and soils near the Brazilian Antarctic Station, King George Island. *Marine Pollution Bulletin, 50*, 185–194.
10. Alongi, D. M., Wattayakorn, G., Boyle, S., Tirendi, F., Payn, C., & Dixon, P., (2004). Influence of roots and climate on mineral and trace element storage and flux in tropical mangrove soils. *Biogeochemistry, 69*, 105–123.
11. De Wolf, H., & Rashid, R., (2008). Heavy metal accumulation in *Littoraria scabra* along polluted and pristine mangrove areas of Tanzania. *Environmental Pollution, 152*, 636–643.
12. Nissenbaum, A., & Swaine, D. J., (1976). Organic–metal interactions in recent sediments: The role of humic substances. *Geochimica et Cosmochimica Acta, 40*, 809–816.
13. Tam, N. F., & Wong, Y. S., (1993). Retention of nutrients and heavy metals in mangrove sediment receiving wastewater of different strengths. *Environmental Technology, 14*, 719–729.
14. Zhou, Y. W., Zhao, B., Peng, Y. S., & Chen, G. Z., (2010). Influence of mangrove reforestation on heavy metal accumulation and speciation in intertidal sediments. *Marine Pollution Bulletin, 60*, 1319–1324.
15. Bayen, S., (2012). Occurrence, bioavailability and toxic effects of trace metals and organic contaminants in mangrove ecosystems: A review. *Environment International, 48*, 84–101.
16. Natesan, U., Kumar, M. M., & Deepthi, K., (2014). Mangrove sediments a sink for heavy metals? An assessment of Muthupet mangroves of Tamil Nadu, southeast coast of India. *Environmental Earth Sciences, 72*, 1255–1270.
17. Lacerda, L. D., Martinelli, L. A., Rezende, C. E., Mozeto, A. A., Ovalle, A. R., Victoria, R. L., Silva, C. A., & Nogueira, F. B., (1988). The fate of trace metals in suspended matter in a mangrove creek during a tidal cycle. *Science of the Total Environment, 75*, 169–180.
18. Harbison, P. A., (1986). Mangrove muds—A sink and a source for trace metals. *Marine Pollution Bulletin, 17*, 246–250.
19. Furukawa, K., Wolanski, E., & Mueller, H., (1997). Currents and sediment transport in mangrove forests. *Estuarine, Coastal and Shelf Science, 44*, 301–310.
20. Donato, D. C., Kauffman, J. B., Murdiyarso, D., Kurnianto, S., Stidham, M., & Kanninen, M., (2011). Mangroves among the most carbon-rich forests in the tropics. *Nature Geoscience, 4*, 293–297.
21. Adame, M. F., Neil, D., Wright, S. F., & Lovelock, C. E., (2010). Sedimentation within and among mangrove forests along a gradient of geomorphological settings. *Estuarine, Coastal and Shelf Science, 86*, 21–30.

22. Spohn, M., & Giani, L., (2010). Water-stable aggregates, glomalin-related soil protein, 686 and carbohydrates in a chronosequence of sandy hydromorphic soils. *Soil Biology and Biochemistry, 42*, 1505–1511.
23. Jia, X., Zhao, Y., Liu, T., Huang, S., & Chang, Y., (2016). Elevated CO_2 increases 642 glomalin-related soil protein (GRSP) in the rhizosphere of *Robinia pseudoacacia* L. seedlings in Pb-and Cd-contaminated soils. *Environmental Pollution, 218*, 349–357.
24. Lacerda, L. D., & Rezende, C. E., (1984). Dinhmica de metals pesados em manguezais. In: *II Simpdsio Brasileiro Sobre Recursos do Mar.*
25. Lacerda, L. D., Rezende, C. E., Jos, D. M. V., & Francisco, M. C. F., (1986). Metallic composition of mangrove leaves from Southeastern Brazil. *Revista Brasileira de Biologia, 46*, 395–399.
26. Hossain, M. A., Hasanuzzaman, M., & Fujita, M., (2010). Up-regulation of antioxidant and glyoxylase systems by exogenous glycinebetaine and proline in mung bean confer tolerance to cadmium stress. *Physiology and Molecular Biology of Plants, 16*, 259–272.
27. Rasico, N., & Navari-Izzo, F., (2011). Heavy metal hyperaccumulating plants: How and why do they do it? And what makes them so interesting? *Plant Science, 180*, 169–181.
28. Villiers, F., Ducrix, C., Hugouviex, V., Jarno, N., Ezan, E., Garin, J., Junot, C., & Bourguignon, J., (2011). Investigating the plant response to cadmium exposure by proteomic and metabolomic approaches. *Proteomics, 11*, 1650–1663.
29. MacFarlane, G. R., & Burchett, M. D., (2002). Toxicity, growth and accumulation relationships of copper, lead and zinc in the grey mangrove *Avicennia marina* (Forsk.) Vierh. *Marine Environmental Research, 54*, 65–84.
30. Cuong, D. T., Bayen, S., Wurl, O., Subramanian, K., Wong, K. K. S., Sivasothi, N., & Obbard, J. P., (2005). Heavy metal contamination in mangrove habitats of Singapore, baseline. *Marine Pollution Bulletin, 50*, 1713–1744.
31. Dalvi, A. A., & Bhalerao, S. A., (2013). Response of plants towards heavy metal toxicity: An overview of avoidance, tolerance and uptake mechanism. *Annals of Plant Sciences, 2*, 362–368.
32. Peer, W. A., Baxter, I. R., Richards, E. L., Freeman, J. L., & Murphy, A. S., (2005). In: Tamas, M. J., & Martinoia, E., (eds.), *Phytoremediation and Hyperaccumulator Plants in Molecular Biology of Metal Homeostasis and Detoxification* (pp. 299–340). Berlin: Springer.
33. Ali, H., Khan, E., & Sajad, M. A., (2013). Phytoremediation of heavy metals-concepts and applications. *Chemosphere, 91*, 869–881.
34. Berti, W. R., & Cunningham, S. D., (2000). In: Raskin, I., & Ensley, B. D., (eds.), *Phytostabilization of Metals in Phytoremediation of Toxic Metals: Using Plants to Clean-up the Environment* (pp. 71–88). New York, NY: John Wiley & Sons Inc.
35. Yan, A., Wang, Y., Tan, S. N., Mohd, Y. M. L., Ghosh, S., & Chen, Z., (2020). Phytoremediation: A promising approach for revegetation of heavy metal-polluted land. *Frontiers in Plant Science, 11*, 359.
36. Aken, B. V., Correa, P. A., & Schnoor, J. L., (2009). Phytoremediation of polychlorinated biphenyls: New trends and promises. *Environmental Science & Technology, 44*, 2767–2776.
37. Jacob, J. M., Karthik, C., Saratale, R. G., Kumar, S. S., Prabakar, D., Kadirvelu, K., et al., (2018). Biological approaches to tackle heavy metal pollution: A survey of literature. *Journal of Environmental Management, 217*, 56–70.

38. Wuana, R. A., & Okieimen, F. E., (2011). Heavy metals in contaminated soils: A review of sources, chemistry, risks and best available strategies for remediation. *ISRN Ecology, 2011*, 402647.
39. DalCorso, G., Fasani, E., Manara, A., Visioli, G., & Furini, A., (2019). Heavy metal pollutions: State of the art and innovation in phytoremediation. *International Journal of Molecular Sciences, 20*, 3412.
40. Hall, J., (2002). Cellular mechanisms for heavy metal detoxification and tolerance. *Journal of Experimental Botany, 53*, 1–11.
41. Ernst, W. H., (2005). Phytoextraction of mine wastes–options and impossibilities. *Chemie Der Erde Geochemistry, 65*, 29–42.
42. Salt, D. E., Blaylock, M., Kumar, N. P. B. A., Dushenkov, V., Ensley, B. D., Chet, I., et al., (1995). Phytoremediation: A novel strategy for the removal of toxic metals from the environment using plants. *Nature Biotechnology, 13*, 468–474.
43. Marques, A. P., Rangel, A. O., & Castro, P. M., (2009). Remediation of heavy metal contaminated soils: Phytoremediation as a potentially promising clean-up technology. *Critical Reviews in Environmental Science and Technology, 39*, 622–654.
44. Palutoglu, M., Akgul, B., Suyarko, V., Yakovenko, M., Kryuchenko, N., & Sasmaz, A., (2018). Phytoremediation of cadmium by native plants grown on mining soil. *Bulletin of Environmental Contamination and Toxicology, 100*, 293–297.
45. Galal, T. M., & Shehata, H. S., (2015). Bioaccumulation and translocation of heavy metals by *Plantago major* L. grown in contaminated soils under the effect of traffic pollution. *Ecological Indicators, 48*, 244–251.
46. Gong, X., Huang, D., Liu, Y., Zeng, G., Wang, R., Wan, J., Zhang, C., Cheng, M., Qin, X., & Xue, W., (2017). Stabilized nanoscale zerovalent iron mediated cadmium accumulation and oxidative damage of *Boehmeria nivea* (L.) Gaudich cultivated in cadmium contaminated sediments. *Environmental Science Technology, 51*, 11308–11316.
47. Ganeshkumar, A., Arun, G., Vinothkumar, S., & Rajaram, R., (2019). Bioaccumulation and translocation efficacy of heavy metals by *Rhizophora mucronata* from tropical mangrove ecosystem, Southeast coast of India. *Ecohydrology and Hydrobiology, 19*, 66–74.
48. Peters, E. C., Gassman, N. J., Firman, J. C., Richmond, R. H., & Power, E. A., (1997). Ecotoxicology of tropical marine ecosystems. *Environmental Toxicology and Chemistry. 16*, 12–40.
49. Pahalawattaarachchi, V., Purushothaman, C. S., & Vennila, A., (2009). Metal phytoremediation potential of *Rhizophora mucronata* (Lam.). *Indian Journal of Marine Sciences, 38*, 178–183.
50. Sing, L. Y., (2010). *Impact of Sonneratia Caseolaris, an Exotic Mangrove Species, on Macrobenthic Community* (p. 574). City University of Hong Kong.
51. Richter, O., Nguyen, H. A., Nguyen, K. L., Nguyen, V. P., Biester, H., & Schmidt, P., (2016). Phytoremediation by mangrove trees: Experimental studies and model development. *Chemical Engineering Journal, 294*, 389–399.
52. Paz-Alberto, A. M., & Sigua, G. C., (2013). Phytoremediation: A green technology to remove environmental pollutants. *American Journal of Climate Change, 2*, 71–86.
53. Khan, W. R., Rasheed, F., Zulkifli, S. Z., Kasim, M. R., Zimmer, M., Pazi, A. M., Kamrudin, N. A., et al., (2020). Phytoextraction potential of *Rhizophora apiculata*:

A case study in matang mangrove forest reserve, Malaysia. *Tropical Conservation Science, 13*, 1–8.
54. Qiu, Y. W., Yu, K. F., Zhang, G., & Wang, W. X., (2011). Accumulation and partitioning of seven trace metals in mangroves and sediment cores from three estuarine wetlands of Hainan Island. *Journal of Hazardous Materials, 190*, 631–638.
55. Usman, A. R., Lee, S. S., Awad, Y. M., Lim, K. J., Yang, J. E., & Ok, Y. S., (2012). Soil pollution assessment and identification of hyper-accumulating plants in chromated copper arsenate (CCA) contaminated sites, Korea. *Chemosphere, 87*, 872–878.
56. Branquinho, C., Serrano, H. C., Pinto, M. J., & Martins-Loucao, M. A., (2007). Revisiting the plant hyperaccumulation criteria to rare plants and earth abundant elements. *Environmental Pollution, 14*, 437–443.
57. Rezvani, M., & Zaefarian, F., (2011). Bioaccumulation and translocation factors of cadmium and lead in *Aeluropus littoralis*. *Australian Journal Agricultural Engineering, 2*, 114–119.
58. Mellem, J. J., Baijnath, H., & Odhav, B., (2012). Bio-accumulation of Cr, Hg, As, Pb, Cu and Ni with the ability for hyperaccumulation by *Amaranthus dubius*. *African Journal of Agricultural Research, 7*, 591–596.
59. Kamari, A., Yusoff, M., Najiah, S., Putra, W. P., Ishak, C. F., Hashim, N., Mohamed, A., & Phillip, E., (2014). Metal uptake in water spinach grown on contaminated soil amended with chicken manure and coconut tree sawdust. *Environmental Engineering and Management Journal, 13*, 2219–2228.
60. Baker, A. J., (1981). Accumulators and excluders-strategies in the response of plants to heavy metals. *Journal of Plant Nutrition, 3*, 643–654.
61. Haque, N., Peralta-Videa, J. R., Jones, G. L., Gill, T. E., & Gardea-Torresdey, J. L., (2008). Screening the phytoremediation potential of desert broom (*Baccharis sarothroides* gray) growing on mine tailings in Arizona, USA. *Environmental Pollution, 153*, 362–368.
62. Hussain, S., Akram, M., Abbas, G., Murtaza, B., Shahid, M., Shah, N. S., Bibi, I., & Niazi, N. K., (2017). Arsenic tolerance and phytoremediation potential of *Conocarpus erectus* L. and *Populus deltoides* L. *International Journal of Phytoremediation, 19*, 985–991.
63. Sun, H., Wu, W., & Wang, L., (2009). Phenanthrene partitioning in sediment—Surfactant—Fresh/saline water systems. *Environmental Pollution, 157*, 2520–2528.
64. Takarina, N. D., & Pin, T. G., (2017). Bioconcentration factor (BCF) and translocation factor (TF) of heavy metals in mangrove trees of Blanakan fish farm. *Makara Journal of Science. 7*, 77–81.
65. Banerjee, K., Chakraborty, S., Paul, R., & Mitra, A., (2018). Accumulation of metals by mangrove species and potential for bioremediation. In: *Multifunctional Wetlands* (pp. 275–300). Springer, Cham.
66. Mukherjee, S., & Dutta, M., (2016). Biological oxygen demand in controlling fish production and cost of supplementary feed towards better sustainability of a sewage-fed aquaculture system: A case study of East Kolkata Wetlands, West Bengal, India. *International Journal of Waste Resources, 6*, 209.
67. Mitra, A., & Choudhury, A., (1992). Trace metals in macrobenthic molluscs of the Hooghly estuary, India. *Marine Pollution Bulletin, UK, 26*, 521, 522.
68. UNEP, (1992). *Pollution and the Marine Environment in the Indian Ocean.* UNEP regional seas program activity center, Geneva, Switzerland.

69. Gimeno-García, E., Andreu, V., & Boluda, R., (1996). Heavy metals incidence in the application of inorganic fertilizers and pesticides to rice farming soils. *Environmental Pollution, 92*, 19–25.
70. Mitra, A., (1998). Status of coastal pollution in West Bengal, with special reference to heavy metals. *Journal of Indian Ocean Studies, 5*, 135–138.
71. Chakraborty, D., (2014). *Bio-monitoring of Heavy Metal Pollution of Mangrove Ecosystem Using Biochemical-Markers*. M.Phil Thesis. School of Oceanographic Studies. Jadavpur University. doi: 10.13140/RG.2.1.4146.0327.
72. Hortellani, M. A., Sarkis, J. E. S., Bonetti, J., & Bonetti, C., (2005). Evaluation of mercury contamination in sediments from Santos-São Vicente estuarine system, São Paulo State, Brazil. *Journal of Brazilian Chemical Society, 16*, 1140–1149.
73. Wuana, R. A., & Okieimen, F. E., (2011). *Heavy Metals in Contaminated Soils: A Review of Sources, Chemistry, Risks and Best Available Strategies for Remediation*. International Scholarly Research Notices.
74. Usman, A. R. A., Alkreda, R. S., & Al-Wabel, M. I., (2013). Heavy metal contamination in sediments and mangroves from the coast of Red Sea: *Avicennia marina* as potential metal bioaccumulator. *Ecotoxicology and Environmental Safety, 97*, 263–270.
75. Chowdhury, A., & Maiti, S. K., (2016). Assessing the ecological health risk in a conserved mangrove ecosystem due to heavy metal pollution: A case study from Sundarbans biosphere reserve, India. *Human and Ecological Risk Assessment: An International Journal, 22*, 1519–1541.
76. Bakshi, M., Ghosh, S., Chakraborty, D., Hazra, S., & Chaudhuri, P., (2018). Assessment of potentially toxic metal (PTM) pollution in mangrove habitats using biochemical markers: A case study on *Avicennia officinalis* L. in and around Sundarbans, India. *Marine Pollution Bulletin, 133*, 157–172.
77. Kumar, S., Karmoker, J., Pal, B. K., Luo, C., & Zhao, M., (2019). Trace metals contamination in different compartments of the Sundarbans mangrove: A review. *Marine Pollution Bulletin. 148*, 47–60.
78. Danda, A. A., Ghosh, N., Bandyopadhyay, J., & Hazra, S., (2019). Managed retreat: Adaptation to climate change in the Sundarbans ecoregion in the Bengal delta. *Journal of the Indian Ocean Region, 15*, 317–335.
79. Ghosh, S., Chakraborty, D., Dash, P., Patra, S., Nandy, P., & Mondal, P. P., (2018). Climate risks adaptation strategies for Indian Sundarbans. *Peer J. Preprints*. No. e26963v2.
80. Rahman, M. M., Mandal, B. K., Chowdhury, T. R., Sengupta, M. K., Chowdhury, U. K., Lodh, D., Chanda, C. R., et al., (2003). Arsenic groundwater contamination and sufferings of people in north 24-parganas, one of the nine arsenic affected districts of West Bengal, India. *Journal of Environmental Science Health Part A., 38*, 25–59.
81. Signes-Pastor, A. J., Mitra, K., Sarkhel, S., Hobbes, M., Burló, F., De Groot, W. T., & Carbonell-Barrachina, A. A., (2008). Arsenic speciation in food and estimation of the dietary intake of inorganic arsenic in a rural village of West Bengal, India. *Journal of Agricultural Food Chemistry, 56*, 9469–9474.
82. Chowdhury, A., Naz, A., & Maiti, S. K., (2017). Health risk assessment of 'tiger prawn seed' collectors exposed to heavy metal pollution in the conserved mangrove forest of Indian Sundarbans: A socio-environmental perspective. *Human and Ecological Risk Assessment: An International Journal, 23*, 203–224.

83. Bhattacharya, B. D., Nayak, D. C., Sarkar, S. K., Biswas, S. N., Rakshit, D., & Ahmed, M. K., (2015). Distribution of dissolved trace metals in coastal regions of Indian Sundarban mangrove wetland: A multivariate approach. *Journal of Cleaner Production, 96* 233–243.
84. Gabriel, A. V., & Salmo, S. G., (2014). Assessment of trace metal bioaccumulation by *Avicennia marina* (Forsk.) in the last remaining mangrove stands in Manila Bay, the Philippines. *Bulletin of Environmental Contamination and Toxicology, 93*, 722–727.
85. Rodriguez-Iruretagoiena, A., Trebolazabala, J., Martinez-Arkarazo, I., De Diego, A., & Madariaga, J. M., (2015). Metals and metalloids in fruits of tomatoes (*Solanum lycopersicum*) and their cultivation soils in the Basque Country: Concentrations and accumulation trends. *Food Chemistry, 173*, 1083–1089.
86. Chowdhury, R., Favas, P. J., Pratas, J., Jonathan, M. P., Ganesh, P. S., & Sarkar, S. K., (2015). Accumulation of trace metals by mangrove plants in Indian Sundarban Wetland: Prospects for phytoremediation. *International Journal of Phytoremediation, 17,* 885–894.
87. Rahman, M. S., Saha, N., Ahmed, A. S., Babu, S. O., Islam, A. R., Begum, B. A., Jolly, Y. N., et al., (2021). Depth-related dynamics of physicochemical characteristics and heavy metal accumulation in mangrove sediment and plant: *Acanthus ilicifolius* as a potential phytoextractor. *Marine Pollution Bulletin, 173*, 113160.
88. Radziemska, M., Gusi-atin, Z. M., Holatko, J., Ham-Merschmiedt, T., Głuchowski, A., Mizerski, A., Jaskulska, I., et al., (2021). Nano zero valent iron (nZVI) as an Amendment for phytostabilization of highly multi-PTE contaminated soil. *Materials, 14*, 2559.
89. Lebrun, M., Miard, F., Bucci, A., Fougère, L., Nandillon, R., Naclerio, G., Scippa, G. S., et al., (2021). The rhizosphere of *Salix viminalis* plants after a phytostabilization process assisted by biochar, compost, and iron grit: Chemical and (micro)-biological analyses. *Environmental Science and Pollution Research, 28*, 47447–47462.
90. Du, J., Zhang, L., Ali, A., Li, R., Xiao, R., Guo, D., Liu, X., et al., (2019). Research on thermal disposal of phytoremediation plant waste: Stability of potentially toxic metals (PTMs) and oxidation resistance of biochars. *Process Safety and Environmental Protection, 125*, 260–268.

CHAPTER 7

Plant-Assisted Bioremediation of Heavy Metals

FAIZA KHAN, IQRA IJAZ, MUHAMMAD UZAIR, and NOREEN FATIMA

Department of Plant Breeding and Genetics, University of Agriculture, Faisalabad, Pakistan

ABSTRACT

Environmental pollution is either caused by natural disasters or anthropogenic activities. Pollutants or heavy metals from these sources directly effects agricultural lands and water resources hence threatening to food security. Serious health conditions are caused by the presence of heavy metals in food. Heavy metals like Zn, Cd, Al, As, Co, Cr, Cu, and Ni are becoming abundant in soil and slowly seeps down in water. Bioremediation is the only safest and cheapest method for the reclamation of soil and water. Among bioremediation, (microbial bioremediation, phytoremediation, and mycoremediation) phytoremediation is the most effective way to reduce heavy metals, which in result soil quality is improved. Transport, translocation, transformation, and mineralization are the processes to degrade heavy metals. Growing of hyper accumulated plants successively or more foliage bearing plants like willow are two main approaches for this purpose. Using bacteria in conjunction with plants can enhance the uptake of pollutants. More biodegradable form is designed by these bacteria in the progression of chelation, methylation, redox reactions, leaching, and the production of siderophores. Besides bacteria mycorrhizal fungi is also being used to aid plants in uptake of heavy metals, e.g., plant growth-promoting rhizobacteria (PGPR). Various mechanisms are used for plant assisted bioremediation including phytoextraction, phytovolatilization, and phytostabilization. Wastewater or polluted water

Phytoremediation and Biofortification: Strategies for Sustainable Environmental and Health Management. Nand Kumar Singh, Shadma Afzal, & Tariq Aftab (Eds.)

can also made useful by implanting treatments. Wetlands are less studied in with respect to phytoremediation. So, plant assisted bioremediation is useful approach in improving the land environment.

7.1 INTRODUCTION

Anthropogenic chemicals have been tremendously increased in our environment due to industrialization from last few centuries leading towards contamination of land. Transportation, use of chemicals, fertilizers, and disposal processes are contaminating soil [1]. Human health and biota are affected by harmful effects of waste materials. Among them heavy metals are of main concern depending upon their toxicity [2]. Heavy metals cause toxicity due to its persistent nature in environment. Metals are required for proper functioning of biological entities, but they are harmful at higher levels. They are required by plants at lower levels but as their concentration increase, they cause damage [3]. Heavy metals are toxic (Pb, Cd, Cr, Hg, Zn, Hg, Ur, Se, Ag, Au, Ni, As), they are harmful to plants and affect their growth, activity of essential enzymes and mineral uptake [4]. Some heavy metals are toxic even at low concentrations [3]. These metals accumulate in organs of animals who take them depending upon nature of metal and its reactivity, potential of being carcinogenic. Chromium and Nikal have been studied for their carcinogenic effects in humans [5]. They cause oxidative stress by production of ROS due to free radicals and non-radicals such as O_2 [4]. At acidic pH metal ions replace protons and make hydroxo-metal complexes, they also effect mobility of other metals in soil [6]. So heavy metal pollution is thoughtful environmental issue due to their persistent and non-biodegradable nature.

Bioremediation is cost and environment friendly method for elimination of heavy metals from polluted environments in which innate biological processes are being utilized for elimination of hazardous substances using plants and microorganisms [6]. Classical procedures follow "suck and truck" strategies which is off site treatment of polluted products, but bioremediation and phytoremediation provide opportunity for in situ elimination of contaminated soil and water. For this purpose, combination of different treatments is used as single treatment is not enough for removal of variety of contaminants [1]. Prime rule for bioremediation is lessening of hazardous of chemical from environment by changing in pH, utilization of redox reactions and adsorption of pollutants. It is grounded upon redox reactions that involves modification of biochemistry and work on microbiology by injecting specific reagents to boost the degradation and abstraction of contaminants from environment [4].

Phytoremediation is utilization of plants for removal of contaminants, they have capacity to bind and extract the contaminants [4]. Phytoremediation includes many processes, i.e., Rhizofiltration, Phytostabilization, Phytoextraction, Phytovolatilization, Phytotransformation. Rhizofiltration is used to treat the industrial waste through aquatic and terrestrial plants, in which they accumulate heavy metals in their roots. It is useful for both in-situ and ex-situ treatment. Phytostabilization is used to soil sediments and sludges by reduction in percolation of contaminated water in soil [7]. While phytovolatilization is process by which plants extract contaminants from soil and convert it to gaseous state. It is useful for extraction of Hg, Se, and As [8]. Phytotransformation is degradation of contaminants through plants, i.e., Cr uptake through plants [9].

An important and efficient way of removal of heavy metals is through microorganisms who have capability to degrade or to change them in non-hazardous forms, for this purpose indigenous and genetically modified organisms (GMOs) are used that utilized different approaches, i.e., accumulation of substances and biomethylation/transformation [10]. Microorganisms also change the valency of metals; this phenomenon is useful for detoxification for those heavy metals that have various valence forms and their toxicity differ in different valance forms. It has been studied in different valance forms of Cr. Other mechanisms being used by microorganisms are compartmentalization, metal binding and volatilization [11]. Many microorganisms use processes of biosorption in which bio-sorbent have high affinity for metal ions, i.e., Zn and Cd can be removed by biosorption by *Saccharomyces cerevisiae* by process of ion exchange [12]. Fungi has also proved to be potential source for biodegradation of heavy metals.

Phytoremediation is an effective and cost saving way for the removal of pollutants by process that is chiefly driven by solar energy [13]. Phytoremediation involves in extraction of heavy metals from soil by roots of plants and assimilation into biomass [13]. And their translocation to shoot is desirable [14]. Because of its efficacy and cost efficiency, heavy metal removal using phytoremediation, particularly the use of hyperaccumulators to breakdown and detoxify pollutants, has recently gained a lot of attention [15]. Hyperaccumulation criterion vary depending on the metal, ranging from 100 mg per kg dry mass (DM) for Cd to 1,000 mg per kg for Cu. These values show a shoot-to-soil metal concentration ratio, and the bioaccumulation factor is more than one [16].

Genetic engineering has role in designing modified microorganisms and plants that have enhanced ability to degrade contaminants in environment. It has the advantage of allowing researchers to create microbial strains that can tolerate harsh settings and act as bio-remediators in a variety of environments. Development of "microbial biosensors" through genetic engineering that can swiftly and accurately monitor the degree of contamination in hazardous places. Heavy metal concentrations have been measured using a variety of biosensors [17]. Bacteria that are associated with plants (endophytes and rhizospheric bacteria) has been engineered for elimination of heavy metals [18]. Other microbes can also be explored for this purpose. Recent breakthroughs in omics technologies have aided in the identification of features that optimize the benefits of field remediation methods. The bacterial genes responsible for metal degradation can be inserted into plant tissues using the designer plant technique, allowing metal degradation within the plant tissues [19]. In comparison to physio-chemical based techniques, which are less eco-friendly and more hazardous to human health, the use of genetically engineered (GE) plant-based bioremediation for various heavy metals pollution is in the forefront due to its eco-friendliness and reduced health dangers [20].

Genomics has primarily been studied and applied in microbial genetics and agriculture (e.g., GMO crops), but it is now being employed as a bioremediation method. Understanding genomics in the context of microbe-mediated remediation provides insight into genes related to microbial sensitivity to hazardous metals in the soil. The use of genomics in bioremediation allows researchers to examine microorganisms not only on the basis of biochemical parameters, but also on molecular levels relevant to mechanism. Integration of modern techniques can be proved as remedy of contaminants in our aquatic and soil environments.

7.2 PROCESSES OF PHYTOREMEDIATION

There are different processes by which plants remediate the soil from heavy metal contamination viz., phytoextraction, phytostabilization, rhizofiltration, phytovolatilization, phytodegradation, rhizodegradation, phytodesalination, and hydraulic control.

7.2.1 PHYTOEXTRACTION

Phytoextraction is the adsorption of heavy metals by the roots followed by their translocation to the above ground parts like shoot and leaves and subsequent storing [21]. Hyperaccumulators (plants that store high concentration of heavy metals) are used for this purpose. The criteria for a plant to be a hyperaccumulator varies with the heavy metal in consideration as the threshold of each heavy metal is different, e.g., plants accumulating more than 100 ppm cadmium (Cd) are considered as Cd hyperaccumulators, however lead (Pb) hyperaccumulators uptake more than 1,000 ppm Pb [22]. Phyto mining is an emerging technology that uses plants to extract metals of high or commercial value like nickel, silver, and gold, etc. [23].

7.2.2 PHYTOSTABILIZATION

In contrast to phytoextraction, phytostabilization mainly limits the heavy metals to the roots or root zone. Phytostabilization acts to reduce the bioavailability of heavy metals by their chemical fixation or accumulation mainly in roots. In this way the heavy metal contents of the soil remain the same except the heavy metals are unable to contaminate the ground water as well as the food chain. The chemical fixation of heavy metals in the soil is mainly achieved by physical stabilization of the soil or precipitation in the root zone due to the effect of different organic compounds released by the roots [24]. It has been reported that an amino acid compound, i.e., Phyto siderophore released by the plants of Poaceae family plays an important role in the phytostabilization by forming stable complexes with heavy metals like copper, zinc, cadmium, and iron [25].

7.2.3 RHIZOFILTRATION

In this technique, roots of the plant are used to reclaim contaminated ground water, surface water or wastewater having low heavy metal concentration in large volume. Both terrestrial and aquatic plants are used for this purpose [26]. The roots mainly work by absorbing the heavy metals and sometimes by their precipitation. Plants with roots having good regeneration capability and large surface area are preferred [27].

7.2.4 PHYTOVOLATILIZATION

Phytovolatilization involves the uptake of the heavy metals by the plant, conversion to volatile form and subsequent release into the atmosphere by transpiration [28]. This type of phytoremediation is most controversial as the contaminants released into the atmosphere may be deposited back by precipitation into the ecosystem causing off-site contamination.

7.2.5 PHYTODEGRADATION

Phytodegradation is the breakdown of complex organic compounds into simpler molecule contaminants by plants either by up taking the contaminants followed by their metabolic degradation within the plant or through the release of degrading compounds into the rhizosphere [29]. The enzymes for degradation include dehalogenase and oxygenase [30].

7.2.6 RHIZODEGRADATION

Rhizodegradation is the breakdown of the organic contaminants in the rhizosphere (root zone) by rhizobacteria. It is a slower process than phytodegradation. Different microorganisms like bacteria, fungi, and yeast, etc. Have the ability to degrade different organic contaminants like solvents and fuel [31]. Plants can increase the efficiency of rhizobacteria by exuding promoting substances like flavonoids, amino acids and carbohydrates [30].

7.2.7 PHYTODESALINATION

It is a recent and emerging technology. It is the use of halophytic plants to remove the salts from saline soil. The halophytic plants are naturally more tolerant to heavy metals than glycophytic or salt sensitive plants [30].

7.2.8 HYDRAULIC CONTROL

In hydraulic control trees with high transpiration rate and long roots are used to take in a large volume of surface as well as ground water. As the water is pumped out, the contaminants dissolved in it are also taken up resulting in the rectification of the soil and water table. It also reduces the danger of heavy metal escape from surface water [32].

All of the technologies described above are not mutually exclusive and can be used in combination to improve the efficiency of remediation [31].

7.3 FACTORS AFFECTING BIOREMEDIATION

Control and optimization are the complex procedure in bioremediation structure. Environmental factors including temperature, soil type, nutrients, electron acceptors and oxygen, capability of microbes to convert toxic waste into less toxic substances, how much contaminants are available to microbial population.

➢ **Advantages:**

- Use of natural substances makes it more acceptable by public.
- In the presence of contaminant, the number of bio-agent increases and vice versa.
- By products after biodegradation are usually harmless and are mainly in the form of biomass of cell, water, and CO_2.
- Wide variety of pollutants are destroyed by the use of biodegradation.
- Whole demolition is potential is present instead of moving from soil to water or water to soil.
- Transportation or disposal of byproducts is not required hence Normal human activities are not affected by the use of biodegradation.
- It is less expensive than other treatments for removing contaminants.

➢ **Disadvantages:**

- All toxins are not vulnerable to speedy and widespread dilapidation; only biodegradable compounds can be eliminated from environment.
- Sometime the byproducts after degrading process are more toxic than the present toxin.
- Microbial population which are metabolically active, suitable levels of pollutants and nutrients and best atmosphere for microbial growth are main things to be needed.
- It takes longer to work than other treatments like excavation.
- There are no satisfactory results for bioremediation managements.

7.4 TYPES OF BIOREMEDIATION

There are three types of bioremediations – microbial bioremediation, phytoremediation, and mycoremediation, and each of them is discussed in subsections.

7.4.1 MICROBIAL BIOREMEDIATION

Microbes can be grown in anaerobic conditions, in water, high heat, in condition of dessert and at almost subzero temperature so Isolation of micro-organisms can be done in almost any environmental conditions. Microbes are adapted to any conations this is the main reason they can be used to remove toxicity from wide range of environmental conditions by only using carbon and energy source. They are divided into two groups Aerobic (in which presence of oxygen is required, e.g., are Alcaligenes, Pseudomonas, Rhodococcus, Sphingomonas, and Mycobacterium). These microbes have increasing interest in degrading both polyaromatic and alkanes like hydro-carbons and pesticides. Anaerobic (in this absence of oxygen is not required). DE chlorination of the solvent trichloroethylene (TCE), polychlorinat-ed biphenyls (PCBs) in river sediments, and chloroform are some toxins can be degraded by the use if anaerobic bacteria. Varied kind of environmental impurities are degraded by the use of fungi such as *Phanaerochaete chryso-sporium* (white rot fungus). Saw dust, corn cobs or Straw are the substrates which are commonly used. Methane and carbon are used as an energy source by some anaerobic bacteria called Methylotrophs. Methane monooxygenase is the early enzyme in the aerobic degradation and has a broad substrate range. This enzyme is active against many pollutants including 1,2-dichloroethane and Chlorinated aliphatic trichloroethylene. Constant spread or availability of toxins and bacteria is not easily achieved hence Interaction of pollutants and bacteria is an important step. Chemotactic reaction is shown by some types of bacteria which is known as sensitivity towards chemicals. In the case of fungi, it makes filaments in the direction of pollutants. Surfactants like sodium dodecyl sulfate (SDS) can be used as enhancers for the mobility of contaminants. Activity and growth of microorganisms must be enthused which is mainly done by the use of oxygen and nutrients to aid the local microbes. Toxin break down and formation of essential enzymes require those essential building blocks of life. Carbon is the most basic compound among the phosphorous and nitrogen. Carbon makes 95% of the cell composition among all these elements. The contaminated soil requires pretreatment like excavation [33].

7.4.2 PHYTOREMIDIATION

Microbes have very limited petroleum-based digestion. Substances are metabolized and filtered naturally by plants in natural ecosystem. Contaminants are accumulated metabolizes and converted into low level of toxins by plants. Soil and water contaminants are removed by plantation. In 1991 the term phytoremediation was proposed. Based on contaminant five types of phytoremediation techniques are available phytoextraction, phytotransformation, phytostabilization, phytodegradation, rhizofiltration. Accumulation of toxins into roots and leaves is called phytoaccumulation or phytoextraction. Toxins then entered into body mass of plants which is not degraded like other bioremediation process. Less toxic forms are produced after up taking from soil and water in the process of phytodegradatio and phytotransformation. Trivalent chromium state is attained from hexavalent which is less toxic. A process in which contaminated soil migration is reduced is called phytostabilization. Contaminants that seep down the soil are absorbed and become the part of biomass of the plant then this will never become the part of environment. Activity existing in rhizosphere breaks the contaminants then it's called Phytodegradation or rhizodegradation. Bacteria, yeast, or plants produced proteins or enzymes in the soil enhance the activity. Plants and microbes have symbiotic relation which is known as Rhizodegradation. Microbes deliver healthy soil environment and plant provide nutrient to live prosperously. Through rhizofilterration wetlands are purified by the uptake of contaminants through roots [33].

7.4.3 MYCOREMIDIATION

Coal, metal mining and industrial activities produce lots of water pollution. Accumulation of heavy metals is the main problem these days. Contaminated wastewater is needed to be treated with some clean green and cheaper way. Potential of fungi can be utilized due to accumulation and adsorption capacity. In mycodegradation Wide variety of elements are despoiled by fungi. In various situations fungi can be seen as destructive agent like wood, textiles, stored paper, lather, wrapping material and plastic. *Penicillium simplicissimum* can degenerate polyethylene having molecular weight of 4,000 to 2,800. Weight and mechanical properties of polyethylene bags has changed by the use of enzymes *Aspergillus flavus* and *Mucor rouxii*. Polyethylene degradation is also done by the use of white rot fungus.

Fungus can escape in air from water. Moist substrates like wood are the place where filamentous fungi can colonize and break the air water intrusion to grow into the air. Before escaping water to make Arial hyphae, Surface tension of water is reduced by *Schizophyllum commune.* Large amount of hydrophobins is released in the watery surroundings surface tension is produced, at water air interference stable protein film is formed. Especially hyphae of A. Niger move their direction upward [34].

7.5 FUTURE PROSPECTS

Phytoremediation is a technique that uses plants and their associated microbial populations to degrade, eliminate, convert, or immobilize hazardous chemicals in soils, sediments, ground water, and surface water. Phytoremediation has several advantages over traditional approaches, including low cost, limited environmental disruption, public acceptance, and the ability to treat a variety of toxins. Furthermore, plants, as autotrophic systems with vast biomass, require just a little amount of nutrient input and prevent pollutants from spreading by water and wind erosion.

KEYWORDS

- **bioremediation**
- **mineralization**
- **phytoextraction**
- **phytoremediation**
- **plant growth promoting rhizobacteria**
- **translocation**
- **transport**

REFERENCES

1. Truu, J., Truu, M., Espenberg, M., Nõlvak, H., & Juhanson, J., (2015). Send orders for reprints to Reprints@benthamscience.Ae phytoremediation and plant-assisted bioremediation in soil and treatment wetlands: A review. *Open Biotechnol. J., 9*, 85–92.
2. Ancona, V., Grenni, P., Caracciolo, A. B., & Campanale, C., (2017). Soil biological communities and ecosystem resilience. *Soil Biol. Communities Ecosyst. Resil.*
3. Dixit, R., Wasiullah, Malaviya, D., Pandiyan, K., Singh, U. B., Sahu, A., Shukla, R., Singh, B. P., Rai, J. P., Sharma, P. K., et al., (2015). Bioremediation of heavy metals from soil and aquatic environment: An overview of principles and criteria of fundamental processes. *Sustain., 7*(2), 2189–2212.

4. Ojuederie, O. B., & Babalola, O. O., (2017). Microbial and plant-assisted bioremediation of heavy metal polluted environments: A review. *Int. J. Environ. Res. Public Health, 14*(12).
5. Mahurpawar, M., (2015). Effects of heavy metals on human health effects of heavy metals on human health. *Int. J. Res. -Granthaalayah, 3*(9SE), 1–7.
6. Ayangbenro, A. S., & Babalola, O. O., (2017). A new strategy for heavy metal polluted environments: A review of microbial biosorbents. *Int. J. Environ. Res. Public Health, 14*(1), 1–16.
7. Du, H., Huang, M., Zhang, Z., & Cheng, S., (2014). Genome-wide analysis of the AP2/ERF gene family in maize waterlogging stress response. *Euphytica, 198*(1), 115–126.
8. Sakakibara, M., Watanabe, A., Sano, S., Inoue, M., & Kaise, T., (2007). Phytoextraction and phytovolatili-zation of arsenic from as-contaminated soils by *Pteris vittata*. *Assoc. Environ. Heal. Sci. – 22nd Annu. Int. Conf. Contam. Soils, Sediments Water, 12*, 258–263.
9. Caçador, I., & Duarte, B., (2015). Chromium phyto-transformation in salt marshes: The role of halophytes. In: *Phytoremediation* (pp. 211–217). Springer.
10. Gupta, A., & Joia, J., (2016). Microbes as potential tool for remediation of heavy metals: A review. *J. Microb. Biochem. Technol., 8*(4), 364–372.
11. Wu, G., Kang, H., Zhang, X., Shao, H., Chu, L., & Ruan, C., (2010). A critical review on the bio-removal of hazardous heavy metals from contaminated soils: Issues, progress, eco-environmental concerns and opportunities. *J. Hazard. Mater., 174*(1–3), 1–8.
12. Kilár, F., (2009). Cadmium biosorption on native *Saccharomyces cerevisiae* cells in aqueous suspension. *Agric. Environ., 1*, 20–30.
13. Alkorta, I., Hernández-Allica, J., Becerril, J. M., Amezaga, I., Albizu, I., & Garbisu, C., (2004). Recent findings on the phytoremediation of soils contaminated with environmentally toxic heavy metals and metalloids such as zinc, cadmium, lead, and arsenic. *Rev. Environ. Sci. Biotechnol., 3*(1), 71–90.
14. Sekara, A., Poniedzialeek, M., Ciura, J., & Jedrszczyk, E., (2005). Cadmium and lead accumulation and distribution in the organs of nine crops: Implications for phytoremediation. *Polish J. Environ. Stud., 14*(4), 509–516.
15. Ahemad, M., (2012). Implications of bacterial resistance against heavy metals in bioremediation: A review. *J. Inst. Integr. Omi. Appl. Biotechnol., 3*(3).
16. Baker, A. J. M., McGrath, S. P., Sidoli, C. M. D., & Reeves, R. D., (1994). The possibility of in situ heavy metal decontamination of polluted soils using crops of metal-accumulating plants. *Resour. Conserv. Recycl., 11*(1–4), 41–49.
17. Verma, N., & Singh, M., (2005). Biosensors for heavy metals. *Biometals, 18*(2), 121–129.
18. Divya, B., & Kumar, M. D., (2011). Plant–microbe interaction with enhanced bioremediation. *Res. J. Biotechnol. Vol, 6*, 4.
19. Van, A. B., Tehrani, R., & Schnoor, J. L., (2011). Endophyte-assisted phytoremediation of explosives in poplar trees by methylobacterium populi BJ001 T. In: *Endophytes of Forest Trees* (pp. 217–234). Springer.
20. Singh, B. K., (2011). Emerging and genomic approaches in bioremediation. In: *Proceedings of the 4th International Contaminated Site Remediation Conference* (pp. 11–15). Adelaide, Australia.
21. Yao, Z., Li, J., Xie, H., & Yu, C., (2012). Review on remediation technologies of soil contaminated by heavy metals. *Procedia Environ. Sci., 16*, 722–729.

22. Ullah, A., Heng, S., Munis, M. F. H., Fahad, S., & Yang, X., (2015). Phytoremediation of heavy metals assisted by plant growth promoting (PGP) bacteria: A review. *Environ. Exp. Bot., 117*, 28–40.
23. Gardea-Torresdey, J. L., Peralta-Videa, J. R., De La Rosa, G., & Parsons, J. G., (2005). Phytoremediation of heavy metals and study of the metal coordination by x-ray absorption spectroscopy. *Coord. Chem. Rev., 249*(17, 18), 1797–1810.
24. Mahar, A., Wang, P., Ali, A., Awasthi, M. K., Lahori, A. H., Wang, Q., Li, R., & Zhang, Z., (2016). Challenges and opportunities in the phytoremediation of heavy metals contaminated soils: A review. *Ecotoxicol. Environ. Saf., 126*, 111–121.
25. Shackira, A. M., & Puthur, J. T., (2019). Phytostabilization of heavy metals: Understanding of principles and practices. In: *Plant-Metal Interactions* (pp. 263–282). Springer.
26. Jadia, C. D., & Fulekar, M. H., (2009). Phytoremediation of heavy metals: Recent techniques. *African J. Biotechnol., 8*(6).
27. McIntyre, T., (2003). Phytoremediation of heavy metals from soils. *Phytoremediation,* 97–123.
28. Jabeen, R., Ahmad, A., & Iqbal, M., (2009). Phytoremediation of heavy metals: Physiological and molecular mechanisms. *Bot. Rev., 75*(4), 339–364.
29. Tangahu, B. V., Sheikh, A. S. R., Basri, H., Idris, M., Anuar, N., & Mukhlisin, M., (2011). A review on heavy metals (As, Pb, and Hg) uptake by plants through phytoremediation. *Int. J. Chem. Eng., 2011.*
30. Ali, H., Khan, E., & Sajad, M. A., (2013). Phytoremediation of heavy metals—Concepts and applications. *Chemosphere, 91*(7), 869–881.
31. Ghosh, M., & Singh, S. P., (2005). A review on phytoremediation of heavy metals and utilization of it's by products. *Asian J. Energy Env., 6*(4), 18.
32. Muthusaravanan, S., Sivarajasekar, N., Vivek, J. S., Paramasivan, T., Naushad, M., Prakashmaran, J., Gayathri, V., & Al-Duaij, O. K., (2018). Phytoremediation of heavy metals: Mechanisms, methods and enhancements. *Environ. Chem. Lett., 16*(4), 1339–1359.
33. Kumar, V., & Dwivedi, S. K., (2021). Mycoremediation of heavy metals: Processes, mechanisms, and affecting factors. *Environ. Sci. Pollut. Res.,* 1–38.
34. Singh, H., (2006). *Mycoremediation: Fungal Bioremediation.* John Wiley & Sons.

CHAPTER 8

Molecular and Physiological Attributes Regulating Phytoremediation Potential of Plants

NAIR G. SARATH,[1] C. AKSHAYA PRAKASH,[2] EDAPPAYIL JANEESHMA,[3] K. P. RAJ ASWATHI,[4] A. M. SHACKIRA,[5] DELSE PAREKKATTIL SEBASTIAN,[2] and JOS T. PUTHUR[4]

[1]*Department of Botany, Mar Athanasius College [Autonomous], Kothamangalam, Kerala, India*

[2]*Department of Botany, St. Joseph's College [Autonomous], Devagiri, Kozhikode, Kerala, India*

[3]*Department of Integrated Botany, MES-KVM College, Valanchery, Malappuram, Kerala, India*

[4]*Plant Physiology and Biochemistry Division, Department of Botany, University of Calicut, C.U. Campus P.O. Kerala, India*

[5]*Department of Botany, Sir Syed College, Taliparamba, Kannur, Kerala, India*

ABSTRACT

Phytoremediation is a widely acknowledged technology that utilizes plants for the reclamation of contaminated environments. For the efficient removal of contaminants, the selection of plant candidates should be based on some criteria like having higher biomass with a well-branched and long root system, high pollutant accumulation capacity, and increased tolerance to the pollutants. Plants exhibit various defense mechanisms against inorganic or organic contaminant stress by augmenting the synthesis of antioxidants,

Phytoremediation and Biofortification: Strategies for Sustainable Environmental and Health Management. Nand Kumar Singh, Shadma Afzal, & Tariq Aftab (Eds.)

osmolytes, etc. The molecular and physiological mechanisms functioning in the plants help to identify hazardous consequences of organic as well as inorganic contaminants to plants. The physiological alterations take place as a result of the coordinated responses of several stress induced genes and other underlying molecular attributes. Unraveling the molecular and physiological mechanisms that are concerned with stress tolerance would help in better understanding and improvement of phytoremediation technology. Moreover, distinguishing hyperaccumulators from non-accumulator plants also requires molecular and physiological markers. Understanding the molecular mechanisms upregulated in hyperaccumulator plants can aid in the designing of plants with a higher capacity for heavy metal tolerance, which will be beneficial for phytoremediation and effective soil reclamation.

8.1 INTRODUCTION

Environmental pollution has been a huge issue for the universe since the dawn of the industrial revolution. Mining, foundries, and smelters, as well as other metal-based industrial operations, are the major contributors of various kinds of pollution. Heavy metal pollution comes from a variety of sources, including geogenic, industrial, agricultural, pharmaceutical, and household effluents [1]. Earth's crust naturally contains trace amounts of heavy metals, however rising anthropogenic activities have increased their levels, resulting in toxicity and cause serious difficulties for plants and animals as they pass through the food chain. The heavy metals like iron [Fe], manganese [Mn], copper [Cu], zinc [Zn], nickel [Ni], molybdenum [Mo], and cobalt [Co] are generally required in minute amounts for the proper plant growth and development. On the other hand, it can be injurious to plants at high levels. Other heavy metals, such as lead [Pb], chromium [Cr], cadmium [Cd], and mercury [Hg], have no beneficial effect on plants and are considered as "major hazards" since they are extremely toxic to both plants and animals [2].

Solubility, soil pH, plant growth phases, species types, fertilizers, and soil qualities are all elements that influence metal uptake from soil [3]. Translocation of toxic metal ions to crop grains increases the risk of hyperaccumulation in humans, resulting in toxicity symptoms such as bone mineralization and cancer, as well as damage to the normal functioning of vital organs in the human body such as brain, lungs, kidney, and liver, as well as changes in blood composition [4]. Heavy metal poisoning of natural ecosystems is currently a global environmental concern. Heavy metals have a deleterious impact on plant morphology, physiology, and biochemistry, as well as reducing biomass accumulation at high concentrations [5]. Heavy metal accumulation has a negative impact on important metabolic activities such as photosynthesis, mineral nutrition, and water relations.

Some plants which are tolerant or hyper accumulators of toxic metals which are highly utilized for the purpose of remediation of these toxic metals. They have an adaptive and dynamic physiology to cope up the heavy metal contamination in the growing habitat. The molecular events underlined in the plants helps them to withstand the toxic condition and make it as good agent for phytoremediation.

8.2 PHYTOREMEDIATION

Phytoremediation is a technology that exploits the ability of plants to absorb, degrade, immobilize or stabilize compounds present in their growth medium and thereby removing the pollutants present in the soil, water, etc. [6]. Since this process is driven by solar energy, the cost of application and maintenance is cheap, eco-friendly, and easy to apply [7]. Phytoremediation occurs through two main strategies: stabilization of contamination in the site and cleaning up the site by eradicating the contaminants from the contaminated site. The various strategies like phytoextraction, phytodegradation, phytostabilization or rhizostabilization, phytovolatilization, etc., may fall into one of the above-mentioned main strategies. Phytoextraction, phytodegradation, phytovolatilization clean up the contaminated areas while phytostabilization stabilizes the contaminants within the soil, thereby reducing their bioavailability [8].

8.3 REMEDIATION STRATEGIES FUNCTIONING IN PLANTS

8.3.1 PHYTOEXTRACTION

This phytoremediation strategy refers to the use of plants to remove pollutants from the polluted matrix by absorbing and accumulating these pollutants in the shoots of the plants [8]. The plants used for phytoextraction should satisfy specific criteria like having high biomass, tolerating toxic metals and high accumulation potential of contaminants in the shoots [9]. Growing the plants that satisfy the above criteria, harvesting the contaminated biomass, reducing contaminated biomass by incinerating or thermal decomposition and either disposing the waste safely or recycling the metal contaminants are all different steps to carry out effective phytoextraction [10]. The plants with an inherent capacity to accumulate high levels of toxic contaminants [hyperaccumulators], plants that produce very high biomass, and genetically modified plants can effectively remove contaminants by phytoextraction [8]. The concentration of contaminants, the depth of contamination in the polluted areas and the bioavailable fraction of contaminants also determine the effectiveness of phytoextraction [11].

8.3.2 PHYTOSTABILIZATION

Phytostabilization is the stabilization of contaminants in the root zone of the plants, thereby making the contaminants less available. This prevents the contaminants from leaching into the groundwater, prevents groundwater contamination, and also prevents the entry of contaminants into the food chain [12]. Heavy metal precipitation in the rhizosphere, absorption or adsorption, sequestration of the contaminants in the root cells, etc., are the mechanisms of phytostabilization [13]. Phytostabilization ensures that the contaminants do not leach into groundwater sources and are not dispersed by wind [9]. The main advantage of this strategy is that there is no burden of disposal of contaminated biomass compared to phytoextraction [14].

8.3.3 PHYTOVOLATILIZATION

The strategy of phytovolatilization exploits the capability of contaminant absorption and conversion of these contaminants into a minor toxic volatile compound, which are then released into the atmosphere through the transpiration stream. However, this mechanism is effective for only some volatilizing heavy metals like mercury, arsenic, etc., and organic pollutants [15]. Phytovolatilization is also advantageous because there is no requirement for contaminated biomass disposal. However, in its modified form, the pollutant remains in the atmosphere and is not a complete solution for contaminant removal. Moreover, precipitation may bring pollutants to the soil or water [9].

8.3.4 PHYTOFILTRATION

Phytofiltration is the phytoremediation strategy utilizing plant roots, shoots or seedlings to eliminate contaminants from polluted waters. When roots are used for cleanup, the technique is called rhizofiltration, when cut shoots are used, the process is caulofiltration, and when seedlings are used, the process is blastofiltration [16]. Adsorption or absorption of heavy metals occurs on roots in this case. This is because pH of the root zone changes due to the release of root exudates enabling the contaminants to precipitate on the roots [17]. A dense root system, contaminant tolerance and high biomass are the basic requirements of plants utilized for phytofiltration [18].

8.3.5 PHYTODEGRADATION

Phytodegradation, also called phytotransformation, is when the plant completely breaks down the contaminant due to the metabolic activities occurring inside the plants and by the activity of plant exudates. This strategy is mainly applicable to organic pollutants. It is an attractive strategy because it degrades the contaminants at the site itself, and there is a chance for complete mineralization. Therefore, removal of plants from the contaminated sites is not necessary. The main enzymes responsible for phytodegradation that occurs outside the plant are dehalogenase, peroxidase (POD), nitroreductase, nitrilase, and phosphatase. These enzymes cause the transformation of contaminants in the soil. To degrade contaminants inside the plant, primarily plants should be capable of absorbing the contaminants. Absorption is dependent on hydrophobicity, polarity, and solubility. For example, non-polar substances adsorb onto root surfaces while polar compounds are absorbed by roots and translocated in the plants [19].

8.4 HEAVY METAL DETOXIFICATION MECHANISMS UNDERLYING IN PLANTS

Plants receive heavy metals into their biomass through the roots and shoots [20]. Toxic concentrations of heavy metals in plants disrupt metabolic processes and the proper functioning of plants. Heavy metals disrupt the structure of proteins as they form bonds with sulfhydryl groups [21], influences the cytoplasmic membrane integrity negatively [22], and disturb the activities of various enzymes of photosynthesis, and respiration [23]. Elevated production of reactive oxygen species (ROS) like hydroxyl and superoxide free radicals, non-free radical species like H_2O_2 and singlet oxygen, and cytotoxic compounds like methylglyoxal are the after-effects of heavy metal toxicity. This results in oxidative stress, thereby causing lipid and protein oxidation, leakage of ions, DNA damage, etc., finally leading to apoptosis [23–25].

When plants come across high concentrations of heavy metals, they exhibit various defense mechanisms. The foremost step a plant takes against heavy metal stress is metal avoidance. This strategy eliminates stress by inhibiting the absorption of metal by plant roots [26]. If heavy metals overcome this strategy of plants and enter the plant, then the next mechanism, tolerance, is activated. Tolerance mechanisms mainly include detoxification strategies like sequestration of heavy metals and their compartmentalization in organelles like vacuoles [20], metal ion chelation, binding of metal ions to the cell wall, synthesis of osmoprotectants, etc. [27–31]. If these strategies are unsuccessful, then comes the role of antioxidant defense mechanisms [30].

8.4.1 AVOIDANCE

The avoidance mechanism restricts the entry of heavy metals into the plant tissues through the root. This strategy works at the extracellular level. The main mechanisms restricting metal uptake are exudation of organic acids or amino acids from root and immobilization of heavy metals by mycorrhiza [31]. Plant root exudates include organic acids like citric acid, oxalic acids, etc., amino acids like methionine, lysine, etc. When the plants face heavy metal stress, the production of these root exudates increases, these compounds form complexes with heavy metals, resulting in non-toxic, stable substances that the roots cannot absorb. Certain root exudates increase the rhizospheric pH, which helps to precipitate the heavy metals, decreasing their bioavailability [31].

Moreover, these exudates can serve as nutrient sources for the rhizospheric microbes. The secretions of these microbes can also help to restrict the uptake of heavy metals by the roots [32, 33]. The activity of mycorrhizal fungi also plays a vital role in restricting metal entry into the plants. This is done by the release of phenolic compounds, chelating agents, siderophores, phosphate ions by mycorrhizal fungi which leads to the formation of insoluble metal salts, by the adsorption of metals on the fungal cell surface, which occurs by hydrated iron oxides and sulfides precipitation, by the accumulation of metal ions in fungal arbuscules, hyphae or vesicles or by the detoxification of metal ions in fungal cells due to the activity of metallothioneins (MTs) and vacuole polyphosphates [34].

8.4.2 TOLERANCE

Metal tolerance through chelating agents like various organic acids, cell wall binding molecules, and amino acids, including phytochelatins (PCs) and MTs, is a well-defined defense strategy in various plants against the toxic level of metal ions. For instance, in crop plants like rice, maize, and wheat, organic acids like citrate, malate, lactate, and oxalate plays a crucial role in the detoxification of metal ions, specifically Al, Cd, Pb, and Zn, by forming strong bonds with the carboxyl group of the acids and the metal ions [35–39]. Organic acids refer to the critical metabolites of various pathways operating in plants and may include the anabolic pathways like C4, CAM, and catabolic pathways like Krebs cycle, glyoxylate cycle, etc. [39, 40].

Like organic acids, the carboxylic group of pectic components in the cell wall, such as uronic acids, polygalacturonic acids, etc., act as the binding agent for metal ions, thereby limiting the entry into the plant cell [39, 41, 42]. In *Medicago sativa,* Cd exposure induced the enhanced synthesis of xylo-galacturonan in the cell wall, which serves as a potent tolerance acquisition mechanism of the plant as it limits the cell wall binding of Cd [43]. Methy-lesterified pectins are found exceptionally suitable for the complexation of Pb and minimizing the toxicity to a greater extent even in the lower group of plants, as evidenced by Krzesłowska et al. [44] in *Funaria hygrometrica.* Hence, like higher plants in lower groups like algae, specific components such as alginates also take part in the complexation of toxic metal ions, thereby aiding the detoxification process [45].

In addition to the above components, recently, two primary classes of amino acids [cysteine-rich], namely PCs and MTs, were widely characterized in several metal tolerant plants, including the lower groups like algae. The PC can form stable complexes with many toxic metal ions, which may include Cd, Ag, Cu, and As, and the biosynthesis of PC inside plant cells is carried out by the enzyme phytochelatin synthase (PCS), which is activated by the metal ions [46]. Like PCs, MTs are also endogenously synthesized cysteine-rich residues and are highly efficient in complexing transition metal ions like Cu, Zn, Fe, Mn, etc. These MT-metal complexes bind the metals firmly but can be easily exchanged with other protein molecules due to their low kinetic stability [47]. The gene pool of PC and MT characterized from a number of plants till date will be the future focus of research as it will reveal the metal tolerance pathways in plants at the cellular level.

8.5 PHYSIOLOGICAL FUNCTIONALITY OF PLANTS TOWARDS HEAVY METALS

As soon as heavy metals overcome all the avoidance strategies and enter the plants, the plants try to reduce the effects of heavy metal stress by initiating various cellular defense mechanisms. An important mechanism to nullify metal toxicity effects is by the synthesis of several molecules like nicotianamine (NA), mugineic acids (MAs), spermine, organic acids, MTs, PCs, heat shock proteins, flavonoids, and phenolic compounds, amino acids like proline, hormones like jasmonic acid, salicylic acid, etc. [26, 31, 48]. When these are not sufficient to counteract the production of ROS caused due to heavy metal stress, the antioxidant defense mechanism comes into play. The antioxidant enzymes that scavenge the free radicals include catalase (CAT), superoxide dismutase (SOD), guaiacol POD, ascorbate peroxidase (APX), etc. [enzymatic antioxidants], ascorbate, glutathione (GSH), carotenoids, etc. [24, 49–51].

8.5.1 EFFECTS ON PLANT GROWTH AND PHOTOSYNTHESIS

Growth reduction is the main physiological symptom of heavy metal toxicity in plants [52]. Photosynthetic efficiency and respiration also decrease, affecting metabolism and energy production. Transpiration is also affected, which negatively affects material transportation in plants [53]. Nutrient and water uptake efficiency decreases when plants face heavy metal stress [54]. The altered functioning of leaf and root due to metal toxicity negatively affects flowering, embryo formation, and seed production. Al toxicity in plants inhibited root growth which may be attributed to the interaction of Al with cell division and elongation of root meristem [55, 56]. However, a study conducted on an As hyperaccumulator, *Isatis cappadocica* showed that plant growth was enhanced with 50 μM AsV. This may be due to the detoxification of As when As forms complexes with thiol compounds present in the plant [57].

Photosynthesis is a complex energy-producing process that requires the coordinated working of several cellular and molecular structures. Heavy metals change the morphology of leaves and affect the ultrastructure of chloroplast, mainly the membranes of thylakoids. Heavy metals break down the lipid bilayers of thylakoid membranes by liberating lipids, proteins as well as other constituents of photosynthetic membranes [58]. Leaf growth is retarded due to heavy metal toxicity. Heavy metals decrease plants' chlorophyll and carotenoid content [59]. The composition and structure of Rubisco and other proteins involved in photosynthesis are also affected by heavy metals [60], which affects the proper functioning of light and dark photosynthetic reactions. The transport of electrons between PSI and PSII is inhibited due to the damage caused to the membrane lipids by heavy metal toxicity [61]. Heavy metals also hamper the conductance of stomata by changing guard cell turgidity, which is caused as a result of changes in the morphology and water imbalance in stomata. The increased production of ROS under the influence of stressed conditions also decreases the rate of photosynthesis [58]. Vernay et al. [62] found that *Datura innoxia* exhibited decreased biosynthesis of chlorophyll and carbon dioxide (CO_2) assimilation due to chromium stress, mainly due to stomatal closure. Therefore, this heavy metal decreases the photosynthetic rate, stomatal conductance, and transpiration. The inhibition of CO_2 assimilation may be attributed to decreased utilization of ATP and NADPH, non-photochemical quenching, and a reduced electron transport rate in PS II, which leads to decreased photosynthesis in plants. Zn toxicity changed the guard cell and mesophyll tissue morphology, further decreasing the stomatal conductance in *Beta vulgaris* [63]. The biosynthesis of Chl *a* and Chl *b* was inhibited under Cd, Pb, and Mn stress in *Suaeda glauca* [Bunge] Bunge and *Arabidopsis thaliana* [L.] Heynh [64]. High concentrations of Al in the soil retarded photosynthesis in *Cacao* genotypes [65].

Plants have mechanisms that are inherent in them to cope up with heavy metal toxicity. There are mechanisms to regulate photosynthesis, including scavenging ROS, signal transduction, antioxidant enzyme activity, detoxification, and sequestration of heavy metals, etc. When plants face heavy metal stress, the antioxidant defense system is activated. The increased production of H_2O_2 in the chloroplast is a signal. Interacting H_2O_2 regulates gene expression with proteins containing thiol groups, cell signaling pathways, and transcription factors. The role of phytohormones in regulating photosynthesis is also a major one. The concentration of abscisic acid, ethylene, etc., increased under various abiotic stresses, including heavy metal stress depicting their role in stress tolerance [58]. Chloroplast small heat shock proteins were reported to protect photosynthesis during Zn, Pb, Ni, and Cu stress in *Zea mays* [66].

8.5.2 SUPERIORITY IN ANTI-OXIDATIVE MECHANISMS OF HYPERACCUMULATORS

The accumulation of heavy metals in plant cells augments the production of ROS, which are toxic to plants. ROS induces oxidative damage, which affects photosynthesis and respiration [67], breaks the integrity of nucleic acids, proteins, membranes, etc. [68]. However, plants have inherent defense systems to oppose the oxidative damage caused due to ROS, of which the antioxidant defense system is the main strategy to detoxify ROS [69]. Hyperaccumulator plants can accumulate sufficiently high concentrations of heavy metals in their biomass. They are capable of doing so because of the high expression of various antioxidant enzymes, which is attributed to the upregulation of genes encoding these enzymes [70]. For example, *Stanleya pinnata,* a Se hyperaccumulator plant, produced about 1.5-fold greater concentrations of antioxidant enzymes than the non-accumulator *S. albescens* [71]. The genes regulating the synthesis of GSH, GSH POD, APX was all highly expressed/upregulated in the hyperaccumulator *S. pinnata* which was the cause of reduced ROS in it [70]. *Isatis cappadocica,* an As hyperaccumulator, exhibited high antioxidant enzymes SOD, CAT, and GSH reductase under As stress [72]. *Pteris vittata* also exhibited increased activity of antioxidant enzymes and lesser thiobarbituric acid content, which may be the reasons for the categorization of this plant as an As hyperaccumulator [73]. *Tetraena qataranse*, a Pb hyperaccumulator, exhibited enhanced activity of the antioxidant enzymes SOD, CAT, APX, and GSH reductase [74]. Sidhu et al. [75] also found that Pb augmented the antioxidant enzyme activity of the plant *Coronopus didymus* which helped the plant to tolerate Pb stress.

8.5.3 MAINTENANCE OF OSMOTICUM

Plants produce various organic osmolytes when encountering stressed conditions like drought, heavy metals, temperature, high light, etc. Osmolytes are compounds with high solubility and low molecular weight (LMW). They play a significant role in protecting plants from abiotic stresses through various mechanisms like maintaining proper cellular osmoticum, ROS detoxification, membrane integrity maintenance, enzymes or protein stabilization, protect components of cells from dehydration [76]. Examples of various osmolytes are glucose, sucrose, fructose, proline, sugar alcohols, glycine betaine, polyamines, etc. [77].

The concentration of heavy metals in polluted soils does not significantly impact the osmotic potential of soil solution. However, the plants face physiological drought leading to secondary water stress due to heavy metals present in the soil. Studies conducted by Drazkiewicz & Baszyńki [78] and Vaillant et al. [79] reported that heavy metal toxicity affected transpiration, water content of leaves, water potential, and stomatal conductance of plants negatively. Lower water content in leaves may be attributed to the accumulation or deposition of heavy metals in the vessels [80]. The study conducted by Adejumo et al. [81] on the metallophytic plants *Gomphrena celosioides* and *Sporobolus pyramidalis* reported a high concentration of the osmolytes proline, glycine betaine, and phenolics under heavy metal stress. These compounds play the role of osmoregulators, thereby mitigating oxidative stress. Lutts & Lefèvre [82] suggested xerohalophytes as promising candidates for landscaping and desert greenification as they can grow underwater stress efficiently. The drought resistance capability of plants can also provide heavy metal stress resistance to plants to some extent. In this view, xerohalophytes can be possible candidates for phytoremediation [82].

Besides the inherent potential of plants to produce an increased concentration of osmolytes under heavy metal stress, exogenous application of certain compounds has also been found to be effective in maintaining osmoticum in plants by enhancing the production of osmolytes. The application of Se in tomato plants further enhanced the osmolyte content [proline and glycine betaine] in the presence of Cd [83]. Ahmad et al. [84] has already reported that Se can modulate the production of osmolytes, thereby maintaining the cellular water balance. The increased production of proline and glycine betaine in tomato plants due to the application of Se increased the water extraction efficiency from soil which helped in the protection of plant metabolism [69].

8.6 MOLECULAR ATTRIBUTES UNDERLYING IN PLANTS IN RESPONSE TO HEAVY METALS

8.6.1 SIGNALING RESPONSE IN PLANTS AGAINST HEAVY METALS

Heavy metal-induced signaling cascades aid the plants in tolerating the extreme oxidation stress and ionic imbalance induced by metal ions. This signaling induced dynamism in the molecular and metabolic responses of plants determining the tolerance level of plants. There are three major signaling pathways working in plants in response to heavy metal stress; they are: (a) mitogen-activated protein kinase [MAPK or MAP kinase] signaling [85]; (b) calcium signaling [86]; and (c) hormone signaling [87].

MAP kinase transfers the stress related signals for the up-regulation or down-regulation of different cellular events and metals like Cd, Cu, and

As activates these signaling pathways in different crop plants like rice, wheat, and maize [88]. Similarly, this gene was over-expressed in hyperaccumulators such as *B. juncea* and *Arabidopsis* on the exposure of different heavy metals [89]. The role of MAPK in ROS scavenging and heavy metal chelation was evidenced in rice roots when treated with different antioxidants and GSH, where the pretreated seeds alleviated the impact of metal stress. Thus, the activation of MAPK became limited. MAPK activation depends upon the accumulation of potent ROS molecules, including NO [89]. In maize leaves, excess Cu content activated MAPK cascades, which positively correlated with the activity of antioxidant enzymes such as APX, CAT, and SOD [90].

The ionic concentration of Calcium [Ca^{2+}] which is a secondary messenger that regulates normal metabolism and stress-induced responses in plants cells [91]. Different Ca sensitive proteins like Calcineurin B-like proteins [CBLs], Ca^{2+} dependent protein kinases [CDPKs], Calmodulins [CaMs], and CaM like proteins [CMLs] plays specific roles in metal tolerance [92]. Moreover, the external application of Ca^{2+} alleviated the negative effect of Cu in *Brassica juncea* and Arabidopsis and thus reduced the accumulation of ROS molecules [93]. Similarly, when *Cucurbita pepo* is pretreated with $CaCl_2$ prior to the exposure of Ni, toxicity aid in mitigating the toxic impacts of Ni on the plant metabolism [94]. The Ca treatment reduced the alteration in ion homeostasis and proline accumulation in relation to the reduced uptake of Ni to the cells of *C. pepo*. This study also proved that Ca mediates the activation of different antioxidant enzymes, which should contribute heavy metal tolerance potential to plants [94].

According to the varying environmental cues, different hormones modulate the responses of plants. Auxin, ethylene, cytokinin, brassinosteroids, jasmonic acid, and abscisic acid [ABA] are the different plant hormones that exhibit prominent activities under metal toxicity [95]. The role of auxin homeostasis in hydrogen peroxide accumulation and the potential of ethylene to regulate ROS accumulation were reviewed by Emamverdian et al. [95]. Jasmonic acid and salicylic acid signals the excess heavy metal in the environment, and these growth regulators are indirectly involved in the synthesis of PCs [96]. The exogenous application of cytokinin mitigates the effect of metal toxicity on plant growth [95]. The application of 6-Benzylaminopurine alleviated the toxic effects of Cu stress in *Ricinus*, especially this cytokinin help to maintain the photosynthetic responses of plants by safeguarding the photosystem activities (Figure 8.1) [97].

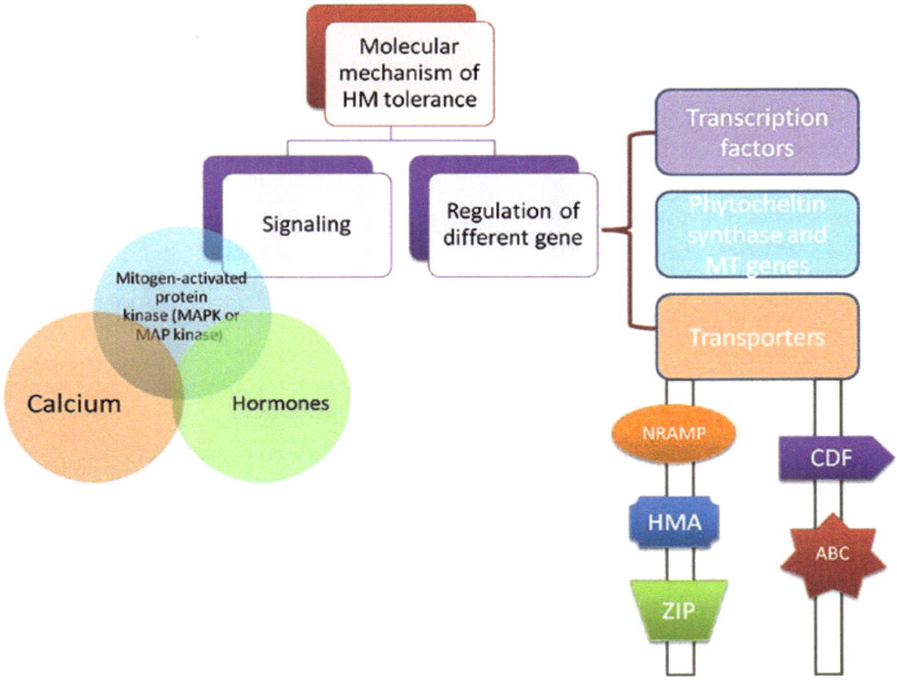

FIGURE 8.1 Molecular mechanism of heavy metal tolerance.

8.6.2 REGULATION OF GENE EXPRESSION

Under heavy metal stress, plants are forced to prevent the excess influx of metal ions in the cytoplasm by activating different transporters in the root cell membrane. Similarly, plant metabolites efficiently chelate the metal ions and transport this complex to the vacuole through tonoplast with activation of different transcription factors [91]. Different transporters play a significant role in determining the metal stress tolerance of the plant. Zinc–iron permease [ZIP], natural resistant associated macrophage protein [NRAMP], heavy metal transport ATPase [CPx- and P1B-ATPase], cation diffusion facilitator [CDF], an ATP-binding cassette [ABC] transporters are the different transporters that aid the conduction of metals inside the plant. In rice plants, excess Zn, Cu, and Cd accumulation elicited the *OsZIP1* transporter gene, and it has a prominent role in the detoxification of metal [98]. The importance of this transporter in Cd stress tolerance was proved with the help of an *OSZIP1* mutant [plant with suppressed *OSZIP1* gene], where this plant showed high sensitivity towards Cd compared to the wild plant [98].

In a comparative study conducted between *Brassica rapa* and *Solanum nigrum,* the *S. nigurum* showed the maximum Cd and Zn accumulation. Differential expression of *the NRAMP* gene was found in these plants, which made the difference in the metal accumulation potential of *S. nigurum* and *B. rapa* [99]. Cd and Zn induced the *NRAMP* gene expression, and overexpression of this gene was found in root cells than the shoot cells [99]. Analysis of Cd tolerance in *Theobroma cacao* showed the significance of *NRAMP* in the uptake of Cd, and overexpression of this gene influences the rate of uptake of metal ions [100]. The detailed analysis of *NRAMP*-structure showed the importance of N-terminus in the functionality of this protein, and AtNRAMP3 of *Arabidopsis thaliana* plays a significant role in the metal tolerance [101]. Higher expression levels of transporter genes *NRAMP* and *ZIP* in tomato genotype YSL189 increase Cd accumulation efficiency, and thus this genotype accumulates more Cd than the other one, HZ903 [102].

The role of HMA-family proteins in plant heavy metal tolerance was examined in *Theobroma cacao,* and it contributes to Cd sequestration [100]. CDF transporters have specific roles in metal tolerance; MTP11 cation diffusion facilitators of *Arabidopsis* and poplar plants impart a high Mn tolerance to a yeast mutant [manganese-hypersensitive]. Moreover, this transporter localized in the Golgi-apparatus could help in the vesicular trafficking and exocytosis of excess manganese [103]. A study conducted in the 18 candidate *CDF* genes from the turnip genome proved the importance of this transporter in heavy metal tolerance and treatment of different metals such as Zn, Cu, Mn, Fe, Co, Na, and Cd were increased the expression level of *BrrMTP* gene [104]. Moreover, this study insight the insignificant role *BrrMTP* gene in mineral homeostasis [104].

Similar to different transporters, genes involved in the biosynthesis of different metabolites get overexpressed under heavy metal stress. The overexpression of the PCS gene, pyrroline-5-carboxylate synthetase gene [P5CS-involved in proline biosynthesis] was observed in tomato plants exposed to Pb and Cd [105]. As PCs have a prominent role in the complexation of metals, the PCS expression directly affects the PC accumulation and PC-metal complex formation [106]. Similarly, metallothionein genes were also overexpressed under metal stress, where these metabolites were involved in the compartmentalization of toxic metal ions in the cytoplasm [107]. Class I MT genes identified from the rice genome have a crucial role in detoxifying As [V], and this gene is actively expressed in arsenic-tolerant rice cultivars [108]. In a mangrove species, *Bruguiera gymnorrhiza*, the *BgMT2 [*class 2 metallothionein] gene exhibited a high *expression level under Cu, Zn, and Pb stress [109]. The same gene was overexpressed in Azolla filiculoides* on exposure to Cd, Cu, Zn, and Ni toxicity. Current investigation in the specificity in the functioning of different genes involved in heavy metal stress will accelerate the research to model new organisms with high heavy metal tolerance potential.

8.6.3 TRANSGENIC PLANTS IN HEAVY METAL PHYTOREMEDIATION

Decontamination of polluted soil by balancing the ecosociological and economical attributes has turned out to be a need of the hour. It has an immediate effect on the existence of life on earth. The successful laboratory stories of phytoremediation have also been practiced at the field level [even in controlled conditions], at least in some cases. For the successful establishment of a nature-based remediation technique, features like depth and density of polluted soil, contaminant bioavailability, the influence of chelating agents, and the candidate plant to be selected should be considered for designing the field trials [110]. In general, native hyperaccumulator plants or transgenically modified non-hyperaccumulator plants are chosen for designing the field trials as it will ensure the efficiency of the process to a certain extend. Genetic alterations may precisely aim to improve the resistance to a particular pollutant, enhance the ability to degrade/volatilize the pollutant, increase the rate of accumulation/translocation, augment the biomass yield, etc. [110, 111].

For instance, the genetically altered Indian mustard varieties were evaluated under field conditions to assess the level of decontamination of Se polluted soils. It was successfully reported that the transgenic varieties exhibited an increased tolerance over the wild type, which was substantiated by the overexpression of genes like adenosine triphosphate (ATP) sulfurylase [APS] gamma-glutamyl-cysteine synthetase [ECS], and GSH synthetase [GS] [112]. Similarly, the polar plants were genetically modified with enhanced capacity to decontaminate trichloroethylene [TCE], and the efficiency was analyzed in field conditions and laboratory experiments. The results were satisfying with the enhanced rate of TCE removal in the transgenic lines over the control ones, as reported by Andrew et al. [113]. Recently, a transgenic variety of Arabidopsis has been produced by incorporating the ABC transporter gene from Poplar, namely PtABC1, which confer tolerance up to 76% towards Hg [111]. Unfortunately, behind these successful stories of field-level implementations of phytoremediation, from an ecological point of view, there is still a potential risk of horizontal gene transfer from the transgenic plants, which needs to be appropriately addressed. In addition, the actual cycle of remediation under field trials also turns to be a challenge since, in most cases, a long duration for the complete decontamination seems to be a drawback [110, 114].

8.7 FUTURE PERSPECTIVES AND CONCLUSION

For the past few decades, the feasibility of plant-based technology like phytoremediation has drawn much attention as it offers a new green alternative to the costly, complex, and less efficient decontamination technique. A number of works have been carried out to unravel the defense strategy underlying the detoxification of pollutants, including accumulation, degradation, stabilization, and volatilization. The ever-increasing database regarding the number of plants exploited as phytoremediating agents has shed light on the specific criteria for an appropriate candidate. However, there are still controversies on the field level implementation of the technique, especially in transgenic plants, which need to be addressed swiftly. The chances of horizontal gene transfer from the transgenic lines to the wild plants challenge biosafety concerns. Hence, a more futuristic approach is good enough to overcome the issues of field trials and the issues raised against the genetically modified agents of phytoremediation.

KEYWORDS

- **antioxidants**
- **heavy metals**
- **hyperaccumulators**
- **phytoextraction**
- **phytoremediation**
- **phytostabilization**
- **tolerance**

REFERENCES

1. Tchounwou, P. B., Yedjou, C. G., Patlolla, A. K., & Sutton, D. J., (2012). Heavy metal toxicity and the environment. In: Luch, A., (ed.), *Molecular, Clinical and Environmental Toxicology* (pp. 133–164). Springer, Basel.
2. Chibuike, G. U., & Obiora, S. C., (2014). Heavy metal polluted soils: Effect on plants and bioremediation methods. *Applied and Environmental Soil Science, 2014.*
3. Sharma, R. K., Agrawal, M., & Marshall, F., (2006). Heavy metal contamination in vegetables grown in wastewater irrigated areas of Varanasi, India. *Bulletin of Environmental Contamination & Toxicology, 77*(2).
4. Ma, C., White, J. C., Dhankher, O. P., & Xing, B., (2015). Metal-based nanotoxicity and detoxification pathways in higher plants. *Environmental Science & Technology, 49*(12), 7109–7122.
5. Sarath, N. G., Shackira, A. M., El-Serehy, H. A., Hefft, D. I., & Puthur, J. T., (2022). Phytostabilization of arsenic and associated physio-anatomical changes in *Acanthus ilicifolius* L. *Environmental Pollution,* 118828.
6. Berti, W. R., & Cunningham, S. D., (2000). Phytostabilization of metals. In: Raskin, I., & Ensley, B. D., (eds.), *Phytoremediation of Toxic Metals: Using Plants to Clean-up the Environment* (pp. 71–88). New York, NY: John Wiley & Sons, Inc.
7. Jacob, J. M., Karthik, C., Saratale, R. G., Kumar, S. S., Prabakar, D., Kadirvelu, K., et al., (2018). Biological approaches to tackle heavy metal pollution: A survey of literature. *J. Environ. Manage., 217*, 56–70. doi: 10.1016/j.jenvman.2018.03.077.
8. Suman, J., Uhlik, O., Viktorova, J., & Macek, T., (2018). Phytoextraction of heavy metals: A promising tool for clean-up of polluted environment? *Front Plant Sci., 9*, 1476. doi: 10.3389/fpls.2018.01476.
9. Vangronsveld, J., Herzig, R., Weyens, N., Boulet, J., Adriaensen, K., Ruttens, A., et al., (2009). Phytoremediation of contaminated soils and groundwater: Lessons from the field. *Environ. Sci. Pollut. R., 16*, 765–794. doi: 10.1007/s11356-009-0213-6.
10. McGrath, S. P., & Zhao, F. J., (2003). Phytoextraction of metals and metalloids from contaminated soils. *Curr. Opin. Biotechnol., 14*, 277–282. doi: 10.1016/S0958-1669(03)00060-0.
11. Sheoran, V., Sheoran, A., & Poonia, P., (2009). Phytomining: A review. *Miner. Eng., 22*, 1007–1019. doi: 10.1016/j.mineng.2009.04.001.
12. Marques, A. P., Rangel, A. O., & Castro, P. M., (2009). Remediation of heavy metal contaminated soils: Phytoremediation as a potentially promising clean-up technology. *Crit. Rev. Env. Sci. Technol., 39*, 622–654. doi: 10.1080/10643380701798272.
13. Gerhardt, K. E., Gerwing, P. D., & Greenberg, B. M., (2017). Opinion: Taking phytoremediation from proven technology to accepted practice. *Plant Sci., 256*, 170–185. doi: 10.1016/j.plantsci.2016.11.016.
14. Wuana, R. A., & Okieimen, F. E., (2011). Heavy metals in contaminated soils: A review of sources, chemistry, risks and best available strategies for remediation. *ISRN Ecology, 2011*, 402647. doi: 10.5402/2011/402647.
15. Mahar, A., Wang, P., Ali, A., Awasthi, M. K., Lahori, A. H., Wang, Q., et al., (2016). Challenges and opportunities in the phytoremediation of heavy metals contaminated soils: A review. *Ecotox. Environ. Safe., 126*, 111–121. doi: 10.1016/j.ecoenv.2015.12.023.
16. Mesjasz-Przybyłowicz, J., Nakonieczny, M., Migula, P., Augustyniak, M., Tarnawska, M., Reimold, U., et al., (2004). Uptake of cadmium, lead nickel and zinc from soil

and water solutions by the nickel hyperaccumulator *Berkheya coddii. Acta Biol. Cracoviensia Ser. Bot., 46*, 75–85.
17. Javed, M. T., Tanwir, K., Akram, M. S., Shahid, M., Niazi, N. K., & Lindberg, S., (2019). Chapter 20 – phytoremediation of cadmium-polluted water/sediment by aquatic macrophytes: Role of plant-induced pH changes. In: Hasanuzzaman, M., Prasad, M. N. V., & Fujita, M., (eds.), *Cadmium Toxicity and Tolerance in Plants* (pp. 495–529). London: Academic Press. doi: 10.1016/B978-0-12-814864-8.00020-6.
18. Hooda, V., (2007). Phytoremediation of toxic metals from soil and waste water. *J. Environ. Biol., 28*(2 Suppl.) 367–376.
19. Parker, D. R., Feist, L. J., Varvel, T. W., et al., (2003). Selenium phytoremediation potential of *Stanleya pinnata. Plant and Soil,* ***249*,** 157–165. https://doi.org/10.1023/A:1022545629940.
20. Patra, M., Bhowmik, N., Bandopadhyay, B., & Sharma, A., (2004). Comparison of mercury, lead and arsenic with respect to genotoxic effects on plant systems and the development of genetic tolerance. *Environmental and Experimental Botany, 52*, 199–223.
21. Hall, J. L., (2002). Cellular mechanisms for heavy metal detoxification and tolerance. *Journal of Experimental Botany, 53,* 1–11.
22. Farid, M., Shakoor, M. B., Ehsan, A., Ali, S., Zubair, M., & Hanif, M. S., (2013). Morphological, physiological and biochemical responses of different plant species to Cd stress, *International Journal of Chemical and Biochemical Sciences, 3,* 53–60.
23. Hossain, M. A., Piyatida, P., Da Silva, J. A. T., & Fujita, M., (2012). Molecular mechanism of heavy metal toxicity and tolerance in plants: Central role of glutathione in detoxification of reactive oxygen species and methylglyoxal and in heavy metal chelation. *Journal of Botany*.
24. Zengin, F. K., & Munzuroglu, O., (2005). Effects of some heavy metals on content of chlorophyll, proline and some antioxidant chemicals in bean (*Phaseolus vulgaris* L.) seedlings. *Acta Biologica Cracoviensia Series Botanica, 47*, 157–164.
25. Rellán-Álvarez, R., Ortega-Villasante, C., Álvarez-Fernández, A., Campo, F. F. D., & Hernández, L. E., (2006). Stress responses of *Zea mays* to cadmium and mercury. *Plant and Soil, 279*, 41–50.
26. 26.Viehweger, K., (2014). How plants cope with heavy metals. *Botanical Studies, 55*, 1–12.
27. Memon, A. R., Aktoprakligil, D., Zdemur, A., & Vertii, A., (2001). Heavy metal accumulation and detoxification mechanisms in plants. *Turkish Journal of Botany, 25,* 111–121.
28. Prasad, M. N. V., (2004). Phytoremediation of metals in the environment for sustainable development. *Proceedings of the Indian National Science Academy, 70,* 71–98.
29. John, R., Ahmad, P., Gadgil, K., & Sharma, S., (2009). Heavy metal toxicity: Effect on plant growth, biochemical parameters and metal accumulation by *Brassica juncea* L. *International Journal of Plant Production, 3,* 65–76.
30. Manara, A., (2012). Plant responses to heavy metal toxicity. In: Furini, A., (ed.), *Plants and Heavy Metals* (pp. 27–53). Springer Briefs in Molecular Science, Springer, Dordrecht, Netherlands.
31. Dalvi, A. A., & Bhalerao, S. A., (2013). Response of plants towards heavy metal toxicity: An overview of avoidance, tolerance and uptake mechanism, *Annals of Plant Sciences, 2,* 362–368.

32. Wang, Q., Chen, L., He, L. Y., et al., (2016). Increased biomass and reduced heavy metal accumulation of edible tissues of vegetable crops in the presence of plant growth-promoting *Neorhizobium huautlense* T1-17 and biochar[J]. *Agriculture Ecosystems & Environment, 228*, 9–18.
33. Sallah-Ud-Din, R., Farid, M., Saeed, R., et al., (2017). Citric acid enhanced the antioxidant defense system and chromium uptake by *Lemna minor* L. grown in hydroponics under Cr stress[J]. *Environmental Science & Pollution Research, 24,* 1–10.
34. Małachowska-Jutsz, A., & Gnida, A., (2015). Mechanisms of stress avoidance and tolerance by plants used in phytoremediation of heavy metals. *Archives of Environmental Protection, 41,* 104–114.
35. Yang, Y. Y., Jung, J. Y., Song, W. Y., Suh, H. S., & Lee, Y., (2000). Identification of rice varieties with high tolerance or sensitivity to lead and characterization of the mechanism of tolerance. *Plant Physiology, 124*(3), 1019–1026. https:// doi.org/10.1104/pp.124.3.1019.
36. Zhu, X. F., Zheng, C., Hu, Y. T., Jiang, T., Liu, Y., Dong, N. Y., Yang, J. L., & Zheng, S. J., (2011). Cadmium-induced oxalate secretion from root apex is associated with cadmium exclusion and resistance in *Lycopersicon esulentum. Plant, Cell and Environment, 34*(7), 1055–1064. https://doi. org/10.1111/j.1365-3040.2011.02304.x.
37. Li, X., Chen, X., & Cui, X., (2012). Zinc chemical forms and organic acid exudation in non-heading Chinese cabbages under zinc stress. *Agricultural Sciences, 3*(4), 562–566. https://doi.org/10.4236/as.2012.34067.
38. Anjum, N. A., Hasanuzzaman, M., Hossain, M. A., Thangavel, P., Roychoudhury, A., Gill, S. S., Rodrigo, M. A. M., et al., (2015). Jacks of metal/metalloid chelation trade in plants — an overview. *Frontiers in Plant Science, 6*, 192. https://doi.org/10.3389/fpls.2015.00192.
39. Osmolovskaya, N., Dung, V. V., & Kuchaeva, L., (2018). The role of organic acids in heavy metal tolerance in plants. *Bio. Comm., 63*(1), 9–16. https://doi. org/10.21638/spbu03.2018.103.
40. Igamberdiev, A. U., & Eprintsev, A. T., (2016). Organic acids: The pools of fixed carbon involved in redox regulation and energy balance in higher plants. *Frontiers in Plant Science, 7*, 1042.
41. Meychik, N. R., Nikolaeva, Y. I., Komarynets, O. V., & Ermakov, I. P., (2011). Barrier function of the cell wall during uptake of nickel ions. *Russian Journal of Plant Physiology, 58*(3), 409–414. https://doi.org/10.1134/S1021443711030137.
42. Sytar, O., Kumar, A., Latowski, D., Kuczynska, P., Strzałka, K., & Prasad, M. N. V., (2013). Heavy metal-induced oxidative damage, defense reactions, and detoxification mechanisms in plants. *Acta Physiologiae Plantarum, 35*, 985–999.
43. Gutsch, A., Sergeant, K., Keunen, E., et al., (2019). Does long-term cadmium exposure influence the composition of pectic polysaccharides in the cell wall of *Medicago sativa* stems? *BMC Plant Biol., **19***, 271. https://doi.org/10.1186/s12870-019-1859-y.
44. Krzesłowska, M., Lenartowska, M., Mellerowicz, E. J., Samardakiewicz, S., & Woźny, A., (2009). Pectinous cell wall thickenings formation—A response of moss protonemata cells to Pb. *Environ. Exp. Bot., 65*, 119–131.
45. Davis, T. A., Volesky, B., & Mucci, A., (2003). A review of the biochemistry of heavy metal biosorption by brown algae. *Water Res., 37*, 4311–4330.
46. Shoujing, Z., Wenjuan, S., & Yucheng, J., (2021). Overexpression of BnPCS1, a novel phytochelatin synthase gene from ramie (*Boehmeria nivea*), Enhanced Cd tolerance,

accumulation, and translocation in *Arabidopsis thaliana. Frontiers in Plant Science, 12*, 1169.
47. Hasan, M. K., Cheng, Y., Kanwar, M. K., Xian-Yao, C., Ahammed, G. J., & Zhen-Yu, C., (2017). Responses of plant proteins to heavy metal stress—A review. *Frontiers in Plant Science, 8,* 1492.
48. Sharma, S. S., & Dietz, K. J., (2006). The significance of amino acids and amino acid-derived molecules in plant responses and adaptation to heavy metal stress. *The Journal of Experimental Botany, 57,* 711–726.
49. Michalak, A., (2006). Phenolic compounds and their antioxidant activity in plants growing under heavy metal stress. *Polish Journal of Environmental Studies, 15,* 523–530.
50. Rastgoo, L., Alemzadeh, A., & Afsharifar, A., (2011). Isolation of two novel isoforms encoding zinc- and copper-transporting P1B-ATPase from Gouan (*Aeluropus littoralis*). *Plant Omics Journal, 4*, 377–383.
51. Mourato, M., Reis, R., & Martins, L. L., (2012). Characterization of plant antioxidative system in response to abiotic stresses: A focus on heavy metal toxicity. In: Montanaro, G., & Dichio, D., (eds.), *Advances in Selected Plant Physiology Aspects* (pp. 23–44). InTech, Vienna, Austria.
52. Hu, Y. F., Zhou, G., Na, X. F., Yang, L., Nan, W. B., Liu, X., Zhang, Y. Q., Li, J. L., & Bi, Y. R., (2013). Cadmium interferes with maintenance of auxin homeostasis in Arabidopsis seedlings. *J. Plant Physiol., 170,* 965–975.
53. Ying, R. R., Qiu, R. L., Tang, Y. T., Hu, P. J., Qiu, H., Chen, H. R., Shi, T. H., & Morel, J. L., (2010). Cadmium tolerance of carbon assimilation enzymes and chloroplast in Zn/Cd hyperaccumulator *Picris divaricata. J. Plant Physiol., 167,* 81–87.
54. Poschenrieder, C., & Barcelo, J., (2004). Water relations in heavy metal stressed plants. In: Poschenrieder, C., & Barcelo, J., (eds.), *Heavy Metal Stress in Plants* (pp. 207–299). Springer, Berlin.
55. Ciamporova, M., (2002). Morphological and structural responses of plant roots to aluminum at organ, tissue and cellular levels. *Biology of Plants, 45,* 161–171.
56. Dipierro, N., Mondelli, D., Paciolla, C., Brunetti, G., & Dipierro, S., (2005). Changes in the ascorbate system in the response of pumpkin (*Cucurbita pepo* L.) roots to aluminum stress. *Journal of Plant Physiology, 162*, 529–536.
57. Karimi, N., Ghaderian, S. M., Marofi, H., & Schat, H., (2011). Analysis of arsenic in soil and vegetation of a contaminated area in Zarshuran, Iran, identifies two angiosperm arsenic hyperaccumulators. *Int. J. Phytoremediation, 12,* 159–173.
58. Khan, M., Nawaz, N., Ali, I., Azam, M., Rizwan, M., Ahmad, P., & Ali, S., (2019). Regulation of photosynthesis under metal stress. *Photosynthesis, Productivity and Environmental Stress,* 95–105.
59. Kuzminov, F. I., Brown, C. M., Fadeev, V. V., & Gorbunov, M. Y., (2013). Effects of metal toxicity on photosynthetic processes in coral symbionts, *Symbiodinium* spp. *J. Exp. Marine Biol. Ecol., 446,* 216–227.
60. Franco, E., Alessandrelli, S., Masojidek, J., Margonelli, A., & Giardi, M. T., (1999). Modulation of D1 protein turnover under cadmium and heat stresses monitored by [35S] methionine incorporation. *Plant Sci., 144,* 53–61.
61. Rama, D. S., & Prasad, M. N. V., (2004). Membrane lipid alterations in heavy metal exposed plants. *Heavy Metal Stress Plants,* 127–181.

62. Vernay, P., Gauthier-Moussard, C., Jean, L., Bordas, F., Faure, O., Ledoigt, G., & Hitmi, A., (2008). Effect of chromium species on phytochemical and physiological parameters in *Datura innoxia*. *Chemosphere, 72*, 763–771. doi: 10.1016/j.chemosphere.2008.03.018.
63. Sagardoy, R., et al., (2010). Stomatal and mesophyll conductances to CO_2 are the main limitations to photosynthesis in sugar beet (*Beta vulgaris*) plants grown with excess zinc. *New Phytol., 187,* 145–158.
64. Zhang, X., Li, M., Yang, H., et al., (2018). Physiological responses of *Suaeda glauca* and *Arabidopsis thaliana* in phytoremediation of heavy metals. *J. Environ. Manage., 223,* 132–139.
65. Ribeiro, M. A. Q., De Almeida, A. A. F., Mielke, M. S., et al., (2013). Aluminum effects on growth, photosynthesis, and mineral nutrition of cacao genotypes. *J Plant Nutr., 36,* 1161–1179.
66. Heckathorn, S. A., Mueller, J. K., LaGuidice, S., Zhu, B., Barrett, T., Blair, B., & Dong, Y., (2004). Chloroplast small heat-shock proteins protect photosynthesis during heavy metal stress. *American Journal of Botany, 91*, 1312–1318.
67. Ahmad, P., Sarwat, M., Bhat, N. A., Wani, M. R., Kazi, A. G., & Tran, L. S., (2015). Alleviation of cadmium toxicity in *Brassica juncea* L. (Czern. & Coss.) by calcium application involves various physiological and biochemical strategies. *PLoS One, 10,* e0114571.
68. Tuteja, N., Ahmad, P., Panda, B. B., & Tuteja, R., (2009). Genotoxic stress in plants: Shedding light on DNA damage, repair and DNA repair helicases. *Mutat Res., 681,* 134–149.
69. Ahanger, M. A., Agarwal, R. M., Tomar, N. S., & Shrivastava, M., (2015). Potassium induces positive changes in nitrogen metabolism and antioxidant system of oat (*Avena sativa* L. cultivar Kent). *J. Plant Int., 10*, 211–223.
70. Wang, J., Cappa, J. J., Harris, J. P., Edger, P. P., Zhou, W., Pires, J. C., Adair, M., Unruh, S. A., et al., (2018). Transcriptome-wide comparison of selenium hyperaccumulator and non-accumulator *Stanleya* species provides new insight into key processes mediating the hyperaccumulation syndrome. *Plant Biotechnol. J.,* 1–13.
71. Freeman, J. L., Tamaoki, M., Stushnoff, C., Quinn, C. F., Cappa, J. J., Devonshire, J., Fakra, S. C., et al., (2010). Molecular mechanisms of selenium tolerance and hyperaccumulation in *Stanleya pinnata*. *Plant Physiol., 153*, 1630–1652.
72. Souri, Z., Naser, K.; Letúzia, M. De. O., (2017). Antioxidant enzymes responses in shoots of arsenic hyperaccumulator, *Isatis cappadocica* Desv., under interaction of arsenate and phosphate. *Environmental Technology*.
73. Srivastava, M., Ma, L. Q., Singh, N., & Singh, S., (2005). Antioxidant responses of hyper-accumulator and sensitive fern species to arsenic. *J. Exp. Bot., 56*, 1335–1342.
74. Usman, K., Abu-Dieyeh, M. H., Zouari, N., & Al-Ghouti, M. A., (2020). Lead (Pb) bioaccumulation and antioxidative responses in *Tetraena qataranse*. *Sci. Rep., 10,* 17070 https://doi.org/10.1038/s41598-020-73621-z.
75. Sidhu, G. P. S., Singh, H. P., Batish, D. R., & Kohli, R. K., (2016). Effect of lead on oxidative status, antioxidative response and metal accumulation in *Coronopus didymus*. *Plant Physiol. Biochem., 105,* 290–296.
76. Ashraf, M. F., & Foolad, M. R., (2007). Roles of glycine betaine and proline in improving plant abiotic stress resistance. *Environmental and Experimental Botany, 59*, 206–216.
77. Mudgal, V., Madaan, N., & Mudgal, A., (2010). Biochemical mechanisms of salt tolerance in plants: A review. *International Journal of Botany, 6*, 136–143.

78. Drazkiewicz, M., & Baszyńki, T., (2005). Growth parameters and photosynthetic pigments in leaf segments of *Zea mays* exposed to cadmium, as related to protection mechanisms. *Journal of Plant Physiology, 162,* 1013–1021.
79. Vaillant, N., Monnet, F., Hitmi, A., Sallanon, H., & Coudret, A., (2005). Comparative study of responses in four *Datura* species to a zinc stress. *Chemosphere, 59,* 1005–1013.
80. Barceló, J., & Poschenrieder, C., (1990). Plant water relations as affected by heavy metal stress: A review. *Journal of Plant Nutrition, 13*, 1–37.
81. Adejumo, S. A., Oniosun, B., Akpoilih, O. A., et al., (2021). Anatomical changes, osmolytes accumulation and distribution in the native plants growing on Pb-contaminated sites. *Environ Geochem Health, 43*.
82. Lutts, S., & Isabelle, L., (2015). How can we take advantage of halophyte properties to cope with heavy metal toxicity in salt-affected areas? *Annals of Botany, 115,* 509–528, https://doi.org/10.1093/aob/mcu264.
83. Alyemeni, M. N., Ahanger, M. A., Wijaya, L., et al., (2018). Selenium mitigates cadmium-induced oxidative stress in tomato (*Solanum lycopersicum* L.) plants by modulating chlorophyll fluorescence, osmolyte accumulation, and antioxidant system. *Protoplasma., 255,* 459–469 https://doi.org/10.1007/s00709-017-1162-4.
84. Ahmad, P., Abd, A. E. F., Hashem, A., Sarwat, M., & Gucel, S., (2016). Exogenous application of selenium mitigates cadmium toxicity in *Brassica juncea* l. (Czern & Coss) by up-regulating antioxidative system and secondary metabolites. *J. Plant Growth Regul., 35*, 936–950.
85. Heavy metal stress. Activation of distinct mitogen-activated protein kinase pathways by copper and cadmium | *Plant Physiology* | Oxford Academic Available online: https://academic.oup.com/plphys/article/136/2/3276/6112494?login=true (accessed on 22 May 2023).
86. Chakrabarty, D., Trivedi, P. K., Misra, P., Tiwari, M., Shri, M., Shukla, D., Kumar, S., Rai, A., Pandey, A., Nigam, D., et al., (2009). Comparative transcriptome analysis of arsenate and arsenite stresses in rice seedlings. *Chemosphere, 74*, 688–702. doi: 10.1016/j.chemosphere.2008.09.082.
87. Wang, R., Wang, J., Zhao, L., Yang, S., & Song, Y., (2015). Impact of heavy metal stresses on the growth and auxin homeostasis of *Arabidopsis* seedlings. *Biometals, 28*, 123–132. doi: 10.1007/s10534-014-9808-6.
88. Tian, Q., Zhang, X., Ramesh, S., Gilliham, M., Tyerman, S. D., & Zhang, W. H., (2014). Ethylene negatively regulates aluminum-induced malate efflux from wheat roots and tobacco cells transformed with TaALMT1. *Journal of Experimental Botany, 65*, 2415–2426. doi: 10.1093/jxb/eru123.
89. *Arsenic Stress Activates MAP Kinase in Rice Roots and Leaves*. ScienceDirect. Available online: https://www.sciencedirect.com/science/article/abs/pii/S0003986110004662 (accessed on 22 May 2023).
90. Liu, J., Wang, J., Lee, S., & Wen, R., (2018). Copper-caused oxidative stress triggers the activation of antioxidant enzymes via ZmMPK3 in maize leaves. *Plos One, 13*, e0203612. doi: 10.1371/journal.pone.0203612.
91. Jalmi, S. K., Bhagat, P. K., Verma, D., Noryang, S., Tayyeba, S., Singh, K., Sharma, D., & Sinha, A. K., (2018). Traversing the links between heavy metal stress and plant signaling. *Frontiers in Plant Science, 9*, 12. doi: 10.3389/fpls.2018.00012.

92. *Calmodulins and Calcineurin B-like Proteins: Calcium Sensors for Specific Signal Response Coupling in Plants*. PubMed. Available online: https://pubmed.ncbi.nlm.nih.gov/12045290/ (accessed on 22 May 2023).
93. Malik, Z., Afzal, S., Danish, M., Abbasi, G. H., Bukhari, S. A. H., Khan, M. I., Dawood, M., Kamran, M., Soliman, M. H., Rizwan, M., et al., (2020). Role of nitric oxide and calcium signaling in abiotic stress tolerance in plants. In: *Protective Chemical Agents in the Amelioration of Plant Abiotic Stress* (pp. 563–581). John Wiley & Sons, Ltd, ISBN 978-1-119-55215-4.
94. Valivand, M., & Amooaghaie, R., (2021). Calcium signaling confers nickel tolerance in *Cucurbita pepo* L. *International Journal of Phytoremediation, 23*, 362–373. doi: 10.1080/15226514.2020.1814992.
95. Emamverdian, A., Ding, Y., Mokhberdoran, F., & Ahmad, Z., (2020). Mechanisms of selected plant hormones under heavy metal stress. *Polish Journal of Environmental Studies, 30*, 497–507. doi: 10.15244/pjoes/122809.
96. Pál, M., Janda, T., & Szalai, G., (2018). Interactions between plant hormones and thiol-related heavy metal chelators. *Plant Growth Regul., 85*, 173–185. doi: 10.1007/s10725-018-0391-7.
97. Sameena, P. P., Kalaji, H. M., Żuk-Gołaszewska, K., Horaczek, T., Sierka, E., & Puthur, J. T., (2021). 6-benzylaminopurine alleviates the impact of Cu^{2+} toxicity on photosynthetic performance of *Ricinus communis* L. seedlings. *International Journal of Molecular Sciences, 22*, 13349. doi: 10.3390/ijms222413349.
98. Liu, X., Shengjun, F., Zhang, B., Wang, M., Cao, H., Rono, J., Chen, X., & Yang, Z. M., (2019). OsZIP1 functions as a metal efflux transporter limiting excess zinc, copper and cadmium accumulation in rice. *BMC Plant Biology, 19*. doi: 10.1186/s12870-019-1899-3.
99. Song, Y., Hudek, L., Freestone, D., Puhui, J., Michalczyk, A. A., Senlin, Z., Ackland, M. L., et al., (2014). Comparative analyses of cadmium and zinc uptake correlated with changes in natural resistance-associated macrophage protein (NRAMP) Expression in *Solanum nigrum* L. and *Brassica rapa*. *Environ. Chem., 11*, 653–660. doi: 10.1071/EN14078.
100. Moore, R. E. T., Ullah, I., De Oliveira, V. H., Hammond, S. J., Strekopytov, S., Tibbett, M., Dunwell, J. M., & Rehkämper, M., (2020). Cadmium isotope fractionation reveals genetic variation in Cd uptake and translocation by *Theobroma cacao* and role of natural resistance-associated macrophage protein 5 and heavy metal ATPase-family transporters. *Hortic Res., 7*, 1–11. doi: 10.1038/s41438-020-0292-6.
101. Li, J., Wang, L., Zheng, L., Wang, Y., Chen, X., & Zhang, W., (2018). A functional study identifying critical residues involving metal transport activity and selectivity in natural resistance-associated macrophage protein 3 in *Arabidopsis thaliana*. *International Journal of Molecular Sciences, 19*, 1430. doi: 10.3390/ijms19051430.
102. Zhao, S., Zhang, Y., Ye, X., Zhang, Q., & Xiao, W., (2015). Responses to cadmium stress in two tomato genotypes differing in heavy metal accumulation. *Turk J. Bot., 39*, 615–624.
103. Peiter, E., Montanini, B., Gobert, A., Pedas, P., Husted, S., Maathuis, F. J. M., Blaudez, D., et al., (2007). A secretory pathway-localized cation diffusion facilitator confers plant manganese tolerance. *PNAS, 104*, 8532–8537. doi: 10.1073/pnas.0609507104.

104. Li, X., Wu, Y., Li, B., He, W., Yang, Y., & Yang, Y., (2018). Genome-wide identification and expression analysis of the cation diffusion facilitator gene family in turnip under diverse metal ion stresses. *Frontiers in Genetics, 9*, 103. doi: 10.3389/fgene.2018.00103.
105. Kısa, D., (2019). Responses of phytochelatin and proline-related genes expression associated with heavy metal stress in *Solanum lycopersicum. Acta Botanica Croatica, 78*, 9–16. doi: 10.2478/botcro-2018-0023.
106. Gupta, D. K., Vandenhove, H., & Inouhe, M., (2013). Role of phytochelatins in heavy metal stress and detoxification mechanisms in plants. In: Gupta, D. K., Corpas, F. J., & Palma, J. M., (eds.), *Heavy Metal Stress in Plants* (pp. 73–94). Springer: Berlin, Heidelberg. ISBN 978-3-642-38469-1.
107. Sharma, R., Bhardwaj, R., Handa, N., Gautam, V., Kohli, S. K., Bali, S., Kaur, P., Thukral, A. K., Arora, S., Ohri, P., et al., (2016). Chapter 10 – responses of phytochelatins and metallothioneins in alleviation of heavy metal stress in plants: An overview. In: Ahmad, P., (ed.), *Plant Metal Interaction* (pp. 263–283). Elsevier. ISBN 978-0-12-803158-2.
108. Gautam, N., Verma, P. K., Verma, S., Tripathi, R. D., Trivedi, P. K., Adhikari, B., & Chakrabarty, D., (2012). Genome-wide identification of rice class I metallothionein gene: Tissue expression patterns and induction in response to heavy metal stress. *Funct. Integr. Genomics, 12*, 635–647. doi: 10.1007/s10142-012-0297-9.
109. Huang, G. Y., & Wang, Y. S., (2009). Expression analysis of type 2 metallothionein gene in mangrove species (*Bruguiera gymnorrhiza*) under heavy metal stress. *Chemosphere, 77*, 1026–1029. doi: 10.1016/j.chemosphere.2009.07.073.
110. Wang, L., Deyi, H., Zhengtao, S., Jin, Z., Xiyue, J., Yong, S. O., Filip, M. G. T., & Jörg, R., (2019). Field trials of phytomining and phytoremediation: A critical review of influencing factors and effects of additives. *Critical Reviews in Environmental Science and Technology.* doi: 10.1080/10643389.2019.1705724.
111. Sun, L. P., Ma, Y. F., Wang, H. H., Huang, W. P., Wang, X. Z., Han, L., & Wang, B. J., (2018). Overexpression of PtABCC1 contributes to mercury tolerance and accumulation in Arabidopsis and poplar. *Biochemical and Biophysical Research Communications, 497*(4), 997–1002.
112. Bañuelos, G., Terry, N., Danika, L. L., Pilon-Smits, E. A. H., & Mackey, B., (2005). Field trial of transgenic Indian mustard plants shows enhanced phytoremediation of selenium-contaminated sediment. *Environ. Sci. Technol., 39*(6), 1771–1777.
113. Andrew, J. C., Legault, E. K., James, C., Keith, S., Indulis, M., Doty, S. L., & Strand, S. E., (2017). A field trial of TCE phytoremediation by genetically modified poplars expressing cytochrome P450 2E1. *Environ. Sci. Technol., 51*, 6090–6099.
114. Fladung, M., & Hoenicka, H., (2012). Fifteen years of forest tree biosafety research in Germany. *Iforest, 5*, 126–130.

CHAPTER 9

Role of Engineered Nanoparticles in Fortification of Plant Nutrients

RICHA PARASHAR, SHADMA AFZAL, and NAND K. SINGH

Department of Biotechnology, Motilal Nehru National Institute of Technology, Allahabad, Prayagraj, Uttar Pradesh, India

ABSTRACT

Micronutrients are those vital nutrients that are required in a balanced form for both humans and plant growth. Nonetheless today, soil and crops are deficient in such micronutrients, causing food insecurity and malnutrition worldwide. An initiative or alternative has been proposed to unravel this issue. Micronutrients can also be used as fertilizers to improve crop production when conventional fertilizers are incompetent. Studies illustrate that metal and metal oxide nanoparticles (NPs) possess great potential for the amplification of a crop's nutrient content resulting in the multiplication of crop yield and biomass production. It can also provide resistance to drought, disease, and pests even in the absence of NPK fertilizers. Different methods have been adopted like dietary diversification, drugs, and industrial fortification but the tenability and supportability of these methods are inadequate. Plants directly get their nutrients from the soil and indirectly from conventional fertilizers but later one yields low efficiency and huge environmental effects. Engineered nanofertilizers are comparatively more competent and target oriented. These engineered NPs have certain advantages over chemical fertilizers like plant nutrients can be supplemented and easily absorbed by plants. Moreover, biological synthesis of metal and metal oxide NPs should be preferred. This would help to eliminate the harmful effects related to NPs for the environment and public health. This chapter also highlights the biosafety of NPs in general.

Phytoremediation and Biofortification: Strategies for Sustainable Environmental and Health Management. Nand Kumar Singh, Shadma Afzal, & Tariq Aftab (Eds.)

9.1 INTRODUCTION

Nutrient deficiency in food taken by humans is one of the factors that contribute to health retardation and this problem majorly dominates in rural areas. Also, according to Ref. [1], the world population could be around 9.6 billion by the end of 2050, with growing population demands for more food. Hence there's a need for fortification of nutrients in food crops. Changes in human lifestyle and environment are some of the reasons that add to the loss of fertility in farmlands. There are certain risk factors that are linked with the use of chemicals in fertilizers like they may emerge from soils, water, or air [2], the supreme aim of applying nanomaterials in agriculture is to reduce hazard chemicals, pest control, nutrient losses, and crop yield improvement [3].

Nutrients that are required by plants are provided through fertilizers and these nutrients get absorbed through the soil ecosystem. A condition that deficit of these micronutrients causes abnormal plant growth. But there is a special condition where soil may not have a shortfall of these micronutrients instead plant is incompetent to absorb and then translocate these nutrients from source to sink might be due to tiny root pore size. Hence it is important to explore more techniques and methods that manifest nutrient loss and improve crop quality in agriculture practices. There is a need to develop sophisticated materials that mediate the precise release of nutrients to target areas only without any adverse environmental non-target effects. It is shown that [4] graphene can be a good carrier of nutrients to plants. It can mediate slow and definite target delivery of nutrients for plant growth and have low environmental risks. There are certain ways that can counter nutrient deficiency like dietary diversification, the use of drugs in industrial fortification [5]. But as these methods are not completely sustainable or affordable.

Nanotechnology is the handling of matter at atomic and molecular levels in the production of specialized microscale-based products. Nanotechnology, as applied to agriculture, helps in minimizing nutrient loss and improves the fortification of crops. These days, sustainable agriculture is required, the development of nano-chemicals has emerged as auspicious agents for plant growth, fertilizers, and pesticides. Nanotechnology has the potential to increase agricultural production either by, nano-formulations of agrochemicals for applying pesticides or fertilizers for crop improvement.

In this regard another method that helps in precision agriculture is Nanoscale coating fertilizers, they are accepted fertilizers loaded or coated with nanoparticles. The encapsulation of nutrients most commonly produces nano-fertilizers with nanomaterials. These Porous nanomaterials remarkably reduce nitrogen loss by modulating demand-based release, and by amplifying the plant uptake process. In this chapter, we detailed the use and functioning of conventional fertilizers (NPK), their alternative zinc and iron oxide-based nanoparticles, and their biosafety aspects.

9.2 INORGANIC FERTILIZERS

Soil is the main source of all plant nutrients, hence deficiency of these essential nutrients in soil may hold back plant growth. In natural environment condition nutrients get recycle between soil and plants. But agricultural crops require more nutrients than natural vegetation for a better yield. Harvested crops also withdrawal some amount of nutrients from soil. Hence there's a need for inorganic fertilizers, animal manures, green manures and legume management. These inorganic fertilizers can ne single nutrient or multi-nutrient. In need of all essential nutrients single nutrient fertilizers required a greater number of applications in soil, but on the other hand multi-nutrient fertilizers have certain advantages like they can fulfill all the nutritional requirements of soil that ultimately absorbed by plant. They are nitrogen, potassium, phosphorous, copper, ferrous, zinc, calcium, magnesium, and manganese. These nutrients have an essential role in plant lifecycle and regulation of metabolic pathways. Multi-nutrient fertilizer provides these nutrients in an easily digestible form ad revive the population of beneficial microorganisms (BMOs) which will help in soil fertility. A general multi-nutrient fertilizer contains the following composition of nutrients N:2–4%, P:0.3–1%, K:1.5–5%, S:0.15–0.8%, Ca:0.2–1.5%, Mg:0.15–1%, Zn:10–100 ppm, Fe:20–500 ppm, Mn:15–250 ppm, Cl:4–50 ppm, Co:2.5–50 ppm, Cu:5–75 ppm and Mo:0.03–10 ppm [6]. The multi-nutrient fertilizer is considered to be very vital for the plant life, owing to the improvement in the soil fertility. Also, multi-nutrient fertilizers are cost-effective and easy-to-use. These properties create a demand among the growers for multi-nutrient fertilizer.

Apart from NPK when other nutrients in the fertilizers are in same ratio then it is called balanced and when they are in particular formula then called complete. A proper soil test will help to determine the application of fertilizers to meet the requirement of crop while taking advantage of nutrients already present in the soil. It will also determine lime requirement and help in diagnosis of problematic soil.

9.2.1 NITROGEN

Nitrogen is the nutrient that is prominent in plant growth, and it is possible to buy fertilizers that contain only nitrogen as the active ingredient. It can be available in both the solid as well as liquid form. Fertilizers that supply most of the nitrogen includes *urea, ammonia,* urea–ammonium nitrates and *ammonium sulfate* [7]. Due to its high nitrogen content, urea can damage plants when applied neat, so it is often sold mixed with other agents. To be wholesome, nitrogen must be present in fixed form. Some nitrogen fixing bacteria and their host plants (notably legumes) have the ability to fix atmospheric nitrogen (N_2) by converting it to ammonia. Applying excess of these fertilizers can lead to destruction of soil nature and fertility and can cause eutrophication. Nitrogen loss is one of the concerns associated with conventional fertilizers. When nitrogen is applied it may loss in environment majorly due to microbial transformation. Nitrate (NO_3^-) and nitrous oxide (N_2O) the two forms having negative environmental impacts. Nitrification inhibitors (NIs) based on dimethylpyrazole phosphate (DMPP), NBPT [N-(n-butyl) thiophosphoric triamide] and boric acid are used globally to minimize nitrogen losses. These compounds inhibit the growth of ammonia oxidizing bacteria (AOB) and ultimately delay ammonium (NH_4^+) oxidation [8]. Manure is the major content of nitrogen, and nitrogen inhibitors works in a way to limit the working of bacteria responsible for converting the ammonium portion of nitrogen from manure into nitrate. This can help to reduce denitrification and nitrate leaching; these both can remove nitrogen from crop root zone. However, similar to how a pesticide wears off over time, NIs only last so long in the soil. Hence the application timing is important. Dicyandiamide (DCD) and 3,4-dimethypyrazole phosphate (DMPP) has been approved by US environmental protection agency as a nitrification inhibitor [9]. One should be careful while applying these fertilizers as they are most suitable when applied in inner part of turn up soil with sufficient amount of water. Surface applications have certain limitations as they converted to NH_3 easily and escape to the environment. Urea is mostly used as nitrogen fertilizer because it is well suited with other nutrients and easily to handle.

9.2.2 PHOSPHORUS

Phosphorus is important from protein synthesis aspect in plants and a component part of nucleic acid molecule. Phosphorus is a part of many plant functions like it involves in nutrient movement from source to sink, transfer of genetic information from one generation to another, energy transferring unit, photosynthesis, and metabolism of sugars and starches [10]. Optimum concentration of phosphorus in soil encourages the root growth and maturity. When there is lack of phosphorus in soil it results into accumulation of sugar in plant that display a reddish-purple color due to anthocyanin pigment. Phosphorus is essential for the plants from seeding stage through to maturity. According to [10] cereals take 75% of the total phosphorus in initial 5 to 6 weeks only. Major organic sources of phosphorus include, certain manures, as well as bone meal and pulverized rock phosphate. Example, blood meal, bone meal, seaweeds, and wood ashes release their nutrients into the soil relatively quickly, while leather dust and pulverized rocks release their nutrients slowly.

Some phosphatic fertilizers used globally are diammonium phosphate (DAP), Monoammonium Phosphate (MAP), NPKs, and SSP. Among all these DAP is the most popular one. It is widely used because it is rich in nutrient content and having magnificent physical properties.

9.2.3 POTASSIUM

Potassium is one of required plant nutrient. It has an important role in plant reproduction and growth. Physical parameters of plant like, size, shape, and color greatly depend on potassium content and availability. Plants absorb potassium in its ionic form that is K^+, synchronization of water in plants that is osmoregulation has been mediated by potassium. It plays major part in ATP production and activation of certain enzymes. During photosynthesis, the opening and closing of stomata is also regulated by potassium, and hence maintain uptake of CO_2. Role of potassium in plants known to improve drought resistance and during the process of starch synthesis, enzyme that act as a catalyst has been activated by potassium. Protein synthesis is also mediated by potassium [11]. Potassium deficiency in soil may cause several symptoms like Deficiency of potassium in soil may causes Symptoms like, chlorosis, slow and dwarf growth, poor resistance to temperature changes, drought and pests, defoliation, weak, unhealthy roots, and uneven ripening of fruits.

Potassium fertilizers are available in organic as well as in inorganic form. The organic potassium fertilizer includes Compost, made from food byproducts is a great source of potassium. Also, banana peels are very high in potassium. Wood ash can also be used, but excess of wood ash can burn the plants. Inorganic potassium fertilizers commercially available can be obtained from potassium chloride (KCl), potassium sulfate, potassium nitrate, and potassium–magnesium salts.

9.3 BIOFERTILIZERS

Back-to-back farming over a fertile land can lead to exhaustion of important nutrients, so it is essential to revive those nutrients with the help of external fertilizers. If soil is deficient in certain nutrients that are important from growth aspects can lead to poor crop productivity, poor harvest, hunger, and malnutrition. Chemical fertilizers assure the resuscitation of land, but they are associated with environmental pollution and non-target effects. Biofertilizers are the natural ways to revive soil nutrients. These are fertilizers that contain microorganisms that enhance the fertility of the soil and provide a finer crop yield when they are added to the soil. When applied to seed, plant surfaces, or soil, promote growth by different mechanisms such as increasing the supply of nutrients, increasing root biomass or root area and increasing nutrient uptake capacity of the plant. It includes living organisms such as mycorrhizal fungi, blue-green algae, and bacteria. As a result, nitrogen becomes available for plants. Rhizobium is considerably more effective and mostly used biofertilizer. Rhizobium, in association with legumes, fixes atmospheric nitrogen. The legumes and their symbiotic association with the rhizobium bacterium result in the formation of root nodules that fix atmospheric N. They also protect the plant from many pathogenic attacks.

9.4 NANOFERTILIZERS

According to the recent studies nanotechnology have the capacity to transform traditional agriculture practices [12–14]. It provides easy and target-based delivery system which is safe. High surface area to volume ratio is that excellent property of nanofertilizers which prioritize it over other conventional fertilizers. As they fulfill complete nutrients requirement of crop by slow release that channel through tiny nanoporous surface of plant. These nanofertilizers can be FeO NPs, ZnO NPs, CuO NPs, TiO_2 NPs, CeO NPs, Se NPs, carbon-based nanoparticles and zeolite based nanofertilizers [50] (Figure 9.1). Application of nanoparticles (NPs) in form of nanofertilizers in agriculture system depends upon certain factors such as size and concentration of NPs, their composition, chemical properties and also on plant species [4]. Table 9.1 depicts some approved nanofertilizers and their constituents used in different parts of the world.

TABLE 9.1 Some Approved Nanofertilizers and Their Constituents Used Globally

SL. No.	Fertilizer	Manufacturer	Country	Constituents	References
1.	Nano-max NPK fertilizer	JU Agri Sciences Pvt. Ltd.	India	Multiple organic acids chelated with major nutrients, amino acids, organic carbon, organic micro-nutrient elements, vitamins, and probiotic.	[15]
2.	TAG NANO	Tropical Agrosystem India (P) Ltd.	India	Protein-lacto-gluconate chelated with micronutrients, vitamins, probiotics, seaweed extracts, and humic acid.	[16]
3.	Nano-green	Nano-Green Sciences, Inc.	India	Extracts of corn, grain, soybeans, potatoes, coconut, and palm.	[21]
4.	Nano-micro-nutrient (EcoStar)	Shan Maw Myae Trading Co., Ltd.	India	Zn (6%), B (2%), Cu (1%), Fe (6%), EDTA Mo (0.05%), Mn (5%+), AMINOS (5%)	[21]
5.	Nano-calcium (magic green)	AC International Network Co., Ltd.	Germany	$CaCO_3$ (77.9%), $MgCO_3$ (7.4%), SiO_2 (7.47%), K (0.2%), Na (0.03%), P (0.02%), Fe (7.4 ppm), Al_2O_3 (6.3 ppm), Sr (804 ppm), sulfate (278 ppm), Ba (174 ppm), Mn (172 ppm), Zn (10 ppm)	[33]
6.	PPC nano	WAI International Development Co., Ltd.	Malaysia	M protein (19.6%), Na_2O (0.3%), K_2O (2.1%), $(NH_4)2SO_4$ (1.7%), diluent (76%)	[21]
7.	Nano-capsule	The Best International Network Co., Ltd.	Thailand	N (0.5%), P_2O_5 (0.7%), K_2O (3.9%), Ca (2.0%) Mg (0.2%), S (0.8%), Fe (2.0%), Mn (0.004%), Cu (0.007%), Zn (0.004%)	[21]
8.	Nano-ultra-fertilizer	SMTET Eco-technologies Co., Ltd.	Taiwan	Organic matter (5.5%), N (10%), P_2O_5 (9%), K_2O (14%), P_2O_5 (8%), K_2O (14%), MgO (3%)	[21]
9.	Biozar nano-fertilizer	Fanavar Nano-Pazhoohesh Markazi Company	Iran	Combination of organic materials, micronutrients, and macro-molecules.	[44]

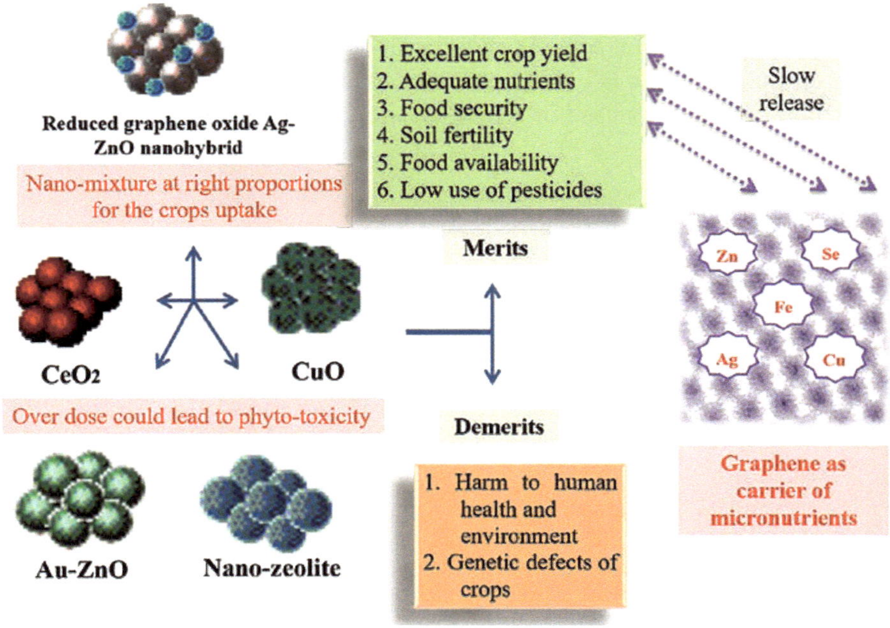

FIGURE 9.1 Improving soil fertility by using metal and metal oxide nanoparticles.

9.5 GENERAL SYNTHESIS OF NANOPARTICLES (NPS)-BASED FERTILIZERS

Nanomaterials can be synthesized by two ways one is top-down approaches and another is bottom-up approaches (Figure 9.2). Top-down approaches includes all the physical methods that are high cost and take time and energy. Where, as bottom-up approaches includes chemical and biological methods, top-down approaches include mechanical milling, laser ablation, sputtering, and bottom-up approaches lists solid state method, liquid state, synthesis method, gas phase methods and biological methods [53]. Biological methods of nanoparticle synthesis have certain advantages over other methods like they have least environmental impacts, ecofriendly material nature, less toxic, inexpensive, economical, and saves time and energy. The main aim with the green synthesis of nanoparticle is to reduce non target effects on environment and public health. Plant extract and microbial synthesis are two major approaches comes under green nanotechnology. Researchers are more focused on plants methods for nanoparticle fabrication as each and every part of plant can be utilized for nanoparticle synthesis for various purpose.

Carbohydrate, protein, flavonoids, phenolics, terpenoids, and saponins are some of important components from plants that have immense role in reducing, stabilizing, and capping metal and metal oxide NPs. A functional group in these biocomponents purpose as organic ligand, they are electron donors and help these NPs to be in compact form. Biological synthesis edges the characteristics of NPs like surface morphologies, close packed nature, biocompatibility, and ecofriendly behavior. In case of selenium NPs, synthesis utilizes plant substrate from *Vitis vinifera* fruits, *Bougainvillea spectabilis* wildflower with sodium selenite, selenous acid or selenium oxide as precursor [54, 55]. However, there are many examples of plant extract-based NPs, but still there are certain unclarities on their applications in agriculture industry.

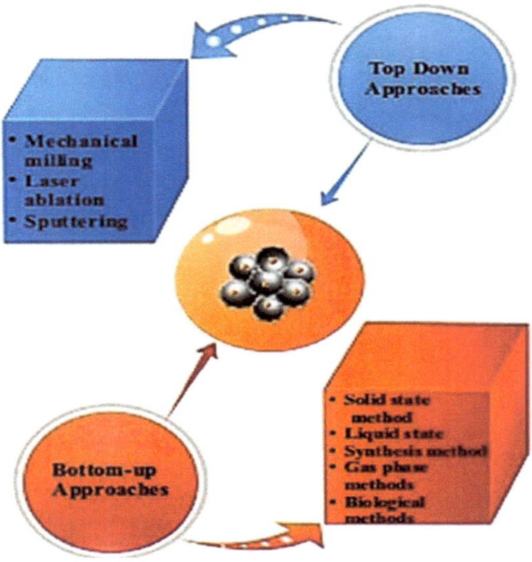

FIGURE 9.2 Top-down and bottom-up approaches.

9.6 ZINC OXIDE (ZNO) NANOPARTICLES (NPS) AS FERTILIZER

Zinc is one of those micronutrients that are required in small quantity in plant. It generally supplied in form of zinc oxide (ZnO) and zinc sulfates ($ZnSO_4H_2O$) in fertilizers. Zinc is important from many regulatory aspects like it mediates most of the enzymes that are essential for hormones and chlorophyll metabolism. In a form of NPs zinc efficiently absorbed by plant system. Either the excess or deficiency of Zn can obstruct the growth pattern and overall development of plant [17]. Malnutrition in food crops is the condition that arises due to the lower absorption of zinc from soil because of its scarcity. According to a survey by WHO (World Health Organization) around one third of population is facing deficiency of zinc in cereals or food crops [18–20]. It can cause hidden hunger diseases. As we know oxides can't be solubilize in water therefore a defined ratio of ZnO and zinc sulfate is used in fertilizers [19]. Alkaline soil is mostly zinc deficient that alters crop productivity.

As uniform concentration of these micronutrient in soil is difficult to achieve due to its small quantity. Carriers are those macronutrients that applied in combination with these micronutrients, but ZnO and zinc sulfate do not require any carriers, they individually act as a micronutrient, but solitary zinc can't. State of these carriers is an important factor that decides the dissolution and absorption of micronutrients in soil. Anhydrous ammonium nitrate (AAN) and orthophosphate are some carriers present in liquid state [52]. Certain soil properties like pH, mineral configuration, organic matter (OM) play a significant role in absorption of micronutrients. Zinc absorption in calcareous soil is affected by the calcium carbonate ($CaCO_3$) that turns its pH to be high. As the reaction occurs between zinc and $CaCO_3$ that precipitate zinc carbonate, that affects crop productivity.

Nature of the soil is also a parameter that defines the absorption of nutrients. Sandy soil find it difficult to hold on nutrients and they effortlessly leached out. Micronutrient's availability and their absorption also depends upon pH of soil, with increasing pH micronutrients absorption decreases, molybdenum is an exception case here. A soil with high pH conditions bearing other clay and organic materials allows zinc to absorb on the surface and later precipitate $Zn(OH)_2$, $ZnCO_3$ and Zn_2SiO_4 [22]. With all these conditions absorption, availability, and distribution of nutrients is severely compromised in plant biosphere. ZnO NPs application has revolutionized the traditional agriculture system as they have the capability to block the excess UV radiations harmful for plant growth and also used in genetic modification of crop variety like nanofoods [23].

9.7 IRON OXIDE NANOPARTICLES (NPS) AS FERTILIZER

Iron is one of the amplest elements on earth's crust, it majorly leached out from ores like magnetite (Fe_3O_4), maghemite (g-Fe_2O_3), and hematite (α-Fe_2O_3). This element has many applications because it is cheap and easily available. Iron is an important element for NPs synthesis due to its various properties like less toxicity and superparamagnetic behavior. These NPs have immense role in improving productivity in agriculture sector. But on the other hand, there are certain limitations also that are associated with iron NPs like reactive nature and being pyrophoric [51]. Soil deficient in iron content make it difficult for plants to absorb it from soil biosphere ad that leads to iron deficiency in food crops results to malnutrition. Iron fertilizers applied to field may react with other compounds and transformed, that cause low absorption by plants, use of iron chelated fertilizers can improve this drawback. γ-Fe_2O_3 NP's can serve as excellent iron fertilizer. It reduces iron deficiency in soil and crops. They promote seed germination, also have significant role in root growth and increases water capacity in chlorophyll. These NPs are dynamic in nature, so when soil is treated with γ-Fe_2O_3 NPs, iron get deliverance and distribute to different parts of plants means source to sink. Table 9.2 mentions the positive and negative effect of some of the NPs on food crops.

9.8 BIOSAFETY OF NANOMATERIALS

Nanomaterial safety examination is always tedious due to many reasons like, type of raw material or salt used during preparation, processing method, size, shape and structure of nanomaterial, reaction in medium, etc., while examining toxicity NPs may get cluster or pile up, rises its unstable nature that changes administered doses. Such NPs modify from their unique properties [34]. NPs from different sources like Zn, Fe, Cu, Ag, etc., may show different end results with plant nutrient fortification and soil absorptivity [35]. NPs may be categorized into three groups on the basis of their method of synthesis, chemically, physically, and biologically synthesized. Among them biological synthesis of NPs considered to be safe as they have least harmful impacts on environment. Properties of NPs depends upon their source material, size formation, method of synthesis and reaction conditions also these factors influence the toxicity of NPs. These external factors may cause changes in reaction chemistry that turns NPs to be toxic. Biological methods of nanoparticle synthesis are preferred over chemical methods.

Environmental conditions can affect the behavior chemistry of NPs to a great extent, one of the examples is Fullerene (C_{60}), it has been studied that fullerene NPs show nontoxic effects for the microorganisms in soil but are toxic in an aquatic biosphere [36, 37]. Toxicology of biomaterial is still a

TABLE 9.2 Required Doses, Positive and Negative Effect of Some of the Nanomaterials on Food Crops

SL. No.	Nanomaterials	Concentration	Positive Impact in Crop Production	Negative Impact in Crop Production	References
1.	ZnO	2,000 mg/L	Increases crop yield in *Zea mays*.	–	[24, 25]
2.	ZnO	<200 mg/kg	–	Lower root no., reduce rice seedling length, obstruct photosynthesis.	[24, 25]
3.	Cu/CuO	≤10 mg/L	Magnify plant growth	Growth retarded, cell death and loss of leaf coloration.	[26]
4.	FeO	250 mg/L	Root elongation, depollute chemicals in root environment.	Reduction in germination in *Hordeum vulgare* and *Linum usitatissimum*.	[27]
5.	Ag	150 ppm	In potato plant it enhances chlorophyll content and improves catalase activities.	Cause sulfidation, hamper absorption of nutrients, delay seed germination and destructs root and shoot growth in mung bean and Chinese cabbage.	[28, 29]
6.	TiO_2	–	Amplifies water and oxygen movement in capsule that response in fast germination and upgrade seed stress resistance.	High rate of movement sometimes harm seed embryo and hamper germination.	[30]
7.	Se	0.05–0.1 mg/kg	Improves different plant metabolites, chlorophyll, and root shoot length.	As a prooxidant in higher concentration can be harmful for plants.	[31, 32]
8.	CeO	2,000 mg/L	Reduced germination in *Medicago sativa*, *Zea mays* and *Cucumis sativus*.	Root elongation, can strike fruit flavor and other nutritional contents.	[27]

topic of research. There is vast source of materials available for NP synthesis and different reactions with the medium of application. So somehow it becomes difficult to study individual behavior of NPs. According to a study it is demonstrated that use of ZnO as nanofertilizer is safe. It fulfills the requirement of zinc micronutrient [38, 39]. But it shows toxic effects in acidic soil in comparison with calcareous soil [40]. In short ZnO toxicity depends upon the pH if soil. Different plants have different kind of behavior in exposure to ZnO NPs. It has been studied that high of roots by these NPs [41–43]. However, this kind of problem can't be seen with zinc ions because sorption of zinc is more efficient in NP form rather than Zn^{+} form, also NPs have poor mobility than ions [25].

If talk about soil biosphere than metal or metal oxide NPs are not favorable for the nutrient fixing microbes present in soil. Some of these microbes are azotobacter, phosphorus, and potassium solubilizing bacteria [45, 46]. It has been also reported that excess of CuO NPs can affect many metabolic processes in wheat plant such as photosynthesis [47, 48]. When these NPs get accumulate in plants, they may get transfer through food chain and indirectly effects human population [49]. Toxicity with any metal or metal oxide NP greatly depends upon its concentration as said above, like a concentration >60–100 microgram/ml can be toxic for Ag NPs and below 30 microgram/ml it is considered to be safe because at this concentration particles are unable to penetrate root cells [26]. However, nanotechnology still improving and finding new methods that contributes towards unparallel development in the area of agriculture.

9.9 CONCLUSION

Certain studies have defined the toxicity and other associated risk factors regarding metal NPs, but still, it's a revolution in modern agriculture practices. Nanotechnology applications can be higher than nuclear energy. The grounds for this are their manipulative properties that alter physiochemical behavior, bioavailability, easy processing, and engineering. Compared to other compounds they are less toxic, and great carrier system use. The biological synthesis of nanomaterials should be preferred for agriculture applications. There must be regulation bodies and norms that counter the effects of nanoproducts on the environment and consumer health. Industries that participate in nanoproduct synthesis should provide complete information regarding their product. After laboratory testing and examination nanoproduct trials should be taken out in the field. They must be initiated at nursery stage for proper tracking. They should be provided during soil preparation so that nutrient deficiency and biosensing can be monitored. Such kind of technology can greatly improve agricultural practices, creates a balance in the environment and sustainability in public health.

KEYWORDS

- **biofertilizers**
- **biofortification**
- **inorganic fertilizers**
- **micronutrients**
- **nanofertilizers**
- **nanoparticles**
- **NPK fertilizers**
- **zinc oxide nanoparticles**

REFERENCES

1. United Nation Department of Economic and Social Affairs, (2015). *World Population Projected to Reach 9.6 Billionby 2050*. United Nation Department of Economic and Social Affairs: New York, NY, USA.
2. Xiong, T., Dumat, C., Dappe, V., Vezin, H., Schreck, E., Shahid, M., Pierart, A., & Sobanska, S., (2017). Copper oxide nanoparticle foliar uptake, phytotoxicity, and consequences for sustainable urban agriculture. *Environ. Sci. Technol., 51*, 5242–5251. [CrossRef].
3. Thakur, S., Thakur, T., & Kumar, R., (2018). Bionanotechnology and its role in agriculture and food industry. *J. Mol. Genet. Med., 12*, 1–5.
4. Kabiri, S., Degryse, F., Tran, D. N. H., Da-Silva, R. C., Mclaughlin, M. J., & Losic, D., (2017). Graphene oxide; A new carrier for slow release of plant micronutrients. *ACS Appl. Mater. Interfaces, 9*, 43325. [CrossRef] [PubMed].
5. Datta, M., & Vitolins, M. Z., (2016). Food fortification and supplement use-are there health implications? *Crit. Rev. Food Sci. Nutr., 56*, 2149–2159. [CrossRef][PubMed].
6. McKenzie, R., (1998). Crop nutrition and fertilizer requirements essential plant nutrients. *Agri-Facts, 540*, 1–7.
7. Monreal, C. M., Derosa, M., Mallubhotla, S. C., Bindraban, P. S., & Dimkpa, C., (2016). Nanotechnologies for increasing the crop use efficiency of fertilizer-micronutrients. *Biol. Fertil. Soils, 52*, 423–437. [CrossRef].
8. Rech, I., Polidoro, J. C., & Pavinato, P. S., (2017). Additives incorporated into urea to reduce nitrogen losses after application to the soil. *Pesqui Agropecu Bras, 52*, 194–204. [CrossRef].
9. Yang, M., Fang, Y.,Sun, D., & Shi, Y., (2016). Efficiency of two nitrification inhibitors (dicyandiamide and 3, 4-dimethypyrazole phosphate) on soil nitrogen transformations and plant productivity: A meta-analysis. *Sci. Rep., 6*, 1–10. [CrossRef].
10. Monostori, T., (2014). *Crop Production* (pp. 15–20). University of Szeged, Faculty of Agriculture Andrássyút15: Hódmezo″vásárhely, Hungary.

11. Hagin, J., & Tucker, B., (1982). *Fertilization of Dryland and Irrigated Soils: Advanced Series in Agricultural Sciences* (pp. 120–140). Springer: Berlin/Heidelberg, Germany.
12. Dimkpa, C. O., & Bindraban, P. S., (2016). Fortification of micronutrients for efficient agronomic production: A review. *Agron. Sustain. Dev., 36*, 7. [CrossRef].
13. Manjunatha, S. B., Biradar, D. P., & Aladakatti, Y. R., (2016). Nanotechnology and its application in agriculture; A review. *J. Farm Sci., 29*, 1–3.
14. Rai, A., Rai, S., & Rakshit, A., (2013). Mycorrhiza-mediated phosphorus use efficiency in plants. *Environ. Exp. Biol., 11*, 107–117.
15. Prasad, R., Bhattacharyya, A., & Nguyen, Q. D., (2017). Nanotechnologyin sustainable agriculture; Recent developments, challenges, and perspective. *Front Microbiol., 8*, 1–13.
16. Diatta, J., Borowiak, K., & Szczepaniak, W., (2018). Evaluation of fertilizers solubility and phosphate release in slightly acidic arable soil. *Arch. Agron. Soil Sci., 64*, 1131–1141. [CrossRef].
17. Singh, N. B., Amist, N., Yadav, K., Singh, D., Pandey, J. K., & Singh, S. C., (2013). Zinc oxide nanoparticlesas fertilizer for the germination, growth and metabolism of vegetable crops. *J. Nanoeng. Nanomanuf., 3*, 353–364. [CrossRef].
18. Biesalski, H. K., (2013). *Hidden Hunger*. Springer: Berlin, Germany.
19. Sadeghzadeh, B., (2013). A review of zinc nutrition and plant breeding. *J. Soil Sci. Plant Nutr., 13*, 905–927. [CrossRef].
20. Guilbert, J. J., (2003). Theworldhealthreport2002=Reducing risks, promoting healthy life. *Educ. Health, 16*, 230.
21. Dimkpa, C., & Bindraban, P., (2017). Nanofertilizers: New products for the industry? *Journal of Agricultural Food Chemistry*. doi: 10.1021/acs.jafc.7b02150.
22. Mortvedt, J. J., & Giordano, P. M., (1967). Crop response to zinc oxide applied in liquid and granular fertilizers. *J. Agric. Food Chem., 15*, 118–122. [CrossRef].
23. Kamran, A., Haroon, Z. K., Muhammad, Z., Imdad, H., & Zeeshan, A., (2016). *Nano-Zinc Oxide as a Future Fertilizer.* Weekly Technology Times.
24. Raddy, R., (2014). *Efficacy of Nano Zinc Particle on Growth.* Master's Thesis, University of Agricultural Sciences, Bangalore, India.
25. Rajput, V. D., Minkina, T. M., Behal, A., Sushkova, S. N., Mandzhieva, S., Singh, R., Gorovtsov, A., et al., (2018). Effects of zinc-oxide nanoparticles on soil, plants, animals and soil organisms: A review. *Environ. Nanotechnol. Monit. Manag., 9*, 76–84. [CrossRef].
26. Rastogi, A., Zivcak, M., Sytar, O., Kalaji, H. M., He, X., Mbarki, S., & Brestic, M., (2017). Impact of metal and metal oxide nanoparticles on plant: A critical review. *Front Chem., 5*, 1–16. [CrossRef][PubMed].
27. Deng, Y., White, J. C., & Xing, B., (2014). Interactions between engineered nanomaterials and agricultural crops: Implications for food safety. *J. Zhejiang Univ. Sci. A, 15*, 552–572. [CrossRef].
28. Sidkey, N. M., Ismail, A. A., Arafa, R. A., & Fathy, R. M., (2016). Impact of silver and selenium nanoparticles synthesized by gamma irradiation and their physiological response on early blight disease of potato. *J. Chem. Pharm. Res., 8*, 934–951.
29. Aslani, F., Bagheri, S., Julkapli, N. M., Juraimi, A. S., Sadat, F., Hashemi, G., & Baghdadi, A., (2014). Effects of engineered nanomaterials on plants growth: An overview. *Sci. World, J.,* 1–28. [CrossRef].

30. Lyu, S., Wei, X., Chen, J., Wang, C., Wang, X., & Pan, D., (2017). Titaniumasa beneficial element for crop production. *Front. Plant Sci., 8*, 1–19. [CrossRef] [PubMed].
31. Sztrik, A., (2016). *Production of Nano-Size Elemental Selenium Particles and Its Investigation in the Soil–Plant–Animal System* (pp. 1–26). Ph.D. Thesis, School of Crop Production, Horticulture and Food Science, University of Debrecen, Debrecen, Hungary.
32. Ragavan, P., Ananth, A., & Rajan, M. R., (2017). Impact of selenium nanoparticles on growth, biochemical characteristics and yield of cluster bean *Cyamopsis tetragonoloba*. *J. Environ. Agric. Biotechnol., 2*, 2917–2926. [CrossRef].
33. Janmohammadi, M., & Amazadeh, T., (2016). Impact of foliar application of nano micronutrient fertilizers and titanium dioxide nanoparticles on the growth and yield components of barley under supplemental irrigation. *Acta Agriculture Slovenica, 107*(2), 265–276.
34. Barrena, R., Casals, E., Colón, J., Font, X., Sánchez, A., & Puntes, V., (2009). Evaluation of the ecotoxicity of model nanoparticles. *Chemosphere, 75*, 850–857. [CrossRef] [PubMed].
35. Rico, C. M., Majumdar, S., Duarte-Gardea, M., Peralta-Videa, J. R., & Gardea-Torresdey, J. L., (2011). Interaction of nanoparticles with edible plants and their possible implications in the food chain. *J. Agric. Food Chem., 59*, 3485–3498.[CrossRef].
36. Masrahi, A., VandeVoort, A. R., & Arai, Y., (2014). Effects of silver nanoparticle on soil-nitrification processes.*Arch. Environ. Contam. Toxicol., 66*, 504–513. [CrossRef] [PubMed].
37. Lyon, D. Y.,Fortner, J. D.,Sayes, C. M.,Colvin, V. L., & Hughe, J. B., (2005). Bacterial cell association and antimicrobial activity of a C60 water suspension. *Environ. Tecxicol. Cam., 24*, 2757–2762.
38. AG ChemGroup, (2017). *Study Finds Potential Use for Zinc Oxide Nanoparticles for Fertilizer Feedstock*. AG Chemgroup: Praha, Czech Republic.
39. De Madrid, U. P., (2017). *Will we be Able to Use Zinc Oxide Nanoparticles as Fertilizers*. Science Daily.
40. García-Gómez, C., Obrador, A., González, D., Babín, M. and Fernández, M.D., (2017) Comparative effect of ZnO NPs, ZnO bulk and $ZnSO_4$ in the antioxidant defences of two plant species growing in two agricultural soils under greenhouse conditions. *Sci. Total Environ., 589*, 11–24.
41. Nair, P. M. G., & Chung, I. M., (2017). Regulation of morphological, molecular and nutrient status in *Arabidopsis thaliana* seedlings in response to ZnO nanoparticles and Zn ion exposure. *Sci. Total Environ., 575*, 187–198. [CrossRef].
42. Schützendübel, A., & Polle, A., (2002). Plant responses to abiotic stresses: Heavy metal-induced oxidative stress and protection by mycorrhization. *J. Exp. Bot., 53*, 1351–1365. [CrossRef].
43. Reddy, P. V. L., Hernandez-Viezcas, J. A., Peralta-Videa, J. R., & Gardea-Torresdey, J. L., (2016). Lessons learned: Are engineered nanomaterials toxic to terrestrial plants? *Sci. Total Environ., 568*, 470–479. [CrossRef].
44. Sidocrowicz, A., & Maqbool, Q., (2019). *Future of Nanofertilizers. Nanotechnology for Agriculture: Crop Production & Protection,* 143–152.
45. Lyu, S., Wei, X., Chen, J., Wang, C., Wang, X., & Pan, D., (2017). Titanium as a beneficial element for crop production. *Front. Plant Sci., 8*, 1–19. [CrossRef] [PubMed].

46. Chai, H., Yao, J., Sun, J., Zhang, C., Liu, W., Zhu, M., & Ceccanti, B., (2015). The effect of metal oxide nanoparticles on functional bacteria and metabolic profiles in agricultural soil. *Bull. Environ. Contam. Toxicol., 94*, 490–495.
47. Dimkpa, C. O., Joan, E., McLean, D. E., & Latta, E. M., (2012). CuO and ZnO nanoparticles: Phytotoxicity, metal speciation and induction of oxidative stress in sand-grown wheat. *Nanoparticle Res., 14*, 1125–1140. [CrossRef].
48. Tang, Y., He, R., Zhao, J., Nie, G., Xu, L., & Xing, B., (2016). Oxidative stress-induced toxicity of CuO nanoparticles and related toxicogenomic responses in *Arabidopsis thaliana. Environ. Pollut., 212*, 605–614. [CrossRef] [PubMed].
49. Lu, C. M., Zhang, C. Y., Wen, J. Q., Wu, G. R., & Tao, M. X., (2002). Research of the effect of nanometer materials on germination and growth enhancement of glycine max and its mechanism. *Soybean Sci., 22*, 168–172.
50. Prasad, R., Bhattacharyya, A., & Nguyen, Q. D., (2017). Nanotechnology in sustainable agriculture: Recent developments, challenges, and perspectives. *Front Microbiol., 8*, 1–13.
51. Zia, M., Phull, A. R., & Ali, J. S., (2016). Synthesis, characterization, applications, and challenges of iron oxide nanoparticles. *Nanotechnol. Sci. Appl., 9*, 49.
52. Mortvedt, J. J., & Giordano, P. M., (1967). Crop response to zinc oxide applied in liquid and granular fertilizers. *J. Agric. Food Chem., 15*, 118–122. [CrossRef].
53. Arole, V. M., & Munde, S. V., (2014). Fabrication of nanomaterials by top-down and bottom-up approaches: An overview. *J. Adv. Appl. Sci. Technol., 1*, 89–93.
54. Sharma, G., Sharma, A. R., Bhavesh, R., Park, J., Ganbold, B., Nam, J. S., & Lee, S. S., (2014). Biomolecule-mediated synthesis of selenium nanoparticles using dried *Vitis vinifera* (raisin) extract. *Molecules, 19*, 2761–2770.
55. Ganesan, V., (2015). Biogenic synthesis and characterization of selenium nanoparticles using the flower of *Bougainvillea spectabilis* willd. *Int. J. Sci. Res., 4*, 690–695.
56. Hu, J., Guo, H., Li, J., Wang, Y., Xiao, L., & Xing, B., (2017). Interaction of -Fe_2O_3 nanoparticles with *Citrus maxima* leaves and the corresponding physiological effects via foliar application. *J. Nanobiotechnol., 15*, 1–12.

CHAPTER 10

Genetic and Agronomic Biofortification of Zinc in Maize and Sweetcorn (*Zea mays* L.)

ZHONG XIANG CHEAH,[1] TIM O'HARE,[2] STEPHEN HARPER,[1] and MICHAEL BELL[1,2]

[1]*The University of Queensland, School of Agriculture and Food Sciences, Gatton, Australia*

[2]*The University of Queensland, Queensland Alliance for Agriculture and Food Innovation, Gatton, Australia*

ABSTRACT

Zinc (Zn) deficiency is a widespread nutritional issue in both developing and developed countries. Various strategies have been implemented to address this issue; however, these strategies were non-sustainable due to logistical and financial challenges. Biofortification of crops is a strategy that utilizes improved plant genetics and better agronomic practices to effectively deliver a high concentration of dietary Zn to consumers to help meet the recommended daily intake of 7–14 mg Zn day^{-1}.

Maize (*Zea mays*) is an important crop for human consumption, animal feed, fuel, and manufacture of pharmaceutical and industrial products. It is currently the third most important crop in the world after rice and wheat in terms of production, and a dietary staple for more than 1.2 billion people. However, despite early identification of maize as an important staple crop for Zn biofortification, there are currently few active maize Zn biofortification breeding program. The main reason for the lack of success in Zn biofortification efforts has been the lack of genotypes with replicable high kernel Zn

Phytoremediation and Biofortification: Strategies for Sustainable Environmental and Health Management.
Nand Kumar Singh, Shadma Afzal, & Tariq Aftab (Eds.)

concentration. Sweetcorn (*Zea mays* ssp. *saccharata*) differs from maize due to genetic mutations that confer a sugary endosperm. It is an important vegetable consumed worldwide and is a promising alternative to maize as a target crop for Zn biofortification. This chapter explores genetic and agronomic approaches to Zn biofortification in maize and sweetcorn.
This chapter will cover six topics, namely:

1. Occurrence of Zn deficiency in the world;
2. Biofortification of crops;
3. The physiology of Zn acquisition by plants;
4. Zn nutrition of maize (*Zea mays* L.) and sweetcorn (*Zea mays* ssp. *saccharata*);
5. Zn biofortification in maize and sweetcorn; and
6. Conclusion: Future directions of Zn biofortification in maize and sweetcorn.

10.1 OCCURRENCE OF ZN DEFICIENCY IN THE WORLD

Prior to the 1990s, agricultural systems worldwide have been focused on increasing yields and productivity to ensure food security and farmer profitability. The more productive cropping systems that were developed resulted in the conversion of many complex intercropped agriculture systems to simplified mono-cropping of cereal grains [1]. Consequently, there was a transition from highly diversified traditional diets that included a larger proportion of vegetables and legumes, to simplified diets dominated by carbohydrate originating from refined and processed cereals. These changes resulted in the emergence of widespread micronutrient deficiencies [2]. Human micronutrient deficiency (minerals and vitamins) is also known as 'hidden hunger,' as the symptoms of this form of malnutrition are often less obvious or visible. It is estimated that one-third of the world's population suffers from micronutrient deficiencies, inflicting long-term consequences on health and productivity [3].

Cereal grains and root tubers constitute a considerable proportion of human diets, representing a major source of calories and mineral nutrients. However, micronutrient concentrations in these crops are often too low to satisfy the recommended daily intake of humans [4]. In communities relying on a subsistence diet dominated by a single staple, micronutrient deficiency symptoms are more prevalent and can be severe. This is further exacerbated when regions with soil mineral deficiencies are inhabited and managed by low-income communities with less financial capacity to fertilize and develop their agricultural systems.

One of the most widespread micronutrient deficiencies affecting global populations is Zn, which is an essential mineral for healthy growth and development in humans [1]. The recommended dietary intake (RDI) of Zn for an adult is 7–14 mg day^{-1} and for breastfeeding women is 16–19 mg day^{-1} [5]. The human body has minimal capacity to store Zn in the blood plasma and continuous intake of Zn is required to maintain sufficient supply to meet demand [6].

Approximately 30% of the world's population and more than half of the children, elderly and pregnant women in developing countries are Zn-deficient [7] where it is the fifth leading risk factor for diseases [8]. In-depth studies of Zn deficiency and its impacts on human health and nutrition have been reviewed [9].

10.2 BIOFORTIFICATION OF CROPS

One approach of providing a micronutrient-rich diet to consumers is biofortification, defined as 'the process of increasing the bioavailable concentrations of nutrients in edible portions of crop plants through genetic selection or agronomic intervention,' or a combination of both. In 2004, the HarvestPlus Challenge Program was launched by the Consultative Group for International Agricultural Research (CGIAR), with the objective of reducing micronutrient deficiency around the world through the biofortification of dietary staples. Research efforts to biofortify crops have primarily focused on staple crops, since these commodities represent a large proportion of the diets in poor households. While biofortified crops cannot match the high micronutrient supply from fortified processed foods or pharmaceutical supplements, they can nonetheless increase the daily intake of minerals and vitamins [10]. Hence, while a successful biofortification effort will not replace the need for dietary diversification and disease reduction programs, it should be implemented complementarily.

10.2.1 GENETIC BIOFORTIFICATION

Genetic biofortification focuses on increasing the efficiency of mineral micronutrient uptake and enhancing accumulation in the edible product or enhancing the genotypic capacity to efficiently synthesize vitamins and antioxidants. Genetic biofortification can also be achieved simultaneously as part of screening for general crop improvement, with examples including selecting for reduced anti-nutrient concentrations and elevated nutrient absorption [11, 12]. In recent years, advances in genomics and molecular technology have opened opportunities to target biofortification research at a molecular level [13]. However, despite intensive research efforts, understanding of genetic by environmental interactions is still incomplete, resulting in a lack of cereal crop genotypes that exhibit consistently enhanced micronutrient accumulation across different environments [14, 15]. Genetic biofortification can also be achieved at the expense of other desirable crop agronomic characteristics including yield, pest, and disease resistance, environmental adaptation, or unintentional accumulation of toxic heavy metals [16, 17]. Genetic biofortification programs therefore need to evaluate genotypes in different environments to be confident that the elevated micronutrient concentrations are consistently expressed, and do not compromise yield, other desirable agronomic traits or food quality or safety.

10.2.2 AGRONOMIC BIOFORTIFICATION

Targeted fertilization strategies and improved soil management are important components of agronomic practices that either achieve or facilitate a desired biofortification outcome [1]. Agronomic biofortification most commonly consists of soil and foliar fertilizer applications. A range of environmental and species or genotype-specific factors can affect the efficacy and efficiency of agronomic biofortification. Therefore, careful consideration is required to make specific adjustments to agronomic strategies suited to new situations to ensure a successful biofortification outcome. An important drawback of agronomic biofortification is the frequent need for regular fertilizer applications to sustain consistently high micronutrient concentrations in crops. This comes at a financial cost, and potentially an environmental cost. A combination of both genetic and agronomic biofortification strategies are likely to be required to biofortify crops without incurring negative environmental impacts [18].

10.3 THE PHYSIOLOGY OF ZN ACQUISITION BY PLANTS

Zinc homeostasis involves the processes of uptake, transport, and translocation, distribution, and storage of Zn to maintain optimal Zn concentrations within physiological limits in different plant organs. A network of ligands, proteins, and transporters, and tightly regulated gene expression control Zn homeostasis, with key variables being the plants internal demand and external supply of Zn [19].

The uptake pathway of Zn from soil to grain consists of multiple steps including acquisition and uptake by root epidermal cells, loading into xylem (apoplastic) or phloem (symplastic) pathways for translocation and unloading for seed filling [20]. Another pathway that can result in elevated grain Zn is through absorption of Zn directly through leaves, which is subsequently remobilized via the phloem, especially from senescing leaves into developing seeds [21]. Plants that express high efficiency in the transport of Zn from roots and leaves to grains are ideal candidates for biofortification. In most cereals, limitations in the transportation and translocation of Zn due to potential restrictions to the efficient transfer from roots to shoots (the "root-shoot barrier"), restricted phloem mobility of Zn and impaired phloem unloading during grain filling may limit the capacity for biofortification [20, 22].

10.3.1 ACQUISITION OF ZN BY ROOTS

Zinc is predominantly taken up from the soil by plant roots as a divalent cation (Zn^{2+}) [23], but also potentially as a monovalent cation ($ZnOH^{+}$) at high soil pH [24] or in chelated form in situations where phytosiderophores are exuded [25]. Zn-efficient plants have been reported as having a longer root length and greater proportion of fine roots that provide a greater surface area for uptake, or the ability to change soil biological and chemical properties to increase the phytoavailability of Zn [26]. An example of the latter is where plants are able to release larger quantities of phytosiderophores to solubilize soil Zn [27].

When inside the roots, Zn can be sequestered within the vacuoles of the root cells or be moved to the vascular cylinder containing the xylem and phloem for transportation to the shoots [23]. This is where the "root-shoot barrier" may occur [28]. Anatomical structures such as the Casparian strip have been proposed as possible barriers at the root-shoot junction that limit Zn transport from roots to shoots [22].

Plants use specific chelators such as metallothioneins (MTs) and phytochelatins (PCs) to avoid potential Zn binding in root cells. These chelators further interact with transport proteins to facilitate Zn transport or sequestration [16]. If the cytosolic Zn concentration is toxic, excess Zn can be detoxified by sequestration in vacuoles.

10.3.2 XYLEM (APOPLASTIC) TRANSPORT OF ZN

The apoplastic pathway is characterized by passive movement through permeable cell walls and free intercellular spaces [20]. For Zn to be transported from root to shoot through the apoplastic pathway, it has to be first exported from the root cells into the xylem. Once loaded into the xylem, transport of Zn then occurs in the transpiration stream. Within the xylem sap, Zn can be in the form of Zn^{2+}, bound with deoxymugineic acid (DMA), nicotianamine (NA), or organic acids such as citrate and malate, etc. [23, 29, 30].

10.3.3 PHLOEM (SYMPLASTIC) TRANSPORT OF ZN

The symplastic pathway is based on active movement through the cytoplasmic network of living plant cells interconnected by plasmodesmata [31]. Loading of Zn into the phloem usually occurs from the xylem or from foliage, however the exact mechanism is yet unknown [20]. To prevent precipitation of Zn within the cytoplasm or non-specific off-target binding with other molecules [23], plants chelate Zn with ligands such as NA, DMA, organic acids, and thiols such as glutathione (GSH) for phloem transport [29, 30, 32].

The remobilization of Zn from vegetative tissue through the phloem to the developing grain is an important pathway for grain Zn accumulation, especially when there is insufficient Zn to support plant uptake post-anthesis [33] and the crop is relying on remobilization of Zn accumulated in vegetative plant parts during early growth. Phloem transport of Zn is of particular importance for effective foliar Zn fertilization, which is dependent on efficient Zn transportation via the phloem and into the grain. Any restrictions to phloem translocation will therefore limit the remobilization of Zn into grains.

10.3.4 LOADING OF ZN FROM VASCULAR TISSUE INTO GRAINS

The final step of grain Zn accumulation is the loading of Zn from phloem into the grain itself. Protein transporters responsible for unloading Zn from the vascular tissues into grains have yet to be identified, and relatively little is known about post-phloem movement of Zn in the developing grain [20]. Nonetheless, studies have reported that increased Zn uptake from roots, and increased root-shoot Zn transport, do not necessarily increase grain Zn concentration [34]. This Zn transport barrier at the shoot-grain junction appears to become increasingly limiting as Zn supply increases [35].

Since different pathways and mechanisms apply to root uptake, xylem, and phloem transport, vegetative remobilization, and grain loading, enhancing one aspect may not translate to biofortification of Zn concentrations in edible crop parts [36]. Successful genetic biofortification will require selection of genotypes with an enhanced capacity for all aspects of Zn transportation, remobilization, and accumulation of Zn into the seed.

10.4 ZN NUTRITION OF MAIZE (*ZEA MAYS* L.) AND SWEETCORN (*ZEA MAYS* SSP. *SACCHARATA*)

Maize is an important crop for human consumption, animal feed, fuel, and manufacture of pharmaceutical and industrial products. It is currently the third most important crop in the world after rice and wheat [37], with annual production estimated to be approximately 832.5 million tons across 163.9 million hectares [38]. Being a dietary staple for more than 1.2 billion people, it provides 20% of the world's calories and 15% of the protein [39]. As such, it has been one of the major target crops of the HarvestPlus biofortification research program [10].

Zinc is a plant essential micronutrient required for growth at all phenological stages, with approximately 50% of the maximum plant total Zn content at physiological maturity accumulated after silking [40]. The highest whole plant Zn uptake rate is reported to occur from the 12th leaf stage through to the silk, blister, and milk stages [41]. Zn deficiency during the reproductive stage causes a delay in the development of anthers and tassels, and also causes pollen sterility [42].

Partitioning of Zn to the ear begins in the early reproductive stages, but there are contrasting reports in the literature regarding the source of Zn for grain filling in maize. Some studies have noted that Zn uptake from soil was the major source of Zn for grain filling, rather than Zn remobilized from vegetative tissues [43], while others have reported that remobilization of Zn from vegetative tissues is very important. The key discriminator would appear to be plant Zn status and soil Zn availability. In soils with adequate Zn availability, strongly positive, near-linear correlations have been found between plant Zn uptake rate from the soil and Zn partitioning into the maize ear [41]. Conversely, Bender et al. [40] reported that as much as 60% of stem Zn was remobilized into grains by full physiological maturity. These authors also reported that there was no minimum threshold for plant Zn uptake rate before partitioning of Zn was initiated to the ear, with these observations collectively indicating that maize and sweetcorn cobs and their kernels are strong sinks for Zn, and so these crops therefore have strong potential for biofortification.

10.5 ZN BIOFORTIFICATION IN MAIZE AND SWEETCORN

The HarvestPlus biofortification target for Zn in maize is 60 mg Zn kg^{-1} (dry mass (DM) basis), with 38–40 mg Zn kg^{-1} set as an interim target value [44]. Currently, kernel Zn concentrations of maize varieties typically range between 15 and 35 mg Zn kg^{-1} [15, 45–48], with a single study reporting the highest mean value of 50 mg Zn kg^{-1} [49]. Despite the identification of maize as an important staple crop for Zn biofortification, there are currently few active maize Zn biofortification breeding programs, with the exception of those run by the International Maize and Wheat Improvement Center (CIMMYT) and the International Institute of Tropical Agriculture (IITA). Although CIMMYT have successfully released Zn-biofortified maize in Colombia & Guatemala [37], there has been limited success in releases adapted for other countries due to a lack of genotypes with consistently high kernel Zn concentrations. This suggests that either genes for efficient Zn uptake and accumulation in kernels are not easily introgressed, or there are strong genetic by environmental effects influencing kernel Zn concentration [15, 44].

An additional constraint for targeting maize for Zn biofortification is that Zn accumulated in the kernel is mostly in the embryo, aleurone layers and pericarp, which are removed during milling into flour [50, 51], thereby decreasing the Zn biofortification value of processed maize. This is compounded by the fact that Zn accumulated within maize kernels is often bound to phytate, and as humans lack the phytase enzyme, the resulting bioavailability of Zn in maize kernels is limited. To date, Zn biofortification research in maize has mainly succeeded in overcoming problems of crop Zn sufficiency for yield improvement but has not yet been able to provide a tested and proven Zn-enriched product for consumers [37].

Sweetcorn differs from maize due to a genetic mutation that confers a sugary endosperm [52]. It is an important vegetable consumed worldwide [53] and is a promising alternative to maize as a target crop for Zn biofortification as the entire kernel, including the Zn-rich embryo and aleurone layer, is consumed at its immature stage. Furthermore, the majority of Zn in sweetcorn kernels is located in the endosperm, where it is bound to S- and N-containing ligands such as cysteine and histidine [54] and so is bioavailable to humans. This novel discovery has only recently been reported and provides a clear incentive for genetic biofortification programs to focus on mechanisms that increase the Zn concentration and content of the endosperm tissue, which could provide clear benefits for increasing dietary Zn intake.

10.5.1 GENETIC ZN BIOFORTIFICATION

10.5.1.1 GENETIC VARIATION AND HERITABILITY OF KERNEL ZN CONCENTRATION

Investigations into the genetic basis for increasing maize kernel Zn concentration found correlations with several QTLs [55–57]. Some studies reported that kernel Zn concentration and accumulation was a result of additive gene action [57, 58], although other studies have reported that kernel Zn concentration is more influenced and controlled by dominant or partially dominant gene actions [56, 59]. Successful biofortification outcomes would likely utilize both additive and dominant gene action to achieve high kernel Zn concentration and content.

Reports concerning the heritability of maize kernel Zn concentration vary. Studying the intermated B73 × Mo17 (IBM) recombinant inbred (R1) population, Baxter et al. [60] identified significant heritability (> 0.60) for kernel Zn concentration. However, other studies have reported transgressive segregants expressing extreme (low to high) kernel Zn concentrations that related to the parental phenotypes [56], suggesting that kernel Zn concentration of inbred parent lines is not necessarily a good indication of hybrid filial generation performance. Other studies found that germplasm did not consistently express elevated kernel Zn concentration when grown across different locations and seasons, and genetic effects accounted for only 34% of the total variation in kernel Zn concentration [61]. These studies suggest the important role played by spatial, geographical, temporal, and environmental differences in determining kernel Zn concentration [15, 55]. It also highlights the importance of agronomic biofortification strategies, as in the absence of plant available Zn for uptake, the biofortification outcome will be poor regardless of superiority in genotypic efficiency for Zn uptake.

10.5.1.2 GRAIN YIELD IMPROVEMENTS AND KERNEL ZN CONCENTRATION

Increases in kernel yield and maize shoot biomass due to improved breeding and agronomic management have typically resulted in a dilution in maize kernel nutrient concentration, such as N, oil, and protein concentrations [62–65]. This dilution may be an effect of having similar total Zn content per kernel but larger kernel DM, similar total cob Zn content but more kernels, or a smaller increase in total cob and kernel Zn content than the relatively larger increase in kernel DM and kernel number.

In sweetcorn, significant differences in kernel Zn concentration were identified due to differences in kernel number per cob. Increase in net Zn accumulation did not match increasing kernel numbers on a cob, resulting in a dilution of kernel Zn concentration [17]. Similar relationships could be predicted for maize, since assimilate accumulation in maize is sink-limited when there are sufficient resources during grain filling. However, there is some evidence that kernel Zn concentration in maize is only weakly correlated with grain yield [66], which is encouraging for future biofortification efforts.

10.5.1.3 EMBRYO-TO-ENDOSPERM SIZE RATIO

The concentration of Zn in the embryos of maize and sweetcorn kernels is higher than in the endosperm [51] and illustrate the importance of Zn accumulation in this part of the kernel to total kernel Zn content. Unfortunately, a large proportion of the kernel phytate is also accumulated within the embryo [67], which results in the accumulation of Zn in the embryo occurring predominantly as Zn-phytate [54, 68]. Unless this characteristic can be overcome, any selection for larger embryos may have limited impacts on bioavailable Zn supply.

Recent studies showed maize embryo size is to a certain extend a genetically controlled trait that could be selected for, and that there are therefore opportunities to breed for kernels with altered embryo sizes [69]. Hence to improve the bioavailable Zn in cereal grains, breeding for a smaller embryo-to-endosperm size ratio could allow the accumulation of more Zn and other micronutrients in the endosperm. Other factors such as heterofertilization could also influence the nutrient transfer between endosperm and embryo, which have implications for biofortification [70].

10.5.1.4 MULTIPLE ALEURONE LAYER (MAL) TRAIT AND QUALITY PROTEIN MAIZE (QPM)

The aleurone layer consists of specialized discrete vacuoles where storage proteins are accumulated and bound by a membrane. It was hypothesized that this trait would increase the ratio of Zn-rich cells to regular starchy endosperm cells. Several maize lines with the MAL trait have been identified [71]. Kernels with the multiple aleurone layer (MAL) trait had Zn concentrations of 31.3 mg kg^{-1}, compared to 22.5 mg kg^{-1} for kernels without MAL [72]. Interestingly, this work was not pursued and no commercial lines with the MAL trait were released (Ross M. Welch, pers.com., 21 July 2017). With sufficient understanding of the genetic control and regulation of aleurone cells development, it might be possible to increase kernel Zn concentration by introducing the MAL trait into maize and sweetcorn kernels [73].

The incorporation of the MAL trait can also potentially further improve the amino acid content of maize genotypes expressing the *opaque-2* (o_2) mutation [71]. The o_2 mutation is the basis of a group of the quality protein maize (QPM) maize varieties that accumulate high quantities of non-zein amino acids such as lysine and tryptophan [74]. This o_2 mutation has also been shown confer higher accumulation of minerals, including Zn, Fe, P, and K [75]. Due to the importance of protein in dietary intake and Fe as a micronutrient for human health, utilization of these parent genotypes that possess both MAL and QPM traits to achieve simultaneous biofortification of multiple nutrients could have multiple health benefits for consumers [72].

10.5.2 AGRONOMIC ZN BIOFORTIFICATION

10.5.2.1 SOIL FERTILIZATION

Soil Zn application is an effective method for supplying large amounts of Zn to crops in a variety of cropping systems and is a commonly used strategy [76]. Inorganic forms of Zn fertilizers are often applied to soil, often as part of compound fertilizer products or soil organic amendments. Due to factors affecting solubility, fixation, diffusion rates and bioavailability in soil environments, Zn-chelates are sometimes used as alternative forms of Zn fertilizer [77]. Though Zn-chelates are more expensive than industrial salts, they are usually more effective in maintaining soil Zn in forms that are available to plants, therefore improving plant uptake [78].

Literature on the effectiveness of soil Zn application to biofortify maize kernel Zn is limited, as many studies have only focused on whole plant Zn uptake. Most studies reported significant differences in kernel Zn concentration after soil Zn application (Table 10.1), whereas one study found no significant increase in most sites evaluated [79]. In using soil Zn fertilizers, placement of the fertilizer is particularly important for maize as the root zone actively involved in Zn uptake is reported to be in the top 30 cm of the soil profile [80]. Interestingly, it has also been reported that improved N status of plants promoted the uptake and accumulation of soil-applied Zn in maize plants [81, 82].

10.5.2.2 FOLIAR FERTILIZATION

Foliar Zn application is often used complementarily to soil Zn application to supply additional Zn to plants. In contrast to soil Zn application, most studies reported that uptake of foliar-applied Zn is more effective when those applications are made as Zn salts compared to chelated forms of Zn [83]. Studies consistently show that foliar Zn application was more effective and efficient than soil Zn applications for cereal grain Zn enrichment [84–86]. Total Zn recovery by grains was also higher for foliar-applied Zn [87].

Foliar Zn application was consistently reported to be effective for increasing maize kernel Zn concentration when used alone or when in combination with soil Zn applications (Table 10.1). It has been reported that maize showed the greatest uptake of Zn at three stages: (i) 7th–9th leaf stage; (ii) tasseling; and (iii) the milk stage of kernel development. This was concluded from analysis of foliage concentrations of plants harvested at various growth stages after a uniform foliar Zn application at the 5th leaf stage. Timing of foliar Zn application is crucial to synchronize supply with the onset of increased Zn demand [88]. Studies have shown that the efficacy of foliar Zn fertilizer in improving kernel Zn concentration depends on the stage of plant development [85, 86, 89]. Interestingly in sweetcorn, kernel Zn concentrations were not significantly different between plants receiving foliar Zn application at various timings after tasseling [90].

A combination of soil- and foliar-Zn application was able to achieve a kernel Zn concentration of 71 mg kg^{-1} in sweetcorn, representing the highest reported value in the literature to date in maize or sweetcorn, and even exceeding the HarvestPlus biofortification target for maize [90]. The lack of impact of different timings of foliar Zn application after tasseling on sweet-corn kernel Zn concentration further demonstrate the flexibility of utilizing this agronomic strategy to achieve a positive biofortification outcome. With the RDI for an adult averaging 10 mg Zn d^{-1}, a kernel Zn concentration of 71 mg kg^{-1} in sweetcorn would be able to meet approximately a quarter of that daily demand from consuming a single cob of sweetcorn kernels (approximately 40 g DM, 160 g fresh mass).

10.6 CONCLUSION: FUTURE DIRECTIONS OF ZN BIOFORTIFICATION IN MAIZE AND SWEETCORN

10.6.1 GENETIC ZN BIOFORTIFICATION

Genetic biofortification should explore strategies that can increase the size of the sink for accumulation of bioavailable Zn. A possible approach could

TABLE 10.1 Effects of Soil and Foliar Zn Application at Different Growth Stages on Grain Zn Concentration of Maize

Application Method	Soil Zn † (mg kg–1)	Application Timing	Fertilizer Form	Soil Rate (kg Zn ha–1)	Foliar Rate (% w/v)	Grain Zn Concentration (mgkg–1DM)	References
Soil: broadcast and incorporated	0.68	Pre-sowing	$ZnSO_4$ powder	0	–	15^{a}	[91]
				2.75	–	18^{b}	
				5.50	–	21^{c}	
Soil	0.72	Pre-sowing	$ZnSO_4$	0	–	22^{a}	[48]
				6	–	27^{b}	
				18	–	30^{b}	
				54	–	31^{b}	
Soil	0.65	Pre-sowing	–	0	–	18^{a}	[92]
			$ZnSO_4$	1.2	–	22^{ab}	
			Zn-lignosulfonate	1.2	–	24^{b}	
Soil	0.58	Sowing	ZnO	0	–	17^{a}	[93]
				1	–	27^{bc}	
				2	–	29^{c}	
Soil	0.38	Sowing	–	0	–	23^{a}	[94]
			Zn-poly$(PO_3)_4$	0.5	–	24^{ab}	
			Zn-poly$(PO_3)_4$	1.0	–	24^{ab}	
			Zn-poly$(PO_3)_4$	2.5	–	26^{c}	
			$ZnSO_4.H_2O$	0.5	–	24^{ab}	
			$ZnSO_4.H_2O$	1.0	–	25^{bc}	
			$ZnSO_4.H_2O$	2.5	–	27^{c}	
			Zn-EDTA	1.0	–	25^{b}	

TABLE 10.1 *(Continued)*

Application Method	Soil Zn † (mg kg–1)	Application Timing	Fertilizer Form	Soil Rate (kg Zn ha–1)	Foliar Rate (% w/v)	Grain Zn Concentration (mgkg–1DM)	References
Foliar	0.98	1 month after sowing	–	–	0	21[a]	[95]
			–	–	1%	24[b]	
			–	–	2%	27[c]	
Control	0.65	–	–	0	0	22[a]	[89]
Soil		Sowing	–	12	–	36[b]	
Foliar		9th leaf stage	$ZnSO_4$	–	1%	43[c]	
Both soil and foliar		Anthesis	$ZnSO_4$	–	1%	38[b]	
Control	0.56	–	–	0	0	15[ab]	[87]
Soil: spray and incorporated		Pre-sowing	Aqueous $ZnSO_4.7H_2O$	11	–	17[b]	
Foliar: spray		Tasseling, 1 week after silking	0.5% (w/v) $ZnSO_4.7H_2O$	–	1 kg Zn ha^{-1} (each)	20[bc]	
Both soil and foliar		Both	Both	11	1 kg Zn ha^{-1} (each)	20[c]	
Control	0.43	–	$ZnSO_4.7H_2O$	0	0	22[a]	[96]
Soil: broadcast		Pre-sowing		16	–	27[b]	
Soil: banding (15 cm)		Sowing		16	–	34[d]	
Foliar: spray		25 days after sowing, tasseling		–	0.5%, 0.25%	30[c]	
Broadcast + Foliar		Both		16	0.5%, 0.25%	37[e]	
Banding (15 cm) + Foliar		Both		16	0.5%, 0.25%	42[f]	

TABLE 10.1 *(Continued)*

Application Method	Soil Zn † (mg kg–1)	Application Timing	Fertilizer Form	Soil Rate (kg Zn ha–1)	Foliar Rate (% w/v)	Grain Zn Concentration (mgkg–1DM)	References
Control	0.38	–	$ZnSO_4 \cdot 7H_2O$	0	–	22^{d}	[97]
Soil		Sowing		37.5	–	27^{c}	
Soil				56.25	–	32^{ab}	
Soil				75	–	32^{ab}	
Foliar		V6 and tasseling		0	0.05% each	28^{c}	
Both soil and foliar		Sowing, V6 and tasseling		37.5	0.05% each	31^{b}	
Both soil and foliar				56.25	0.05% each	33^{ab}	
Both soil and foliar				75	0.05% each	34^{a}	
Control	–	–	$ZnSO_4$	0	0	12^{a}	[98]
Soil		–		5	–	20^{b}	
Soil		–		10	–	20^{bc}	
Foliar		Tasseling, milking		–	0.2%	20^{bc}	
Both soil and foliar		–		5	0.2%	22^{bc}	
Both soil and foliar		–		10	0.2%	22^{c}	

† DTPA-extractable soil Zn concentration before commencement of experiment.

a-f Within each application method category, grain Zn concentration means with different letters are significantly different ($P < 0.05$).

be through investigating genes that regulate the size and mass of embryos, thereby attempting to minimize the embryo-to-endosperm size ratio and maximize Zn accumulation in the endosperm without increasing kernel phytate content. However, there is still likely to be strong environmental influences on expression of embryo size and kernel Zn concentration.

At a plant physiology level, research is required to better understand the specific mechanisms of Zn transport and regulation between vegetative organs (e.g., foliage and stems) and reproductive organs (e.g., cob and kernels). Further understanding of the molecular transport of Zn between the transfer cells, aleurone layer and endosperm tissue within the kernel is also required. Specifically, mechanisms regulating Zn homeostasis in plant cells and subcellular Zn compartmentalization need to be investigated. Insights into these processes could be utilized for genetic modification and transgenic studies to enhance the uptake and accumulation of Zn through the manipulation of ligands and transporters across these tissues.

It has been demonstrated that when grain yield increases, the rate of Zn supply into kernels is unable to match the increased supply of assimilates. This suggests that an improvement in the efficiency of kernel Zn accumulation will be required to deliver high kernel Zn concentration and content in high yielding varieties. Research should focus on strategies that could improve the concentration and mobility of Zn within phloem tissues, and subsequent transfer into the developing kernel.

Finally, genetic biofortification strategies that enhance mineral nutrient uptake from soils could contribute to simultaneous accumulation of toxic heavy metalloids such as cadmium. These potentially deleterious outcomes should be carefully examined to ensure that genetic biofortification efforts do not cause other negative impacts for consumers.

10.6.2 AGRONOMIC BIOFORTIFICATION

It was demonstrated that at high soil Zn supply, large increases in foliage and stem Zn concentration and content were not matched by similarly large increases in the kernels. Therefore, studies are needed to understand the remobilization capacity of Zn from vegetative to reproductive plant tissues. Studies using Zn isotopes could help explain the differential kernel accumulation of soil versus foliage-applied Zn post-anthesis, thus providing mechanistic understanding of the poor efficacy of soil-applied Zn at raising kernel Zn concentration in plants with otherwise adequate Zn status.

From a plant nutrition perspective, studies are required to further our understanding of associations between Zn and other nutrients, especially P, N, and S. Past studies on nutrient interactions commonly investigated how the application of other nutrients (P in particular) affected the uptake efficiency of Zn. Further, it has been shown that uptake, utilization, and accumulation of Zn in cereal crops is often related to N fertilizer management and crop N status. Given the importance of P-, N-, and S-based ligands in the speciation of Zn found in sweetcorn kernels, future studies could potentially examine the interactions between supply of these prominent phytonutrients and the transport and bioavailability of Zn.

KEYWORDS

- **agronomic biofortification**
- **bioavailability**
- **biofortification**
- **genetic**
- **maize**
- **micronutrient**
- **nutrition**
- **sweetcorn**
- **Zea mays**
- **zinc**

REFERENCES

1. Bouis, H. E., & Welch, R. M., (2010). Biofortification – a sustainable agricultural strategy for reducing micronutrient malnutrition in the global South. *Crop Sci., 50*(Supplement 1), S20–S32.
2. Remans, R., Fanzo, J., Palm, C., & DeClerck, F. A., (2011). Human nutrition as an ecological service. In: Clerck, F. A. J., Ingram, J. C., & Rio, C. R., (eds.), *Integrating Ecology Into Poverty Alleviation and International Development Efforts: A Practical Guide*. De Springer: Heidelberg, Germany.
3. Beal, T., Massiot, E., Arsenault, J. E., Smith, M. R., & Hijmans, R. J., (2017). Global trends in dietary micronutrient supplies and estimated prevalence of inadequate intakes. *PLoS One, 12*(4), e0175554.

4. La Frano, M. R., Moura, F. F., Boy, E., Lönnerdal, B., & Burri, B. J., (2014). Bioavailability of iron, zinc, and provitamin A carotenoids in biofortified staple crops. *Nutr. Rev., 72*(5), 289–307.
5. Trumbo, P., Yates, A. A., Schlicker, S., & Poos, M., (2001). Dietary reference intakes: Vitamin A, vitamin K, arsenic, boron, chromium, copper, iodine, iron, manganese, molybdenum, nickel, silicon, vanadium, and zinc. *J. Am. Diet. Assoc., 101*(3), 294–301.
6. Wessells, K., Jorgensen, J., Hess, S., Woodhouse, L., Peerson, J., & Brown, K., (2010). Plasma zinc concentration responds rapidly to the initiation and discontinuation of short-term zinc supplementation in healthy men. *The Journal of Nutrition, 140*(12), 2128–2133.
7. Wessells, K. R., Singh, G. M., & Brown, K. H., (2012). Estimating the global prevalence of inadequate zinc intake from national food balance sheets: Effects of methodological assumptions. *PLoS One, 7*(11), e50565.
8. Maret, W., & Sandstead, H. H., (2006). Zinc requirements and the risks and benefits of zinc supplementation. *J. Trace Elem. Med. Biol., 20*(1), 3–18.
9. Gibson, R., (2006). Zinc: The missing link in combating micronutrient malnutrition in developing countries. *Proc. Nutr. Soc., 65*, 51–60.
10. Bouis, H. E., Hotz, C., McClafferty, B., Meenakshi, W. H., & Pfeiffer, H. E., (2011). Biofortification: A new tool to reduce micronutrient malnutrition. *Food Nutr. Bull., 32*(1), S31–S40.
11. Bouis, H. E., (2003). Micronutrient fortification of plants through plant breeding: Can it improve nutrition in man at low cost? *Proc. Nutr. Soc., 62*(2), 403–411.
12. Cheah, Z. X., Harper, S. M., O'Hare, T. J., & Bell, M. J., (2011). In: Heyes, J., & Aksoy, U., (eds.), *Elevated Zinc Concentration in Zeaxanthin-Biofortified Sweetcorn Kernels Compared with a Commercial Hybrid* (pp. 399–403). International Horticulture Congress, Istanbul, Turkey, International Society for Horticultural Science: Istanbul, Turkey.
13. Zhao, F. J., & Shewry, P. R., (2011). Recent developments in modifying crops and agronomic practice to improve human health. *Food Policy, 36*(1), S94–S101.
14. Joshi, A. K., Crossa, J., Arun, B., Chand, R., Trethowan, R., Vargas, M., & Ortiz-Monasterio, I., (2010). Genotype × environment interaction for zinc and iron concentration of wheat grain in eastern Gangetic plains of India. *Field Crops Res., 116*(3), 268–277.
15. Cheah, Z. X., O'Hare, T. J., Harper, S. M., Kochanek, J., & Bell, M. J., (2020). Zinc biofortification of immature maize and sweetcorn (*Zea mays* L.) kernels for human health. *Scientia Horticulturae, 272*, 109559.
16. Shahzad, Z., Rouached, H., & Rakha, A., (2014). Combating mineral malnutrition through iron and zinc biofortification of cereals. *Comprehensive Reviews in Food Science and Food Safety, 13*(3), 329–346.
17. Cheah, Z. X., O'Hare, T. J., Harper, S. M., & Bell, M. J., (2020). Variation in zinc concentration of sweetcorn kernels reflects source–sink dynamics influenced by kernel number. *J. Exp. Bot., 71*(16), 4985–4992.
18. Carvalho, S. M. P., & Vasconcelos, M. W., (2013). Producing more with less: Strategies and novel technologies for plant-based food biofortification. *Food Res. Int., 54*(1), 961–971.
19. Page, V., & Feller, U., (2015). Heavy metals in crop plants: Transport and redistribution processes on the whole plant level. *Agronomy, 5*(3), 447.

20. Olsen, L. I., & Palmgren, M. G., (2014). Many rivers to cross: The journey of zinc from soil to seed. *Frontiers in Plant Science, 5*, 30.
21. Erenoglu, E. B., Kutman, U. B., Ceylan, Y., Yildiz, B., & Cakmak, I., (2011). Improved nitrogen nutrition enhances root uptake, root-to-shoot translocation and remobilization of zinc (^{65}Zn) in wheat. *New Phytol., 189*(2), 438–448.
22. Swamy, B., Rahman, M., Inabangan-Asilo, M., Amparado, A., Manito, C., Chadha-Mohanty, P., Reinke, R., & Slamet-Loedin, I., (2016). Advances in breeding for high grain zinc in rice. *Rice, 9*, 1–16.
23. Sinclair, S. A., & Kramer, U., (2012). The zinc homeostasis network of land plants. *Biochim. Biophys. Acta, 1823*(9), 1553–1567.
24. Broadley, M., Brown, P., Cakmak, I., Rengel, Z., & Zhao, F., (2012). Function of nutrients: Micronutrients. In: Marschner, P., (ed.), *Marschner's Mineral Nutrition of Higher Plants* (3rd edn., pp. 191–248). Academic Press: London.
25. Widodo, Broadley, M. R., Rose, T., Frei, M., Pariasca-Tanaka, J., Yoshihashi, T., Thomson, M., et al., (2010). Response to zinc deficiency of two rice lines with contrasting tolerance is determined by root growth maintenance and organic acid exudation rates, and not by zinc-transporter activity. *New Phytol., 186*(2), 400–414.
26. Rengel, Z., Römheld, V., & Marschner, H., (1998). Uptake of zinc and iron by wheat genotypes differing in tolerance to zinc deficiency. *J. Plant Physiol., 152*(4, 5), 433–438.
27. Cakmak, I., Öztürk, L., Karanlik, S., Marschner, H., & Ekiz, H., (1996). Zinc-efficient wild grasses enhance release of phytosiderophores under zinc deficiency. *J. Plant Nutr., 19*(3, 4), 551–563.
28. Yamaji, N., Xia, J., Mitani-Ueno, N., Yokosho, K., & Ma, J. F., (2013). Preferential delivery of zinc to developing tissues in rice is mediated by P-type heavy metal ATPase OsHMA2. *Plant Physiol., 162*(2), 927.
29. Haydon, M. J., & Cobbett, C. S., (2007). Transporters of ligands for essential metal ions in plants. *New Phytol., 174*(3), 499–506.
30. Clemens, S., (2019). Metal ligands in micronutrient acquisition and homeostasis. *Plant, Cell Environ., 42*(10), 2902–2912.
31. Van, B. A. J. E., Furch, A. C. U., Hafke, J. B., Knoblauch, M., & Patrick, J. W., (2011). Questions on phloem biology. 2. Mass flow, molecular hopping, distribution patterns and macromolecular signaling. *Plant Sci., 181*(4), 325–330.
32. Yoneyama, T., Ishikawa, S., & Fujimaki, S., (2015). Route and regulation of zinc, cadmium, and iron transport in rice plants (*Oryza sativa* L.) during vegetative growth and grain filling: Metal transporters, metal speciation, grain Cd reduction and Zn and Fe biofortification. *Int. J. Mol. Sci., 16*(8), 19111–19129.
33. Nishiyama, R., Tanoi, K., Yanagisawa, S., & Yoneyama, T., (2013). Quantification of zinc transport via the phloem to the grain in rice plants (*Oryza sativa* L.) at early grain-filling by a combination of mathematical modeling and ^{65}Zn tracing. *Soil Sci. Plant Nutr., 59*, 750–755.
34. Yin, H., Gao, X., Stomph, T., Li, L., Zhang, F., & Zou, C., (2016). Zinc concentration in rice (*Oryza sativa* L.) grains and allocation in plants as affected by different zinc fertilization strategies. *Commun. Soil Sci. Plant Anal., 47*(6), 761–768.
35. Wang, Y. X., Specht, A., & Horst, W. J., (2011). Stable isotope labelling and zinc distribution in grains studied by laser ablation ICP-MS in an ear culture system reveals zinc transport barriers during grain filling in wheat. *New Phytol., 189*(2), 428.

36. Hussain, S., Rengel, Z., Mohammadi, S. A., Ebadi-Segherloo, A., & Maqsood, M. A., (2016). Mapping QTL associated with remobilization of zinc from vegetative tissues into grains of barley (*Hordeum vulgare*). *Plant Soil, 399*(1, 2), 193–208.
37. Maqbool, M. A., & Beshir, A., (2019). Zinc biofortification of maize (*Zea mays* L.): Status and challenges. *Plant Breeding, 138*(1), 1–28.
38. Gwirtz, J. A., & Garcia-Casal, M. N., (2014). Processing maize flour and corn meal food products. *Ann. N. Y. Acad. Sci., 1312*(1), 66–75.
39. Nuss, E., Arscott, S., Bresnahan, K., Pixley, K., Rocheford, T., Hotz, C., Siamusantu, W., et al., (2012). Comparative intake of white- versus orange-colored maize by Zambian children in the context of promotion of biofortified maize. *Food Nutr. Bull., 33*(1), 63–71.
40. Bender, R., Haegele, J., Ruffo, M., & Below, F., (2013). Nutrient uptake, partitioning, and remobilization in modern, transgenic insect-protected maize hybrids. *Agron. J., 105*(1), 161–170.
41. Ciampitti, I., & Vyn, T., (2013). Maize nutrient accumulation and partitioning in response to plant density and nitrogen rate: II. Calcium, magnesium, and micronutrients. *Agron. J., 105*(6), 1645–1657.
42. Drissi, S., Aït, H. A., Bamouh, A., & Benbella, M., (2017). Response of corn silage (*Zea mays* L.) to zinc fertilization on a sandy soil under field and outdoor container conditions. *Journal of the Saudi Society of Agricultural Sciences, 16*(2), 145–153.
43. Grzebisz, W., Wronska, M., Diatta, J., & Szczepaniak, W., (2008). Effect of zinc foliar application at an early stage of maize growth on patterns of nutrients and dry matter accumulation by the canopy. Part II. Nitrogen uptake and dry matter accumulation patterns. *Journal of Elementology, 13*(1), 29–39.
44. Ortiz-Monasterio, J. I., Palacios-Rojas, N., Meng, E., Pixley, K., Trethowan, R., & Peña, R. J., (2007). Enhancing the mineral and vitamin content of wheat and maize through plant breeding. *Journal of Cereal Science, 46*(3), 293–307.
45. Bänziger, M., & Long, J., (2000). The potential for increasing the iron and zinc density of maize through plant-breeding. *Food Nutr. Bull., 21*(4), 397–400.
46. Oikeh, S. O., Menkir, A., Maziya-Dixon, B., Welch, R., & Glahn, R. P., (2003). Genotypic differences in concentration and bioavailability of kernel-iron in tropical maize varieties grown under field conditions. *J. Plant Nutr., 26*(10, 11), 2307–2319.
47. Šimić, D., Sudar, R., Ledenčan, T., Jambrović, A., Zdunić, Z., Brkić, I., & Kovačević, V., (2009). Genetic variation of bioavailable iron and zinc in grain of a maize population. *Journal of Cereal Science, 50*(3), 392–397.
48. Kanwal, S., Ranjha, R., & Ahmad, R., (2010). Zinc partitioning in maize grain after soil fertilization with zinc sulfate. *Int. J. Agric. Biol., 12*(2), 299–302.
49. Maziya-Dixon, B., Kling, J., Menkir, A., & Dixon, A., (2000). Genetic variation in total carotene, iron, and zinc contents of maize and cassava genotypes. *Food Nutr. Bull., 21*(4), 419–422.
50. Suri, D. J., & Tanumihardjo, S. A., (2016). Effects of different processing methods on the micronutrient and phytochemical contents of maize: From A to Z. *Comprehensive Reviews in Food Science and Food Safety, 15*(5), 912–926.
51. Cheah, Z. X., Kopittke, P. M., Harper, S. M., O'Hare, T. J., Wang, P., Paterson, D. J., De Jonge, M. D., & Bell, M. J., (2019). *In situ* analyses of inorganic nutrient distribution in sweetcorn and maize kernels using synchrotron-based X-ray fluorescence microscopy. *Ann. Bot., 123*(3), 543–556.

52. Fritz, V. A., Tong, C. B., Rosen, C. J., & Wright, J. A. (2010). Sweet corn (vegetable crop management). University of Minnesota Extension Newsletter. *Commercial Fruit and Vegetable Production.*
53. Szymanek, M., Dobrzański, Jr. B., Niedziółka, I., & Rybczyński, R., (2005). *Sweet Corn: Harvest and Technology Physical Properties and Quality*. B. Dobrzański Institute of Agrophysics Polish Academy of Sciences.
54. Cheah, Z. X., Kopittke, P. M., Harper, S. M., Meyer, G., O'Hare, T. J., & Bell, M. J., (2019). Speciation and accumulation of Zn in sweetcorn kernels for genetic and agronomic biofortification programs. *Planta, 250*(1), 219–227.
55. Agrawal, P., Jaiswal, S., Prasanna, B., Hossain, F., Saha, S., Guleria, S., & Gupta, H., (2012). Genetic variability and stability for kernel iron and zinc concentration in maize (*Zea mays* L.) genotypes. *Indian J. Genet. Plant Breed., 72*(4), 421–428.
56. Qin, H., Cai, Y., Liu, Z., Wang, G., Wang, J., Guo, Y., & Wang, H., (2012). Identification of QTL for zinc and iron concentration in maize kernel and cob. *Euphytica, 187*(3), 345–358.
57. Simic, D., Mladenovic, D. S., Zdunic, Z., Jambrovic, A., Ledencan, T., Brkic, J., Brkic, A., & Brkic, I., (2012). Quantitative trait loci for biofortification traits in maize grain. *J. Hered., 103*(1), 47–54.
58. Long, J., Bänziger, M., & Smith, M., (2004). Diallel analysis of grain iron and zinc density in southern African-adapted maize inbreds. *Crop Sci., 44*(6), 2019–2026.
59. Chakraborti, M., Prasanna, B. M., Singh, A. M., & Hossain, F., (2010). Generation mean analysis of kernel iron and zinc concentrations in maize (*Zea mays*). *Indian J. Agric. Sci., 80*(11), 956–959.
60. Baxter, I. R., Gustin, J. L., Settles, A. M., & Hoekenga, O. A., (2013). Ionomic characterization of maize kernels in the intermated B73 × Mo17 population. *Crop Sci., 53*(1), 208–220.
61. Oikeh, S., Menkir, A., Maziya-Dixon, B., Welch, R., & Glahn, R., (2003). Assessment of concentrations of iron and zinc and bioavailable iron in grains of early-maturing tropical maize varieties. *J. Agric. Food Chem., 51*(12), 3688–3694.
62. Scott, M. P., Edwards, J. W., Bell, C. P., Schussler, J. R., & Smith, J. S., (2006). Grain composition and amino acid content in maize cultivars representing 80 years of commercial maize varieties. *Maydica, 51*, 417–423.
63. Davis, D. R., (2009). Declining fruit and vegetable nutrient composition: What is the evidence? *HortScience, 44*(1), 15–19.
64. Riedell, W. E., (2010). Mineral-nutrient synergism and dilution responses to nitrogen fertilizer in field-grown maize. *J. Plant Nutr. Soil Sci., 173*(6), 869–874.
65. Abdala, L. J., Vitantonio-Mazzini, L. N., Gerde, J. A., Martí, R. F., Murtagh, G., & Borrás, L., (2018). Dry milling grain quality changes in Argentinean maize genotypes released from 1965 to 2016. *Field Crops Res., 226*, 74–82.
66. Chakraborti, M., Prasanna, B., Hossain, F., Singh, A., & Guleria, S., (2009). Genetic evaluation of kernel Fe and Zn concentrations and yield performance of selected maize (*Zea mays* L.) genotypes. *Range Management and Agroforestry, 30*(2), 109–114.
67. Raboy, V., Gerbasi, P., Young, K., Stoneberg, S., Pickett, S., Bauman, A., Murthy, P., et al., (2000). Origin and seed phenotype of maize low phytic acid 1-1 and low phytic acid 2-1. *Plant Physiol., 124*(1), 355–368.
68. Cheah, Z. X., Kopittke, P. M., Scheckel, K. G., Noerpel, M. R., & Bell, M. J., (2020). Comparison of Zn accumulation and speciation in kernels of sweetcorn and maize differing in maturity. *Ann. Bot., 125*(1), 185–193.

69. Suzuki, M., Sato, Y., Wu, S., Kang, B. H., & McCarty, D. R., (2015). Conserved functions of the MATE transporter *BIG EMBRYO1* in regulation of lateral organ size and initiation rate. *The Plant Cell, 27*(8), 2288–2300.
70. Wu, C. C., Diggle, P. K., & Friedman, W. E., (2013). Kin recognition within a seed and the effect of genetic relatedness of an endosperm to its compatriot embryo on maize seed development. *Proceedings of the National Academy of Sciences, 110*(6), 2217.
71. Nelson, O., & Chang, M., (1973). Effect of multiple aleurone layers on the protein and amino acid content of maize endosperm. *Crop Sci., 14*(3), 374–376.
72. Welch, R., Smith, M., Van, C. D., & Schaefer, S., (1993). Improving the mineral reserves and protein quality of maize (*Zea mays* L.) kernels using unique genes. *Plant Soil, 155, 156*, 215–218.
73. Becraft, P. W., & Yi, G., (2011). Regulation of aleurone development in cereal grains. *J. Exp. Bot., 62*(5), 1669–1675.
74. Nuss, E., & Tanumihardjo, S. A., (2011). Quality protein maize for Africa: Closing the protein inadequacy gap in vulnerable populations. *Adv. Nutr., 2*(3), 217–224.
75. Gupta, H. O., Lodha, M. L., Mehta, S. L., Rastogi, D. K., & Joginder, S., (1980). Changes in minerals, proteins and amino acids in hard endosperm *opaque-2 Zea mays* during development. *Indian J. Exp. Biol., 18*(12), 1419–1422.
76. Cakmak, I., & Kutman, U. B., (2018). Agronomic biofortification of cereals with zinc: A review. *Eur. J. Soil Sci., 69*(1), 172–180.
77. Zhao, F. J., & McGrath, S. P., (2009). Biofortification and phytoremediation. *Curr. Opin. Plant Biol., 12*(3), 373–380.
78. Nowack, B., Schwyzer, I., & Schulin, R., (2008). Uptake of Zn and Fe by wheat (*Triticum aestivum* var. Greina) and transfer to the grains in the presence of chelating agents (ethylenediaminedisuccinic acid and ethylenediaminetetraacetic acid). *J. Agric. Food Chem., 56*(12), 4643–4649.
79. Bickel, A., & Killorn, R., (2007). Response of corn to banded zinc sulfate fertilizer in fields with variable soil pH. *Commun. Soil Sci. Plant Anal., 38*(9, 10), 1317–1329.
80. Zhang, Y. Q., Pang, L. L., Yan, P., Liu, D. Y., Zhang, W., Yost, R., Zhang, F. S., & Zou, C. Q., (2013). Zinc fertilizer placement affects zinc content in maize plant. *Plant Soil, 372*(1), 81–92.
81. Iqbal, J., Khan, R., Wahid, A., Sardar, K., Khan, N., Ali, M., Hussain, M., et al., (2016). Effect of nitrogen and zinc on maize (*Zea mays* L.) yield components and plant concentration. *Advances in Environmental Biology, 10*(10), 203.
82. Xue, Y. F., Yue, S. C., Liu, D. Y., Zhang, W., Chen, X. P., & Zou, C. Q., (2019). Dynamic zinc accumulation and contributions of pre- and/or post-silking zinc uptake to grain zinc of maize as affected by nitrogen supply. *Frontiers in Plant Science, 10*, 1203.
83. Wei, Y., Shohag, M. J., & Yang, X., (2012). Biofortification and bioavailability of rice grain zinc as affected by different forms of foliar zinc fertilization. *PLoS One, 7*(9), e45428.
84. Phattarakul, N., Rerkasem, B., Li, L. J., Wu, L. H., Zou, C. Q., Ram, H., Sohu, V. S., et al., (2012). Biofortification of rice grain with zinc through zinc fertilization in different countries. *Plant Soil, 361*(1, 2), 131–141.
85. Mabesa, R. L., Impa, S. M., Grewal, D., & Johnson-Beebout, S. E., (2013). Contrasting grain-Zn response of biofortification rice (*Oryza sativa* L.) breeding lines to foliar Zn application. *Field Crops Res., 149*, 223–233.

86. Ajiboye, B., Cakmak, I., Paterson, D., De Jonge, M. D., Howard, D. L., Stacey, S. P., Torun, A. A., et al., (2015). X-ray fluorescence microscopy of zinc localization in wheat grains biofortified through foliar zinc applications at different growth stages under field conditions. *Plant Soil, 392*(1, 2), 357–370.
87. Wang, J., Mao, H., Zhao, H., Huang, D., & Wang, Z., (2012). Different increases in maize and wheat grain zinc concentrations caused by soil and foliar applications of zinc in loess plateau, China. *Field Crops Res., 135*, 89–96.
88. Jones, C., Olson-Rutz, K., & Dinkins, C., (2011). *Nutrient Uptake Timing by Crops.* MSU Extension, EB0191.
89. Tariq, A., Anjum, S. A., Randhawa, M. A., Ullah, E., Naeem, M., Qamar, R., Ashraf, U., & Nadeem, M., (2014). Influence of zinc nutrition on growth and yield behavior of maize (*Zea mays* L.) hybrids. *American Journal of Plant Sciences, 5*(18), 2646–2654.
90. Cheah, Z. X., Harper, S. M., O'Hare, T. J., Kopittke, P. M., & Bell, M. J., (2022). Improved agronomic biofortification of sweetcorn achieved using foliar rather than soil Zn applications. *Cereal Chem.* doi: 10.1002/cche.10539.
91. Harris, D., Rashid, A., Miraj, G., Arif, M., & Shah, H., (2007). 'On-farm' seed priming with zinc sulphate solution – a cost-effective way to increase the maize yields of resource-poor farmers. *Field Crops Res., 102*(2), 119–127.
92. Martín-Ortiz, D., Hernández-Apaolaza, L., & Gárate, A., (2009). Efficiency of a NPK fertilizer with adhered zinc lignosulfonate as a zinc source for maize (*Zea mays* L.). *J. Agric. Food Chem., 57*(19), 9071–9078.
93. Hossain, M., Jahiruddin, M., Islam, M., & Mian, M., (2008). The requirement of zinc for improvement of crop yield and mineral nutrition in the maize–mungbean–rice system. *Plant Soil, 306*(1), 13–22.
94. Behera, S. K., Shukla, A. K., Singh, M. V., Wanjari, R. H., & Singh, P., (2015). Yield and zinc, copper, manganese and iron concentration in maize (*Zea mays* L.) grown on vertisol as influenced by zinc application from various zinc fertilizers. *J. Plant Nutr., 38*(10), 1544–1557.
95. Mohsin, A., Ahmad, A., Farooq, M., & Ullah, S., (2014). Influence of zinc application through seed treatment and foliar spray on growth, productivity and grain quality of hybrid maize. *Journal of Animal and Plant Sciences, 24*(5), 1494–1503.
96. Imran, M., & Rehim, A., (2017). Zinc fertilization approaches for agronomic biofortification and estimated human bioavailability of zinc in maize grain. *Archives of Agronomy and Soil Science, 63*(1), 106–116.
97. Soleimani, R., (2012). Cumulative and residual effects of zinc sulfate on grain yield, zinc, iron, and copper concentration in corn and wheat. *J. Plant Nutr., 35*(1), 85–92.
98. Debnath, P., Hemalatha, S., & Bhowmik, S., (2016). Effect on yield, quality parameters and nutrient uptake studies of biofortified corn with soil and foliar applied zinc. *Ecology, Environment and Conservation, 22*, S97–S103.

CHAPTER 11

Biofortification of Crop Plants: The Role of Arbuscular Mycorrhizal Fungi

GISELLE CAMARGO MENDES,[1] CAROLINE MÜLLER,[2] and JULIANA SILVA RODRIGUES CABRAL[3]

[1]*Department of Biotechnology, Instituto Federal de Santa Catarina (ISFC), Lages, SC, Brazil*

[2]*Department of Environmental Science and Technology, Universidade Federal da Fronteira Sul (UFFS), Erechim, RS, Brazil*

[3]*Department of Agronomy, Universidade de Rio Verde (UniRV), Rio Verde, GO, Brazil*

ABSTRACT

Arbuscular mycorrhizal fungi (AMF) are widely distributed in soil and show mutual symbiosis with the roots of higher plants (~80%). This association attributes numerous benefits to plants such as biofortification, being able to improve quality, and increasing crop/agricultural productivity. Biofortification is getting more attention to increase the phytoavailability of macronutrients (P and N), micronutrients (Zn, Cu, and Fe) and Se in the major food crops. The symbiosis occurs in stages, the plants release chemical signals through the roots that favor the establishment of the fungus-plant association and colonization in the host plant roots. Studies have been shown that AMF increases both the concentration and the bioavailability of nutrients in plants. This information indicates the importance of inoculating plants in agricultural areas to minimize the use of chemical fertilizers and to reduce environmental impacts, in addition to improving soil quality. This chapter discusses the symbiosis process and the increased concentration and availability of macro- and micronutrients by crop plants in symbiosis with AMF.

Phytoremediation and Biofortification: Strategies for Sustainable Environmental and Health Management.
Nand Kumar Singh, Shadma Afzal, & Tariq Aftab (Eds.)

11.1 INTRODUCTION

Biofortification is a technique that aims to produce food as a way to improve the nutritional status of the population and ensure food security, particularly in developing countries, through greater availability of nutrients and pro-vitamin A [1–3]. Biofortification can occur through breeding processes, genetic mutation, use of agronomic practices and selection of genotypes with higher content of nutrients of interest [4]. However, for efficient biofortification, the focus should be on increasing the availability of nutrients simultaneously with increase in their concentration in the plant [5]. For this, one of the ways of biofortification is the use of obligatory biotrophic microorganisms, such as arbuscular mycorrhizal fungi (AMF).

The AMF, belonging to the phylum Glomeromycota and the class Glomeromycetes, currently constitute a group with five orders, 11–16 families, 39–49 genera and more than 300 species [6, 7; http://www.amf-phylogeny.com]. AMF are classified according to their interaction with the root cells of the host plant, and can be ectomycorrhizae, arbuscular mycorrhizae, ericoid mycorrhizae and orchid mycorrhizae [8]. In general, ectomycorrhizas form intercellular fungal networks, generating the Hartig network in the root cortex. Arbuscular mycorrhiza, ericoid mycorrhiza, and orchid mycorrhiza develop outside and inside the host plant root cells via hyphae, arbuscules, and/or vesicles [8].

By establishing a mutualistic symbiotic relationship, AMF obtain carbohydrates from plants for their growth and development, and plants benefit from the acquisition of water and nutrients [5, 9, 10], which confer them high economic and ecological importance [2, 11]. The AMF have low plant specificity and can colonize the root cortex of about 80% of land plants [12, 13]. This colonization is mainly characterized by the intracellular growth of hyphae in the cortical tissue and by the differentiation of terminal intracellular hyphae into arbuscules, ephemeral structures, responsible for the bidirectional exchange of nutrients between the symbionts, in addition to the participation of other structures specific to the association, such as vesicles and helper cells related to nutrient storage [14, 15].

11.2 SYMBIOSIS

Symbiosis occurs when the fungus spore germinates in the soil and forms the germ tube, allowing the establishment under favorable conditions of temperature and humidity [16]. The symbiotic relationship occurs through the progressive colonization of fungal hyphae in plant roots, through a reciprocal communication of signals between AMF and plants [9]. Initially, fungi exude different oligosaccharides [9, 17], which generate signals responsible for activating the expression of genes that cause an increase in the level of calcium and the accumulation of starch in the root cells of the host plant [18–20]. On the other hand, plants exude chemical signals through the roots, such as strigolactones (SL), and plant hormones derived from carotenoid metabolism [21, 22]. Low concentrations of SL are sufficient to trigger a response in AMF, such as the increase in the expression of mitochondrial genes [23], such as NADH dehydrogenases [24], and the increase in ATP synthesis by intense mitochondrial ATP synthase activity [25, 26], which favors the growth of hyphae. In addition, plant hormones, such as gibberellic acid, which acts to distend the cell wall in root cells, allowing the entry of hyphae [27], and jasmonic acid [28] and abscisic acid [29, 30] also favor the colonization process. Additionally, there is the expression of genes related to the synthesis of cell wall components (chitin deacetylase, chitin synthase) [23], which promote greater development of hyphae in the apoplast region, favoring the arbuscules penetration into the cortical cells [9, 31, 32]. After the fungi establishment and hyphal branching in the roots of host plants, the formation of vesicles with the function of storing carbon from the photosynthetic process may occur [33]. Carbon is transferred in the form of hexoses and stored in the fungus as glycogen [34, 35]. AMF hyphae form extensive networks that can reach up to 12 cm from the root surface [36], which contributes to a reduction in nutrient losses [37] and soil aggregation, preventing erosion [38, 39].

11.3 MACRONUTRIENTS

Most plants are able to establish symbiotic relationships with mycorrhizal arbuscular fungi, and important benefits have been verified for agricultural crops, since symbiosis promotes, among other factors, a critical role for plant nutrient use, especially regarding phosphate (P) and nitrogen (N). The efficiency of such symbiotic systems depends on the ability of the microbial symbionts to uptake/fix nutrients available under an inorganic and/or an organic form in the soil and to translocate them (or the corresponding metabolites) to the symbiotic root [40], as discussed in subsections.

11.3.1 PHOSPHOROUS

In plants, phosphate can remain as inorganic phosphate (Pi) – as it is preferentially absorbed–, be esterified through a hydroxyl group as a single phosphate ester or attached to another phosphate by the energy-rich pyrophosphate bond, forming, for example, adenosine triphosphate (ATP) [41–43]. ATP represents the metabolic energy of cells and is essential for carrying out processes such as photosynthesis, respiration, glycolysis, starch biosynthesis and ion absorption. P also acts as a structural component of nucleic acids, in addition to being present in phospholipids of biomembranes and phosphoproteins [42].

Although essential for plants, the concentration of Pi in the soil solution, at the root-soil interface, is often present at submicromolar levels, with a low rate of diffusion and mobility [43]. Thus, one of the strategies for acquiring Pi from the soil, in addition to the direct absorption of Pi, involves the mycorrhizal pathway. Pi flux through AM associations involves: (i) uptake of P from the soil through the fungal extraradical hyphae; (ii) long-distance movement of the through extraradical and intraradical hyphae; (iii) P release in fungal arbuscules at the arbuscular-cortical cell interface; and (iv) P uptake by plant cortical cells [43–45].

Soil Pi is taken up by AMF through H^+/Pi and Na^+/Pi symporters driven by H^+-ATPases and Na^+-ATPases, respectively, located in the plasma membrane of extraradical hyphae [46, 47], expressed by phosphate transporter genes such as *GmosPT*, *GiPT*, *GvPT*, *GinPT* [48–50]. Pi is accumulated in polyphosphate (polyP), the linear chain of three to hundreds of Pi residues linked by high energy phosphoanhydride bonds [51–53]. PolyP is synthesized by the vacuole transporter chaperones (VTC) complex consisting of VTC1, VTC2, and VTC4 subunits, using ATP as a substrate and accumulated in tubular vacuoles [45, 52, 54, 55]. Subsequently, long-distance P translocation through extraradical and intraradical hyphae occurs driven by the water potential gradient, mediated by a fungal aquaporin (AQP3), identified in *Rhizophagus clarus* [56]. During translocation in extraradical hyphae, polyphosphate is depolymerized by polyphosphatases, exported to the cytosol by the vacuolar Pi exporter (PHO91) and incorporated into ATP for polymerization [52]. In intraradical hyphae, three pathways of P release from arbuscules to host plant cells are described: (i) hydrolysis of the polyphosphate exported to the cytosol by PHO91 and loaded to the Golgi complex for export to the apoplast by SYG1; (ii) exported from vacuoles directly by SYG1 [45]; and (iii) polyP is exported directly through the VTC complex in the plasma membrane of the apoplast and the polyphosphate is hydrolyzed by plant phosphatase [44]. Pi exported to the apoplast is taken up by mycorrhizal inducible Pi transporters (MPT) located in the plant plasma membrane, such as proteins containing the SPX domain [52].

Recent research has demonstrated an increased expression of phosphate transporters in an environment rich in AMF. Gao et al. [57] found that the expression of 10 of the 14 phosphate transporters analyzed in roots of cotton (*Gossypium hirsutum* L.) were higher in plants exposed to AMF *Rhizophagus irregularis* CD1. According to the authors, the increase in P uptake allowed greater growth, photosynthesis, and a 28.6% increase in seed cotton yield. Different species of AMF have demonstrated beneficial effects on P uptake and growth of crop plants, as shown in Table 11.1. *Rhizophagus intradices* (=*Glomus intraradices* Schenck & Smith [145]), in addition to an increase in P content, promoted higher Zn content, root growth, leaf area, chlorophyll content, and tryptophan concentration in *Zea mays* L. [58]. A mix of AMF species containing *Gigaspora margarita, Funneliformis mosseae, Rhizophagus irregularis, Rhizophagus clarum (=Glomus clarum* [145]), *Glomus deserticola, Glomus monosporum, Glomus brasilianum, Glomus aggregatum* also increased P, Fe, and Zn content of wheat (*Triticum durum*) [59].

11.3.2 NITROGEN

Nitrogen (N) is the element required in greater quantities by plants. It is a constituent of nucleic acids, proteins, coenzymes, chlorophyll, phytohormones, and secondary metabolites [42]. AMF are reported to increase rates of soil N mineralization and total N uptake by plants [60–62]. Wheat plants inoculated with AMF *Rhizophagus irregularis* had a higher concentration of C and N, with higher N utilization efficiency and higher plant biomass compared to uninoculated plants [62]. In rice, Wang et al. [63] observed that the mycorrhizal nitrate (NO_3^-) uptake route contributed to up to 42% of the overall rice N uptake when NO_3^- was made available as a N source, being mainly taken up by the OsNPF4.5 transporter. In addition, AMF can transfer large amounts of N in organic forms, such as free amino acids such as glutamine, arginine, and aspartic acid [64].

NO is absorbed by AMF external hyphae mainly in inorganic forms, such as NO_3^- and ammonium (NH_4^+), but it can also be absorbed as organic compounds, such as amino acids [40, 65–69], regardless of whether or not N is limited by the host roots [70, 71]. Several fungal transporters of

NH_4^+, NO_3^-, and organic N were identified in different AMF species, such as *GintAMT1* and *GintAMt2*, NH_4^+ high affinity gene transporters of *Rhizophagus intraradices* (= *Glomus intraradices* [145]) [72, 73], *GiNT*, a nitrate transporter of *R. intraradices* [74], and also an amino acid permease (*GmosAAP1*) was characterized from extraradical structures of *Funneliformis mosseae* (=*Glomus mosseae* [145]) [75], and a dipeptide transporter (*RiPTR2*) from AM fungus *R. irregularis* [76], as examples.

The NO_3^-, when absorbed by FMA, needs to be reduced to nitrite by the enzyme nitrate reductase (NR), using NADPH as a reductant, and this converted to NH_4^+ by nitrite reductase in the intraradical mycelium [72]. The NH_4^+ is then assimilated into amino acids by AMF fungi predominantly via the glutamine synthetase-glutamate synthase (GS-GOGAT) pathway and metabolized to arginine, which is translocated probably together with polyP through the intraradical hyphae [65, 70, 77]. Arginine is broken down by a fungal arginase (CAR1) and urease (URE) into urea, and then to NH_4^+, which is transferred to the host root via ammonia channels [70, 74]. It should be noted that since arginine can bind to polyP, a possible link between P and N translocation in AMF hyphae is believed to occur [70, 78].

Different AMF species have already demonstrated benefits in the growth and productivity of crop species as shown in Table 11.1. Mensah et al. [79] evaluated 10 AMF morphospecies and found that six of these species (*Acaulospora colombiana, Funneliformis mosseae, A. morrowiae, Paraglomus occultum, A. scrobiculata*, and *Claroideoglomus etunicatum*) showed high performance with increment of up to 170% in the *Medicago sativa* biomass, which was positively correlated with the higher uptake of N and P. Yadav et al. [80] using a commercial biofertilizer Symbion-VAM PlusTM® (composed mainly of *Glomus fasciculatum*) verified an increase in N, Fe, and Zn grain content in Wheat (*Triticum aestivum*) cv. HD-3086, in addition to the increase in superoxide dismutase (SOD) activity, and soil organic carbon at harvest stage, β-glucosidase activity of post-harvest soil. A higher N uptake was also reported in rice (*Oryza sativa*) treated with *Rhizophagus irregularis* [63].

In studies with ^{15}N-labeled fertilizers, Ingraffia et al. [59] found that in soils with sufficient N availability, AMF increased N uptake and N concentrations in durum wheat (*Triticum durum* Desf. cv. Anco Marzio), and total N recovered from the fertilizer, without having an effect on plant biomass. These results are important to increase sustainability in agricultural production since AMF favors plant N-related traits When N is not a limiting factor, in addition to allowing the recovery of N from the fertilizer.

11.4 MICRONUTRIENTS

The homeostasis of micronutrient accumulation in plants is related to the exudation of chemical compounds, morphology of the root system, soil aspects (chemical and physical), soil and climate conditions and the interaction with microbiota present in the soil, such as AMF. Mycorrhizal fungi, through their hyphae, also ensure the efficient supply of low mobility micronutrients; and host plants provide carbon compounds in carbohydrate form for fungal growth [33, 81]. Among the main micronutrients acquired by plants in symbiosis, Zn, Fe, and Cu stand out.

11.4.1 ZINC

Zinc (Zn) is present in plants as a divalent cation (Zn^{+2}) [82]. It is an essential micronutrient in plant metabolic processes, with a crucial role in plant development and productivity [83, 84]. Zn is essential in the formation of the pollen tube [85] and the pollen grain [86], in addition to participating as a cofactor of enzymes such as SOD forming the CuZn-SOD complex [87], actively acting to relieve oxidative stress by eliminating free radicals. Furthermore, Zn participates in the chemical structure of more than 300 enzymes [88] and is involved in the biosynthesis of the hormone auxin, because it activates the enzyme tryptophan synthetase, responsible for the synthesis of tryptophan, precursor of indole acetic acid (IAA) synthesis [82, 83]. Zn deficiency is also related to altered plasma membrane permeability [84, 87].

Zn has low mobility in soil [89], and grain cultivation worldwide is affected by soil zinc deficiency [90]. Plants grown in alkaline calcareous soils, in which this micronutrient is unavailable for uptake, are the main causes of Zn deficiency in cereals [91]. In this sense, the use of AMF is an important strategy for biofortification because, in addition to increasing productivity in plants, symbiosis increases the contact area for nutrient uptake [33, 92].

Studies indicate the beneficial effect of using AMF from the genera *Gigaspora*, *Rhizophagus*, and *Glomus* on the Zn uptake in plants [93–96]. In maize plants, in association with AMF *Rhizophagus intraradices* (=*Glomus intraradices* [145]) and cultivated under different concentrations of Zn, there was an uptake of about 24% more zinc, when compared to non-mycorrhizal plants [97]. According to the authors, the fungi favored the architecture of the root system, and increased the availability of Zn in the soil solution. The use of AMF also showed promise in soils in which zinc-based fertilizers were used [98]; in addition, wheat and barley plants inoculated with AMF *Rhizophagus irregularis* (=*Rhizophagus irregulare*) showed an increase of 25% and 13%, respectively, in Zn uptake [94, 95] as show in Table 11.1. The increase in Zn uptake in mycorrhizal barley plants is related to the increase in the expression of the *HvZIP13* gene in the roots, which encodes Zn^{2+} transport proteins [99].

Wheat plants in symbiosis with AMF *Funneliformis mosseae* (=*Glomus mosseae* [145]) had a greater Zn uptake when grown in calcareous soils [100]. Maize plants inoculated with *Rhizophagus intraradices* (=*Glomus intraradices* [145]) also showed higher absorption of Zn and P [101], with both nutrients being low availability in soil, however, Zn is more easily translocated than P [102].

11.4.2 IRON

Iron (Fe) is an important micronutrient for plants as it is involved in key processes of primary metabolism, such as respiration, photosynthesis, nitrogen assimilation and chlorophyll synthesis [103, 104]. Despite being an abundant element in the earth's crust, Fe has low availability to be absorbed by plant roots [105]. Plants uptake Fe as Fe^{+2} (ferrous) and Fe^{+3} (ferric) form and there are two strategies for acquiring this nutrient [86]. Strategy I occur in all plant species, with the exception of grasses, in which the Fe^{+3} ion is reduced to Fe^{+2} by the action of the *FRO* gene that encodes a reductase enzyme (FCR: Ferric Chelate Reductases) [106] by excreting protons before they are absorbed. The protons accumulation favors the acidification of the apoplast and rhizosphere and increases the exudation of acids from organic, phenolic, and flavonoid compounds [107, 108]. On the other hand, in strategy II, the roots of monocotyledonous grasses excrete mugineic acids (MAs) through a sequence of enzymatic reactions which involve the enzymes: (i) nicothiamine synthase (NAS); (ii) nicothiamine amino-transferase (NAAT); and (iii) deoxymugineic acid synthase (DMAS) generating the precursor of MAs that chelates Fe^{+3} forming the Fe^{+3}-MA complex. This complex is taken up via yellow stripe 1-like (YSL) transporters located in the roots [109–111]. In flooded areas, Fe is found in its most available reduced divalent form (Fe^{+2}) due to anaerobic conditions and low pH [112–114].

Plants in association with AMF can increase or decrease iron acquisition [115, 116], and this difference is related to soil characteristics such as acidity and alkalinity [115, 117, 118]. Sorghum plants grown under controlled conditions with AMF *Funneliformis mosseae* (=*Glomus mosseae* [145]) increased the concentration of Fe in the shoot observed by radioactively labeled ^{59}Fe when compared to non-mycorrhizal plants [116]. These results indicate that AMFs have the possibility of assisting in the mobility or removal of Fe in the soil. Maize plants grown in soils with low Fe content and inoculated with AMF species of the genus *Glomus* showed an increase in Fe uptake, in addition to higher shoot and root dry matter [115]. Fe uptake in maize plants may be related to the induction of the expression of the genes *ZmOPT8a* and *ZmOPT8b*, responsible for the Fe transport through oligopeptide transporters (OPT), as observed in roots colonized with AMF *Rhizophagus irregularis* [119].

Quince plants grown in association with the fungi *Funneliformis mosseae* and *Rhizophagus intraradices* and under Fe deficiency showed an induction of the expression of genes phenylalanine ammonia lyase (*PAL*) and ferric chelate reductases (*FRO2*) in the roots, which favored an increase in enzyme activity for synthesis of phenolic compounds and reduction of Fe^{+3} to Fe^{+2} to favor uptake and accumulation in roots and shoots [120] as show in Table 11.1. Sunflower plants grown in association with AMF species of the genus *Glomus* also under Fe deficiency showed improvement in physiological aspects, such as quantum efficiency of photosystem II (*Fv/Fm*), and increased Fe and Zn concentrations in the root and shoot [121]. According to the authors, the increase of nutrients in plant biomass is related to the induction of the *HaFRO1* gene expression, ensuring the availability of iron in the rhizosphere to be uptake by the roots. In addition, the increase in Fe and Zn in the plant is related to the induction of the expression of the genes *HaIRT1*, *HaNramp1*, and *HaZIP1*, which are responsible for encoding Fe and Zn transporters [121].

AMF associated with growth-promoting bacteria showed synergistic effects and increased levels of micronutrients such as Zn and Fe in wheat grains, in addition to parameters related to yield [122] as show in Table 11.1.

11.4.3 COPPER

Cooper (Cu) is a transition element that can be in Cu^{+} (reduced) or Cu^{+2} (oxidized) form. The transition state determines the ability of this element to bind to different substrates, such as sulfur, nitrogen or oxygen, and play its role by interacting with different macromolecules [123]. The Cu is important in physiological processes in plants, acting as cofactors of various proteins [124], in addition to participating effectively in primary metabolism, such as respiration and photosynthesis, where it is as a constituent of the electron transport chain [86, 123].

Wheat plants in symbiosis with AMF *Claroideoglomus etunicatum* (=*Glomus etunicatum* [145]) and *Funneliformis mosseae* (=*Glomus mosseae* [145]) increased stem Cu content [125]. Maize plants grown in calcareous soil and inoculated with *Funneliformis mosseae* had a concentration of Cu in the roots six times higher than non-mycorrhizal plants [126]. It has also been observed a relationship between the uptake of Cu and P, where the availability of Cu and P in the soil directly influences the uptake of Cu in maize plants [118]. Soils with low P contents have an increase in the growth of hyphae and consequently an increase in the contact area with a greater volume of soil, thus favoring the uptake of micronutrients [127].

A study using *in silico* tools identified in the *Rhizophagus irregularis* genome putative genes that may be related to the influx and transport of Fe, Cu, and Zn, such as *Cu-ATPases*, *VITs*, and *NRAMPs*, *RiFTR1*, *RiZRT1*, *RiCTR2*, and *RiZRT3* [128]. This information is extremely relevant for the functional characterization of these genes, in order to understand the homeostasis of these micronutrients during the symbiosis process.

11.5 OTHER NUTRIENT: SELENIUM

Selenium (Se) has emerged in recent years as an important micronutrient for human health, as it minimizes the effects of oxidative stress and improves cardiovascular aspects [129]. However, some regions suffer with Se deficiency in the population, and one of the alternatives is the use of fertilizers associated with microorganisms, in order to obtain this element in a sustainable way [130, 131].

The inorganic forms selenite and selenate bind to soil constituents and become unavailable to plants [132, 133]. Several studies indicate that the availability of selenium in the soil can be influenced by the presence of microorganisms, such as AMF [134, 135]. Durán et al. [130] verified a synergistic effect in wheat plants co-inoculated with AMF *Claroideoglomus claroideum* (= *Glomus claroideum* [145]) and bacteria of the genus *Stenotrophomonas*, *Enterobacter*, *Bacillus*, and *Pseudomonas*, observing an increase of about 24% of grain Se when compared to plants without AMF inoculation. Rice plants inoculated with both AMF *Funneliformis mosseae* and *Diversispora epigaea* (=*Glomus versiforme* [136]), combined with the use of fertilization such as selenite and/or selenate, promoted a greater accumulation of Se in the rice grain, when compared to plants in which only fertilizers were used [131] as show in Table 11.1. The higher Se content in rice grain was related to a change in the microbial composition in the soil, which favored a greater

TABLE 11.1 Responses of Different Arbuscular Mycorrhizal Fungi in Crop Plants

Arbuscular Mycorrhizal Fungi (AMF)	Host Crop	Responses	References
Rhizophagus irregularis	*Triticum aestivum*	Increases in grain yield, number of kernels spike.	[95]
Rhizophagus irregularis	*Hordeum vulgare*	Increases in grain yield, mean grain weight, grain, and straw Zn content.	[95]
*Rhizophagus clarus** (=*Glomus clarum*)	*Zea mays* varieties	Increased provitamin A.	[3]
F. mosseae (IMA1) and *R. irregularis** (=*Glomus mosseae* and *Glomus intraradices*) (IMA5)	*Cicer arietinum*	Higher shoot dry matter, collar diameter, grain yield, shoot N and P, grain N, Fe, and Zn concentration.	[137]
Commercial biofertilizer Symbion-VAM PlusTM® composed mainly of *Rhizophagus fasciculatus** (=*Glomus fasciculatum*)	*Triticum aestivum* cv. HD-2967	Increased 1,000 grain weight, P grain content, superoxide dismutase activity, soil organic carbon at harvest stage, invertase activity of post-harvest soil.	[80]
	Triticum aestivum cv. HD-3086	Increased number of grains/spikes, 1,000 grain weight, P, Ca, Fe, B, soil organic matter, soil dehydrogenase activity.	[122]
M3 *Rhizophagus intraradices** (=*Glomus intraradices*) TNAU-03–08	*Zea mays*, hybrids NK 6240	Increased grain and stover Zn content.	[138]
*Rhizophagus clarus** (=*Glomus clarum*)	*Glycine max*	Increased number of pods, number of seeds.	[81]
R. irregularis WFVAM10	*Triticum durum* L. var. DBA Aurora	Higher P and phytic acid concentrations.	[139]
Diversispora epigaea (=*Glomus versiforme* [136])	*P. trifoliata* L. Raf	Higher shoot and root dry weight, peroxidase, and catalase activities, Fe leaf content and Fe/P ratio.	[140]
*Rhizophagus irregularis** (=*R. irregulare*)	*Oryza sativa*	Higher N uptake.	[63]
Rhizophagus irregularis CD1	*Gossypium hirsutum* L.	Increased total and inorganic P content in leaves, stem, and roots, photosynthetic rate, plant height, yield.	[57]
*Rhizophagus irregularis** (=*R. irregulare*)	*Hordeum vulgare* cv. Compass	Increased in spike dry biomass, P, Mg, and Mn plant content, and P, Mg, Ca, K, S, and Zn grain content.	[141]
Funneliformis mosseae	Strawberry (*Fragaria* x *ananassa* Festival)	Under high levels of Se, AMF increased total phenolics, total soluble solids, total flavonoids, and the antioxidants 2,2-diphenyl-1-picrylhydrazyl (DPPH) and 2,2′-azino-bis (3-ethylbenzothiazoline-6-sulp acid) (ABTS). Higher accumulation rate and translocation factor for Se were also observed in strawberry plants.	[142]

TABLE 11.1 *(Continued)*

Arbuscular Mycorrhizal Fungi (AMF)	Host Crop	Responses	References
Funneliformis mosseae and *Rhizophagus irregularis*	*Lycopersicon esculentum* cv "Moneymaker"	Increased fresh weight fruit, dry shoot weight, β-carotene in fruits, and also volatile compounds as carotenoid–, lipid–, phenylalanine–, and leucine-derived, and increased P in fruits.	[143]
Rhizophagus irregularis	*Triticum aestivum* L.	Increased leaf dry weight, stem dry weight, root dry weight, total dry weight, N accumulation, and leaf C concentration.	[62]
Rhizophagus irregularis	*Andropogon gerardii*	Increased N and P content and plant biomass.	[67]
Rhizophagus irregularis	*Triticum* spp.	Higher N, Fe, and Zn grain concentration.	[144]
Diversispora epigaea (=*Glomus versiforme* [136])	*Allium sativum* L.	Improved the height and fresh weight of garlic plants, positive effect on garlic growth, increased soluble protein content, allicin, and soluble sugar in garlic bulbs.	[13]
Mix of two AM fungi species *Rhizophagus irregularis and Funneliformis mosseae*	*Triticum durum*	Increased total root length, soil root density and reduced total mineral N residual in soil.	[59]
Rhizophagus irregularis	*Medicago truncatula*	Higher plant dry weight, Zn content.	[145]
Diversispora epigaea (=*Glomus versiforme* [136]) or *Funneliformis mosseae*	*Oryza sativa* L. cv. Xinliangyou 6	Increased shoot and root biomass and grain yield, increased Se concentration in leaves, stem, roots, and grain.	[131]
R. intradices or F. mosseae	Quince (*Cydonia oblonga Mill.*) cv. Isfahan	Increased shoot and root weight, chlorophyll concentration, and also increased F_v/F_m, phenolic content, phenylalanine ammonia-lyase and root ferric chelate reductase activities, leaf, and root Fe content under Fe deficiency.	[120]
Rhizophagus intraradices (=Glomus intraradices)	*Zea mays* L.	Increased root growth, leaf area, chlorophyll content, P and Zn plant concentration, Zn grain, and tryptophan concentration.	[58]

**Nomenclature of the phylum Glomeromycota by Schüßler & Walker [146]; and* http://amf-phylogeny.com/.

Se availability, favoring the uptake of this element [131]. Thus, we can highlight the promising use of AMF to obtain Se in rice plants.

11.6 CONCLUSION

In this review, it was possible to verify that the use of AMF is an interesting alternative for nutrient uptake and plant growth and development, as well as for human nutrition through crop biofortification. Increases in the plant nutrient content through mycorrhizal symbiosis also allows the development of clean technology capable of promoting sustainable agriculture. This leads to a reduction in the use of chemical fertilizers in agricultural soils, by increasing the availability of nutrients by AMF, in addition to improving the use of applied fertilizers, especially in low fertility soils, and also improving the soil quality. This occurs since the complex soil microbiota, including AMF, play important roles in nutrient cycling through decomposition, mineralization of organic material and the transformation of organic nutrients into available forms; in addition to favoring nutrient absorption through chelation, solubilization, and soil oxidation and reduction reactions [147, 148]. It is in the rhizosphere, the region of contact between the soil and the plants roots and their exudates, that microbial activities act, improving, for example, nutrient mineralization and uptake, as discussed throughout the chapter. Despite being naturally in the soil, the crop management methods and cultural practices caused a reduction of AMF populations, being necessary the inoculation process to increase and improve the performance of these organisms until the recomposition of the soil microbiota.

KEYWORDS

- **crop productivity**
- **inoculant**
- **macronutrients**
- **micronutrients**
- **nutrients**
- **phosphorous**
- **symbiosis**

REFERENCES

1. Bouis, H. E., Hotz, C., McClafferty, B., Meenakshi, J. V., & Pfeiffer, W. H., (2011). Biofortification: A new tool to reduce micronutrient malnutrition. *Food Nutri. Bull., 32*, 31–40.
2. Rodriguez, A., & Sanders, I. R., (2015). The role of community and population ecology in applying mycorrhizal fungi for improved food security. *ISME Journal, 9*, 1053–1061.
3. Olawuyi, O. J., Azeez, A. O., & Azeez, A. A., (2021). Improvement of provitamin a in maize varieties using arbuscular mycorrhizal fungus, *Glomus clarum. Bio-Research, 19*, 1227–1236.
4. Hess, S. Y., (2012). Zinc: Deficiency disorders and prevention programs. *Encyclopedia of Human Nutrition* (Vol. 4). Elsevier Inc. doi: 10.1016/B978-0-12-375083-9.00287-7.
5. Jha, A. B., & Warkentin, T. D., (2020). Biofortification of pulse crops: Status and future perspectives. *Plants, 9*. doi: 10.3390/plants9010073.
6. Redecker, D., Schüßler, A., Stockinger, H., Stürmer, S. L., Morton, J. B., & Walker, C., (2013). An evidence-based consensus for the classification of arbuscular mycorrhizal fungi (*Glomeromycota*). *Mycorrhiza, 23*, 515–531.
7. Wijayawardene, N. N., Hyde, K. D., Dai, D. Q., Tang, L. Z., Aptroot, A., CastañedaRuiz, R. F., et al., (2020). Outline of fungi and fungi-like taxa. *Mycosphere, 11*, 1060–1456.
8. Brundrett, M. C., & Tedersoo, L., (2018). Evolutionary history of mycorrhizal symbioses and global host plant diversity. *New Phytol., 220*, 1108–1114.
9. Gutjahr, C., & Parniske, M., (2013). Cell and developmental biology of arbuscular mycorrhiza symbiosis. *Ann. Rev. Cell Dev. Biol., 29*, 593–617.
10. Liu, C., Ye, Y., Liu, J., Pu, Y., & Wu, C., (2021). Iron biofortification of crop food by symbiosis with beneficial microorganisms. *J. Plant Nutr., 44*, 2793–2810.
11. Mergulhão, A. C. E. S., Silva, M. V., Lyra, M. C. C. P., Figueiredo, M. V. B., Silva, M. L. R. B., & Maia, L. C., (2014). Caracterização morfológica e molecular de fungos micorrízicos arbusculares isolados de áreas de mineração de gesso, Araripina, PE, Brasil. *Hoehnea, 41*, 393–400.
12. Sturmer, S. L., & Siqueira, J. O., (2013). Fungos micorrízicos. In: Moreira, F. M. S., Cares, J. E., Zanetti, R., & Stürmer, S. L., (eds.). *O Ecossistema Solo: Componentes, Relações Ecológicas e Efeitos na Produção Vegetal* (pp. 289–311). Lavras: Editora UFLA.
13. Yang, F., Pan, Y., Cheng, Z., Ali, A., Zhang, S., Li, X., Qi, X., et al., (2021). Agronomic biofortification of garlic through selenium and arbuscular mycorrhizal fungi application. *Horticulturae, 7*, 230.
14. Maia, L. C., Silva, F. S. B., & Goto, B. T., (2010). Structure, ultrastructure, and germination of glomerospores. In: Siqueira, J. O., Souza, F.A., Cardoso, E. J. B. N., & Tsai, S. M., (eds.), *Mycorrhizae: 30 Years of Research in Brazil* (pp. 76–118). Lavras: Editora UFLA.
15. Smith, S. E., & Smith, F. A., (2011). Roles of arbuscular mycorrhizas in plant nutrition and growth: New paradigms from cellular to ecosystem scales. *Ann. Rev. Plant Biol., 62*, 227–250.
16. Giovannetti, M., Sbrana, C., Avio, L., Citernesi, A. S., & Logi, C., (1993). Differential hyphal morphogenesis in arbuscular mycorrhizal fungi during pre-infection stages. *New Phytol., 125*, 587–593.

17. Maillet, F., Poinsot, V., André, O., Puech-Pagès, V., Haouy, A., Gueunier, M., Cromer, L., et al., (2011). Fungal lipo-chitooligosaccharide symbiotic signals in arbuscular mycorrhiza. *Nature, 469*, 58–63.
18. Chabaud, M., Genre, A., Sieberer, B. J., Faccio, A., Fournier, J., Novero, M., Barker, D. G., & Bonfante, P., (2011). Arbuscular mycorrhizal hyphopodia and germinated spore exudates trigger Ca^{2+} spiking in the legume and nonlegume root epidermis. *New Phytol., 189*, 347–355.
19. Gutjahr, C., Novero, M., Guether, M., Montanari, O., Udvardi, M., & Bonfante, P., (2009). Presymbiotic factors released by the arbuscular mycorrhizal fungus *Gigaspora margarita* induce starch accumulation in *Lotus japonicus* roots. *New Phytol., 183*, 53–61.
20. Kuhn, H., Küster, H., & Requena, N., (2010). Membrane steroidbinding protein 1 induced by a diffusible fungal signal is critical for mycorrhization in *Medicago truncatula*. *New Phytol., 185*, 716–733.
21. Akiyama, K., Matsuzaki, K., & Hayashi, H., (2005). Plant sesquiterpenes induce hyphal branching in arbuscular mycorrhizal fungi. *Nature, 435*, 824–827.
22. Al-Babili, S., & Bouwmeester, H. J., (2015). Strigolactones, a novel carotenoid-derived plant hormone. *Ann. Rev. Plant Biol., 66*, 161–186.
23. Salvioli, A., Ghignone, S., Novero, M., Navazio, L., Venice, F., Bagnaresi, P., & Bonfante, P., (2016). Symbiosis with an endobacterium increases the fitness of a mycorrhizal fungus, raising its bioenergetic potential. *ISME Journal, 10*, 130–144.
24. Besserer, A., Bécard, G., Jauneau, A., Roux, C., & Séjalon-Delmas, N., (2008). GR24, a synthetic analog of strigolactones, stimulates the mitosis and growth of the arbuscular mycorrhizal fungus *Gigaspora rosea* by boosting its energy metabolism. *Plant Physiol, 148*, 402–413.
25. Bonfante, P., & Genre, A., (2015). Arbuscular mycorrhizal dialogues: Do you speak 'plantish' or 'fungish'? *Trends Plant Sci., 20*, 150–154.
26. Kohler, A., Kuo, A., Nagy, L. G., Morin, E., Barry, K. W., Buscot, F., Canbäck, B., et al., (2015). Convergent losses of decay mechanisms and rapid turnover of symbiosis genes inmycorrhizal mutualists. *Nat. Genet., 47*, 410–415.
27. Takeda, N., Handa, Y., Tsuzuki, S., Kojima, M., Sakakibara, H., & Kawaguchi, M., (2015). Gibberellins interfere with symbiosis signaling and gene expression, and alter colonization by arbuscular mycorrhizal fungi in *Lotus japonicus*. *Plant Physiol., 167*, 545–557.
28. Tejeda-Sartorius, M., De-La-Vega, O. M., & Délano-Frier, J. P., (2008). Jasmonic acid influences mycorrhizal colonization in tomato plants by modifying the expression of genes involved in carbohydrate partitioning. *Physiol. Plant., 133*, 339–353.
29. Charpentier, M., Sun, J., Wen, J., Mysore, K. S., Oldroyd, G. E. D., Charpentier, M., Sun, J., Wen, J., et al., (2014). Abscisic acid promotion of arbuscular mycorrhizal colonization requires a component of the protein phosphatase 2A complex. *Plant Physiol., 166*, 2077–2090.
30. Martín-Rodríguez, J. A., León-Morcillo, R., Vierheilig, H., Ocampo-Bote, J. A., Ludwig-Müller, J., & García-Garrido, J. M., (2010). Mycorrhization of the notabilis and sitiens tomato mutants in relation to abscisic acid and ethylene contents. *J Plant Physiol., 167*, 606–613.
31. Bonfante, P., & Genre, A., (2008). Plants and arbuscular mycorrhizal fungi: An evolutionary–developmental perspective. *Trends Plant Sci., 13*, 492–498.

32. Harrison, M. J., (2012). Cellular programs for arbuscular mycorrhizal symbiosis. *Curr Opin Plant Biol., 15*, 691–698.
33. Smith, S. E., & Read, D. J., (2008). *Mycorrhizal Symbiosis* (3rd edn.) Academic Press.
34. Bago, B., Pfeffer, P. E., Zipfel, W., Lammers, P., & Shachar-Hill, Y., (2002). Tracking metabolism and imaging transport in arbuscular mycorrhizal fungi. Metabolism and transport in AM fungi. *Plant Soil, 244*, 189–197.
35. Solaiman, M. D., & Saito, M., (1997). Use of sugars by intraradical hyphae of arbuscular mycorrhizal fungi revealed by radio respirometry. *New Phytol., 136*, 533–538.
36. Jakobsen, I., Abbott, L. K., & Robson, A. D., (1992). External hyphae of vesicular–arbuscular mycorrhizal fungi associated with *Trifolium subterraneum* L. 1. Spread of hyphae and phosphorus inflow into roots. *New Phytol, 120*, 371–380.
37. Cavagnaro, T. R., Bender, S. F., Asghari, H. R., & Van, D. H. M. G. A., (2015). The role of arbuscular mycorrhizas in reducing soil nutrient loss. *Trends Plant Sci., 20*, 283–290.
38. Mardhiah, U., Caruso, T., Gurnell, A., & Rillig, M. C., (2016). Arbuscular mycorrhizal fungal hyphae reduce soil erosion by surface water flow in a greenhouse experiment. *Appl. Soil Ecol., 99*, 137–140.
39. Rillig, M. C., Sosa-Hernandez, M. A., Roy, J., Aguilar-Trigueros, C. A., Valyi, K., & Lehmann, A., (2016). Towards an integrated mycorrhizal technology: Harnessing mycorrhizae for sustainable intensification in agriculture. *Front. Plant Sci., 7*, 1625.
40. Courty, P. E., Smith, P., Koegel, S., Redecker, D., & Wipf, D., (2015). Inorganic nitrogen uptake and transport in beneficial plant root-microbe interactions. *Crit Rev Plant Sci., 34*, 4–16.
41. Veneklaas, E. J., Lambers, H., Bragg, J., Finnegan, P. M., Lovelock, C. E., Plaxton, W. C., Price, C. A., et al., (2012). Opportunities for improving phosphorus-use efficiency in crop plants. *New Phytologist, 195*, 306–320.
42. Hawkesford, M., Horst, W., Kichey, T., Lambers, H., Schjoerring, J., Møller, I. S., & White, P., (2011). Functions of macronutrients. *Marschner's Mineral Nutrition of Higher Plants* (3rd edn.) Elsevier Inc.
43. Johri, A. K., Oelmüller, R., Dua, M., Yadav, V., Kumar, M., Tutea, N., Varma, A., et al., (2015). Fungal association and utilization of phosphate by plants: Success, limitations, and future prospects. *Front. Microbiol., 6*, 984.
44. Saito, K., & Ezawa, T., (2016). Phosphorus metabolism and transport in arbuscular mycorrhizal symbiosis. In: Martin, F., (ed.) *Molecular Mycorrhizal Symbiosis* (pp. 197–216). John Wiley & Sons, Inc.
45. Ezawa, T., & Saito, K., (2018). How do arbuscular mycorrhizal fungi handle phosphate? New insight into fine-tuning of phosphate metabolism. *New Phytol., 220*, 1116–1121.
46. Harrison, M. J., & Van, B. M. L., (1995). A phosphate transporter from the mycorrhizal fungus *Glomus versiforme*. *Nature, 378*, 626–629.
47. Xie, X., Lin, H., Peng, X., Xu, C., Sun, Z., Jiang, K., Huang, A., et al., (2016). Arbuscular mycorrhizal symbiosis requires a phosphate transceptor in the *Gigaspora margarita* fungal symbiont. *Mol. Plant, 9*, 1583–1608.
48. Maldonado-Mendoza, I. E., Dewbre, G. R., & Harrison, M. J., (2001). A phosphate transporter gene from the extra-radical mycelium of an arbuscular mycorrhizal fungus *Glomus intraradices* is regulated in response to phosphate in the environment. *Mol. Plant-Microbe Interact., 14*, 1140–1148.

49. Benedetto, A., Magurno, F., Bonfante, P., & Lanfranco, L., (2005). Expression profiles of a phosphate transporter gene (*GmosPT*) from the endomycorrhizal fungus *Glomus mosseae*. *Mycorrhiza, 15*, 620–627.
50. Fiorilli, V., Lanfranco, L., & Bonfante, P., (2013). The expression of *GintPT*, the phosphate transporter of *Rhizophagus irregularis*, depends on the symbiotic status and phosphate availability. *Planta, 237*, 1267–1277.
51. Ezawa, T., Cavagnaro, T. R., Smith, S. E., Smith, F. A., & Ohtomo, R., (2004). Rapid accumulation of polyphosphate in extraradical hyphae of an arbuscular mycorrhizal fungus as revealed by histochemistry and a polyphosphate kinase/luciferase system. *New Phytol., 161*, 387–392.
52. Kikuchi, Y., Hijikata, N., Yokoyama, K., Ohtomo, R., Handa, Y., Kawaguchi, M., Saito, K., & Ezawa, T., (2014). Polyphosphate accumulation is driven by transcriptome alterations that lead to near-synchronous and near-equivalent uptake of inorganic cations in an arbuscular mycorrhizal fungus. *New Phytol., 204*, 638–649.
53. Nguyen, C. T., & Saito, K., (2021). Role of cell wall polyphosphates in phosphorus transfer at the arbuscular interface in mycorrhizas. *Front. Plant Sci., 12*, 725939.
54. Tani, C., Ohtomo, R., Osaki, M., Kuga, Y., & Ezawa, T., (2009). ATP-dependent but proton gradient-independent polyphosphate-synthesizing activity in extraradical hyphae of an arbuscular mycorrhizal fungus. *Appl. Environ. Microbiol., 75*, 7044–7050.
55. Tisserant, E., Kohler, A., Dozolme-Seddas, P., Balestrini, R., Benabdellah, K., Colard, A., Croll, D., et al., (2012). The transcriptome of the arbuscular mycorrhizal fungus *Glomus intraradices* (DAOM 197198) reveals functional tradeoffs in an obligate symbiont. *New Phytol., 193*, 755–769.
56. Kikuchi, Y., Hijikata, N., Ohtomo, R., Handa, Y., Kawaguchi, M., Saito, K., Masuta, C., & Ezawa, T., (2016). Aquaporin-mediated long-distance polyphosphate translocation directed towards the host in arbuscular mycorrhizal symbiosis: Application of virus induced gene silencing. *New Phytol., 211*, 1202–1208.
57. Gao, X., Guo, H., Zhang, Q., Guo, H., Zhang, L., Zhang, C., Guo, Z., et al., (2020). Arbuscular mycorrhizal fungi (AMF) enhanced the growth, yield, fiber quality and phosphorus regulation in upland cotton (*Gossypium hirsutum* L.). *Sci. Rep., 10*, 2084.
58. Subramanian, K. S., Tenshia, V., Jayalakshmi, K., & Ramachandran, V., (2018). Role of arbuscular mycorrhizal fungus (*Glomus intraradices*) – (fungus aided) in zinc nutrition of maize. *Adv. Agric. Bio., 2*, 94–103.
59. Ingraffia, R., Amato, G., Frenda, A. S., & Giambalvo, D., (2019). Impacts of arbuscular mycorrhizal fungi on nutrient uptake, N_2 fixation, N transfer, and growth in a wheat/faba bean intercropping system. *PLoS One, 14*, e0213672.
60. Saia, S., Benítez, E., García-Garrido, J. M., Settani, L., Amato, G., & Giambalvo, D., (2014). The effect of arbuscular mycorrhizal fungi on total plant nitrogen uptake and nitrogen recovery from soil organic material. *J. Agric. Sci., 152*, 370–378.
61. Meng, L., Zhang, A., Wang, F., Han, X., Wang, D., & Li, A., (2015). Arbuscular mycorrhizal fungi and rhizobium facilitate nitrogen uptake and transfer in soybean/maize intercropping system. *Front. Plant Sci., 6*, 339.
62. Zhu, X., Song, F., Liu, S., & Liu, F., (2016). Arbuscular mycorrhiza improve growth, nitrogen uptake, and nitrogen use efficiency in wheat grown under elevated CO_2. *Mycorrhiza, 26*, 133–140.

63. Wang, S., Chen, A., Xie, K., Yang, X., Luo, Z., Chen, J., Zeng, D., et al., (2020). Functional analysis of the OsNPF4.5 nitrate transporter reveals a conserved mycorrhizal pathway of nitrogen acquisition in plants. *PNAS, 117*, 16649–16659.
64. Whiteside, M. D., Garcia, M. O., & Treseder, K. K., (2012). Amino acid uptake in arbuscular mycorrhizal plants. *PLoS One, 7*, e47643.
65. Bücking, H., & Kafle, A., (2015). Role of arbuscular mycorrhizal fungi in the nitrogen uptake of plants: Current knowledge and research gaps. *Agronomy, 5*, 587–612.
66. Garcia, K., Doidy, J., Zimmermann, S. D., Wipf, D., & Courty, P. E., (2016). Take a trip through the plant and fungal transportome of mycorrhiza. *Trends Plant Sci., 21*, 937–950.
67. Bukovská, P., Bonkowski, M., Konvalinková, T., Beskid, O., Hujslová, M., Püschel, D., Řezáčová, V., et al., (2018). Utilization of organic nitrogen by arbuscular mycorrhizal fungi—is there a specific role for protists and ammonia oxidizers? *Mycorrhiza, 28*, 269–283.
68. Tegeder, M., & Masclaux-Daubresse, C., (2018). Source and sink mechanisms of nitrogen transport and use. *New Phytol., 217*, 35–53.
69. Feng, H., Fan, X., Miller, A. J., & Guohua, X., (2020). Plant nitrogen uptake and assimilation: Regulation of cellular pH homeostasis. *J Exp. Bot., 71*, 4380–4392.
70. Govindarajulu, M., Pfeffer, P. E., Jin, H., Abubaker, J., Douds, D. D., Allen, J. W., Bücking, H., et al., (2005). Nitrogen transfer in the arbuscular mycorrhizal symbiosis. *Nature, 435*, 819–823.
71. Pérez-Tienda, J., Valderas, A., Camañes, G., García-Agustín, P., & Ferrol, N., (2012). Kinetics of NH^{4+} uptake by the arbuscular mycorrhizal fungus *Rhizophagus irregularis*. *Mycorrhiza, 22*, 485–491.
72. López-Pedrosa, A., González-Guerrero, M., Valderas, A., Azcón-Aguilar, C., & Ferrol, N., (2006). *GintAMT1* encodes a functional high-affinity ammonium transporter that is expressed in the extraradical mycelium of *Glomus intraradices*. *Fungal Genet. Biol., 43*, 102110.
73. Pérez-Tienda, J., Testillano, P. S., Balestrini, R., Fiorilli, V., Azcón-Aguilar, C., & Ferrol, N., (2011). *GintAMT2*, a new member of the ammonium transporter family in the arbuscular mycorrhizal fungus *Glomus intraradices*. *Fungal Genet. Biol., 48*, 1044–1055.
74. Tian, C., Kasiborski, B., Koul, R., Lammers, P. J., Bucking, H., & Shachar-Hill, Y., (2010). Regulation of the nitrogen transfer pathway in the arbuscular mycorrhizal symbiosis: Gene characterization and the coordination of expression with nitrogen flux. *Plant Physiol., 153*, 1175–1187.
75. Cappellazzo, G., Lanfranco, L., Fitz, M., Wipf, D., & Bonfante, P., (2008). Characterization of an amino acid permease from the endomycorrhizal fungus *Glomus mosseae*. *Plant Physiol., 147*, 429–437.
76. Belmondo, S., Fiorilli, V., Pérez-Tienda, J., Ferrol, N., Marmeisse, R., & Lanfranco, L., (2014). A dipeptide transporter from the arbuscular mycorrhizal fungus *Rhizophagus irregularis* is upregulated in the intraradical phase. *Front. Plant Sci., 5*, 436.
77. Pérez-Tienda, J., Corrêa, A., Azcón-Aguilar, C., & Ferrol, N., (2014). Transcriptional regulation of host NH^{4+} transporters and GS/GOGAT pathway in arbuscular mycorrhizal rice roots. *Plant Physiol. Biochem., 75*, 1–8.

78. Jin, H., Pfeffer, P. E., Douds, D. D., Piotrowski, E., Lammers, P. J., & ShacharHill, Y., (2005). The uptake, metabolism, transport and transfer of nitrogen in an arbuscular mycorrhizal symbiosis. *New Phytol., 168*, 687–696.
79. Mensah, J. A., Koch, A. M., Antunes, P. M., Kiers, E. T., Hart, M., & Bücking, H., (2015). High functional diversity within species of arbuscular mycorrhizal fungi is associated with differences in phosphate and nitrogen uptake and fungal phosphate metabolism. *Mycorrhiza, 25*, 533–546.
80. Yadav, R., Ror, P., Rathore, P., Kumar, S., & Ramakrishna, W., (2021a). *Bacillus subtilis* CP4, isolated from native soil in combination with arbuscular mycorrhizal fungi promotes biofortification, yield and metabolite production in wheat under field conditions. *J. Appl. Microbiol., 131*, 339–359.
81. Oliveira, T. C., Uehara, H. M., Silva, L. D., Tavares, G. G., Santana, L. R., Cabral, J. S. R., Souchie, E. L., & Mendes, G. C., (2019). Production of soybean in association with the arbuscular mycorrhizal fungi *Rhizophagus clarus* cultivated in field conditions. *Revi Cienc. Agrovet., 18*, 530–535.
82. Castillo-González, J., Ojeda-Barrios, D., Hernández-Rodríguez, A., González-Franco, A. C., Robles-Hernández, L., & López-Ochoa, G. R., (2018). Zinc metalloenzymes in plants. *Interciencia, 43*, 242–248.
83. Samreen, T., Humaira, Shah, H. U., Ullah, S., & Javid, M., (2017). Zinc effect on growth rate, chlorophyll, protein and mineral contents of hydroponically grown mungbeans plant (*Vigna radiata*). *Arab. J. Chem., 10*, S1802–S1807.
84. Khatun, M. A., Hossain, M. M., Bari, M. A., Abdullahil, K. M., Parvez, M. S., Alam, M. F., & Kabir, A. H., (2018). Zinc deficiency tolerance in maize is associated with the up-regulation of Zn transporter genes and antioxidant activities. *Plant Biol., 20*, 765–770.
85. Pandey, N., Pathak, G. C., & Sharma, C. P., (2006). Zinc is critically required for pollen function and fertilization in lentil. *J. Trace Elem. Med. Biol., 20*, 89–96.
86. Broadley, M., Brown, P., Cakmak, I., Rengel, Z., & Zhao, F., (2012). Function of nutrients: Micronutrients. In: Marschner, P., (ed.) *Marschner's Mineral Nutrition of Higher Plants* (pp. 191–248). Academic Press.
87. Cakmak, I., (2000). Possible roles of zinc in protecting plant cells from damage by reactive oxygen species. *New Phytol., 146*, 185–205.
88. McCall, K. A., Huang, C. C., & Fierke, C. A., (2000). Function and mechanism of zinc metalloenzymes. *J. Nutr., 130*, 1437S–1446S.
89. Broadley, M. R., White, P. J., Hammond, J. P., Zelko, I., & Lux, A., (2007). Zinc in plants. *New Phytol., 173*, 677–702.
90. Cakmak, I., (2008). Enrichment of cereal grains with zinc: Agronomic or genetic biofortification? *Plant Soil, 302*, 1–17.
91. Cakmak, I., & Kutman, U. B., (2018). Agronomic biofortification of cereals with zinc: A review. *Eur. J. Soil Sci., 69*, 172–180.
92. Plassard, C., & Dell, B., (2010). Phosphorus nutrition of mycorrhizal trees. *Tree Physiol., 30*, 1129–1139.
93. Almagrabi, O. A., & Abdelmoneim, T. S., (2012). Using of arbuscular mycorrhizal fungi to reduce the deficiency effect of phosphorous fertilization on maize plants (*Zea mays* L.). *Life Sci. J., 9*, 1648–1654.

94. Watts-Williams, S. J., Smith, F. A., McLaughlin, M. J., Patti, A. F., & Cavagnaro, T. R., (2015). How important is the mycorrhizal pathway for plant Zn uptake? *Plant Soil, 390*, 157–166.
95. Coccina, A., Cavagnaro, T. R., Pellegrino, E., Ercoli, L., McLaughlin, M. J., & Watts-Williams, S. J., (2019). The mycorrhizal pathway of zinc uptake contributes to zinc accumulation in barley and wheat grain. *BMC Plant Biology, 19*, 133.
96. Basu, A., Prasad, P., Das, S. N., Kalam, S., Sayyed, R. Z., Reddy, M. S., & El Enshasy, H., (2021). Plant growth promoting rhizobacteria (PGPR) as green bioinoculants: Recent developments, constraints, and prospects. *Sustainability, 13*, 1140.
97. Subramanian, K. S., Bharathi, C., & Jegan, A., (2008). Response of maize to mycorrhizal colonization at varying levels of zinc and phosphorus. *Biol. Fertil. Soils, 45*, 133–144.
98. Lehmann, A., Veresoglou, S. D., Leifheit, E. F., & Rillig, M. C., (2014). Arbuscular mycorrhizal influence on zinc nutrition in crop plants – a meta-analysis. *Soil Biol. Biochem., 69*, 123–131.
99. Watts-Williams, S. J., & Cavagnaro, T. R., (2018). Arbuscular mycorrhizal fungi increase grain zinc concentration and modify the expression of root ZIP transporter genes in a modern barley (*Hordeum vulgare*) cultivar. *Plant Science, 274*, 163–170.
100. Ghasemi-Fasaei, R., Jarrah, M., & Mayel, S., (2012). Dynamics of manganese adsorption onto highly calcareous soils of Southern Iran. *Int. J. Agric. Crop Sci., 4*, 1676–1680.
101. Jansa, J., Mozafar, A., & Frossard, E., (2003). Long–distance transport of P and Zn through the hyphae of an arbuscular mycorrhizal fungus in symbiosis with maize. *Agronomie, 23*, 481–488.
102. Guo, Y., George, E., & Marschner, H., (1996). Contribution of an arbuscular mycorrhizal fungus to the uptake of cadmium and nickel in bean and maize plants. *Plant Soil, 184*, 195–205.
103. Müller, C., Kuki, K. N., Pinheiro, D. T., De Souza, L. R., Silva, S. A. I., Loureiro, M. E., Oliva, M. A., & Almeida, A. M., (2015). Differential physiological responses in rice upon exposure to excess distinct iron forms. *Plant Soil, 391*, 123–138.
104. Tiwari, A., Mamedov, F., Grieco, M., Suorsa, M., Jajoo, A., Styring, S., Tikkanen, M., & Aro, E. M., (2016). Photodamage of iron-sulfur clusters in photosystem I induces non-photochemical energy dissipation. *Nat Plants, 2*, 16035.
105. Morrissey, J., & Guerinot, M. L., (2009). Iron uptake and transport in plants: The good, the bad, and the ionome. *Chem Rev., 109*, 4553–4567.
106. Jeong, J., & Connolly, E. L., (2009). Iron uptake mechanisms in plants: Functions of the FRO family of ferric reductases. *Plant Sci., 176*, 709–714.
107. McCluskey, J., Herdman, L., & Skene, K. R., (2004). Iron deficiency induces changes in metabolism of citrate in lateral roots and cluster roots of *Lupinus albus*. *Physiol. Plant., 121*, 586–594.
108. Valipour, M., Khoshgoftarmanesh, A. H., & Baninasab, B., (2018). Physiological responses of hawthorn (*Crataegus persica* Pojark.) and quince (*Cydonia oblonga* Mill.) rootstocks to bicarbonate-induced iron deficiency in nutrient solution. *J. Plant Nutr. Soil Sci., 181*, 905–913.
109. Curie, C., Cassin, G., Couch, D., Divol, F., Higuchi, K., Le Jean, M., Misson, J., et al., (2009). Metal movement within the plant: Contribution of nicotianamine and yellow stripe 1-like transporters. *Ann. Bot., 103*, 1–11.
110. Conte, S. S., & Walker, E. L., (2011). Transporters contributing to iron trafficking in plants. *Mol. Plant., 4*, 464–476.

111. Kobayashi, T., & Nishizawa, N. K., (2012). Iron uptake, translocation, and regulation in higher plants. *Annu Rev Plant Biol., 63*, 131–152.
112. Guerinot, M. L., & Yi, Y., (1994). Iron: Nutritious, noxious, and not readily available. *Plant Physiol., 104*, 815–820.
113. Pereira, E. G., Oliva, M. A., Siqueira-Silva, A. I., Rosado-Souza, L., Pinheiro, D. T., & Almeida, A. M., (2014). Tropical rice cultivars from lowland and upland cropping systems differ in iron plaque formation. *J. Plant Nutr., 37*, 1373–1394.
114. Müller, C., Silveira, S. F. D. S., Daloso, D. M., Mendes, G. C., Merchant, A., Kuki, K. N., Oliva, M. A., et al., (2017). Ecophysiological responses to excess iron in lowland and upland rice cultivars. *Chemosphere, 189*, 123–133.
115. Clark, R. B., & Zeto, S. K., (1996). Iron acquisition by mycorrhizal maize grown on alkaline soil. *J. Plant Nutr., 19*, 247–264.
116. Caris, C., Hördt, W., Hawkins, H. J., Römheld, V., & George, E., (1998). Studies of iron transport by arbuscular mycorrhizal hyphae from soil to peanut and sorghum plants. *Mycorrhiza, 8*, 35–39.
117. Medeiros, C. A. B., Clark, R. B., & Ellis, J. R., (1993). Effects of MES [2(N-morpholino)-ethane sulfonic acid] and pH on mineral nutrient uptake by mycorrhizal and nonmycorrhizal maize. *J. Plant Nutr., 16*, 2255–2272.
118. Liu, A., Hamel, C., Hamilton, R. I., Ma, B. L., & Smith, D. L., (2000). Acquisition of Cu, Zn, Mn and Fe by mycorrhizal maize (*Zea mays*) growth in soil at different P and micronutrient levels. *Mycorrhiza, 9*, 331–336.
119. Kobae, Y., Tomioka, R., Tanoi, K., Kobayashi, N. I., Ohmori, Y., Nishida, S., & Fujiwara, T., (2014). Selective induction of putative iron transporters, *OPT8a* and *OPT8b*, in maize by mycorrhizal colonization. *Soil Sci. Plant Nutr., 60*, 843–847.
120. Rahimi, S., Baninasab, B., Talebi, M., Gholami, M., & Zarei, M., (2021). Arbuscular mycorrhizal fungi inoculation improves iron deficiency in quince via alterations in host root phenolic compounds and expression of genes. *Sci. Hortic., 285*, 110165.
121. Kabir, A. H., Debnath, T., Das, U., Prity, S. A., Haque, A., Rahman, M. M., & Parvez, Md. S., (2020). Arbuscular mycorrhizal fungi alleviate Fe-deficiency symptoms in sunflower by increasing iron uptake and its availability along with antioxidant defense. *Plant Physiol. Biochem., 150*, 254–262.
122. Yadav, R., Ror, P., Rathore, P., & Ramakrishna, W., (**2020b**).; Bacteria from native soil in combination with arbuscular mycorrhizal fungi augment wheat yield and biofortification. *Plant Physiol. Biochem., 150*, 222–233.
123. Cohu, C. M., & Pilon, M., (2010). Cell biology of copper. In: Rüdiger, H., & Mendel, R. R., (eds.) *Cell Biology of Metals and Nutrients Plant Cell Monographs* (pp. 55–74). Heidelberg, Springer.
124. Printz, B., Lutts, S., Hausman, J. F., & Sergeant, K., (2016). Copper trafficking in plants and its implication on cell wall dynamics. *Front. Plant Sci., 7*, 601.
125. Al-Karaki, G., McMichael, B., & Zak, J., (2004). Field response of wheat to arbuscular mycorrhizal fungi and drought stress. *Mycorrhiza, 14*, 263–269.
126. Kothari, S. K., Marschner, H., & Römheld, V., (1990). Direct and indirect effects of VA mycorrhizal fungi and rhizosphere microorganisms on acquisition of mineral nutrients by maize (*Zea mays* L.) in a calcareous soil. *New Phytologist, **116***, 637–645.
127. Tisdale, S. L., Nelson, W. L., Beaton, J. D., & Havlin, J. L., (1993). *Soil Fertility and Fertilizers* (5th edn., p. 486). MacMillan Publishing Company, New York.

128. Tamayo, E., Gómez-Gallego, T., Azcón-Aguilar, C., & Ferrol, N., (2014). Genome-wide analysis of copper, iron and zinc transporters in the arbuscular mycorrhizal fungus *Rhizophagus irregularis*. *Front. Plant Sci., 5*, 547.
129. Rahman, M. M., Hossain, K. F. B., Banik, S., Sikder, M. T., Akter, M., Bondad, S. E. C., Rahaman, Md. S., et al., (2019). Selenium and zinc protections against metal-(loids)-induced toxicity and disease manifestations: A review. *Ecotoxicol. Environ. Safety, 168*, 146–163.
130. Durán, P., Acuña, J. J., Jorquera, M. A., Azcón, R., Borie, F., Cornejo, P., & Mora, M. L., (2013). Enhanced selenium content in wheat grain by co-inoculation of selenobacteria and arbuscular mycorrhizal fungi: A preliminary study as a potential Se biofortification strategy. *J. Cereal Sci., 57*, 275–280.
131. Chen, X., Zhang, Z., Gu, M., Li, H., Shohag, M. J. I., Shen, F., Wang, X., & Wei, Y., (2020). Combined use of arbuscular mycorrhizal fungus and selenium fertilizer shapes microbial community structure and enhances organic selenium accumulation in rice grain. *Sci. Total Environ., 748*, 141166.
132. Govasmark, E., & Salbu, B., (2011). Translocation and re-translocation of selenium taken up from nutrient solution during vegetable growth in spring wheat. *J Sci. Food Agr., 91*, 1367e1372.
133. Hawkesford, M., & Zhao, F. J., (2007). Strategies for increasing the selenium content in wheat. *J. Cereal Sci., 46*, 282e292.
134. Dungan, R., Yates, S., & Frankenberger, W., (2003). Transformations of selenate and selenite by *Stenotrophomonas maltophilia* isolated from a seleniferous agricultural drainage pond sediment. *Environ. Microbiol., 5*, 287e295.
135. Acuña, J., Jorquera, M., Barra, P., Crowley, D., & Mora, M. L., (2013). Selenobacteria selected from the rhizosphere as a potential tool for Se biofortification of cereal crop. *Biol. Fertil. Soils, 49*, 175–185.
136. Sun, X., Chen, W., Ivanov, S., Maclean, A. M., Wight, H., Ramaraj, T., Mudge, J., et al., (2019). Genome and evolution of the arbuscular mycorrhizal fungus *Diversispora epigaea* (formely *Glomus versiforme*) and its bacterial endosymbionts. *New Phytol., 221*, 1556–1573.
137. Pellegrino, E., & Bedini, S., (2014). Enhancing ecosystem services in sustainable agriculture: Biofertilization and biofortification of chickpea (*Cicer arietinum* L.) by arbuscular mycorrhizal fungi. *Soil Biology and Biochemistry, 68*, 429–439.
138. Suganya, A., Saravanan, A., Baskar, M., Pandiyarajan, P., & Kavimani, R., (2020). Agronomic biofortification of maize (*Zea mays* L.) with zinc by using of graded levels of zinc in combination with zinc solubilizing bacteria and arbuscular mycorrhizal fungi. *J. Plant Nutr., 44*, 988–994.
139. Tran, B. T. T., Cavagnaro, T. R., & Watts-Williams, S. J., (2019). Arbuscular mycorrhizal fungal inoculation and soil zinc fertilization affect the productivity and the bioavailability of zinc and iron in durum wheat. *Mycorrhiza, 29*, 445–457.
140. Wang, M. Y., Xia, R. X., Hu, L. H., Dong, T., & Wu, Q. S., (2007). Arbuscular mycorrhizal fungi alleviate iron deficient chlorosis in *Poncirus trifoliata* L. Raf under calcium bicarbonate stress. *J. Hortic. Sci. Biotechnol., 82*, 776–780.
141. Watts-William, S. J., & Gilbert, S. E., (2020). Arbuscular mycorrhizal fungi affect the concentration and distribution of nutrients in the grain differently in barley compared with wheat. *New Phytol., 3*, 567–577.

142. Antoniou, O., Chrysargyris, A., Xylia, P., & Tzortzakis, N., (2021). Effects of selenium and/or arbuscular mycorrhizal fungal inoculation on strawberry grown in hydroponic trial. *Agronomy, 11*, 721.
143. Hart, M., Ehret, D. L., Krumbein, A., Leung, C., Murch, S., Turi, C., & Franken, P., (2015). Inoculation with arbuscular mycorrhizal fungi improves the nutritional value of tomatoes. *Mycorrhiza, 25*, 359–376.
144. Ercoli, L., Schüßler, A., Arduini, I., & Pellegrino, E., (2017). Strong increase of durum wheat iron and zinc content by field-inoculation with arbuscular mycorrhizal fungi at different soil nitrogen availabilities. *Plant Soil, 419*, 153–167.
145. Watts-Williams, S. J., Cavagnaro, T. R., & Tyerman, S. D., (2018). Variable effects of arbuscular mycorrhizal fungal inoculation on physiological and molecular measures of root and stomatal conductance of diverse *Medicago truncatula* accessions. *Plant Cell Environ., 42*, 285–294.
146. Schüßler, A., & Walker, C., (2015). *The Glomeromycota: A Species List with New Families and Genera.* The Royal Botanic Garden, Edinburgh and Kew, UK.
147. Kaur, S., & Suseela, V., (2020). Unraveling arbuscular mycorrhiza-induced changes in plant primary and secondary metabolome. *Metabolites, 10*, 335.
148. Omotayo, O. P., & Babalola, O. O., (2021). Resident rhizosphere microbiome's ecological dynamics and conservation: Towards achieving the envisioned sustainable development goals, a review. *Int. Soil Water Conserv. Res., 9*, 127–142.

CHAPTER 12

Stable Crops Biofortified with Increased Protein, Oils, and Vitamins

MAJID ABDOLI[1,2]

[1]*Department of Plant Production and Genetics, Faculty of Agriculture, University of Maragheh, Maragheh, Iran*

[2]*Nomads Affairs Organization of Iran (NAOI), Kermanshah, Iran*

ABSTRACT

People of poor are more exposed to risks of hidden hunger and malnutrition due to intake of carbohydrate-rich but deficient of micronutrient plant-based food. The development of high-yielding cultivars but by poor micronutrients further intensified the undernourishment. Several interventions have been used to increase dietary intake of nutrients such as dietary variation, clinical or pharmaceutical supplementation, industrial fortification (is the deliberately increasing the content of vitamins and essential macro- and micro-nutrients in a food), and crop biofortification (is a method by which the nutritional value of crops can be increased with the help of breeding and agronomic practices) are the major ways to get rid of these macro- and micro-nutrient malnutrition. Among them, crop biofortification is only one of the most appropriate, sustainable, and cost-effective ways to cater to the nutrient-deficient population of developing countries. Also, biofortification of staple crops has been proposed as a strategy to address micronutrient undernourishment, particularly with respect to insufficient intake of iron (Fe), zinc (Zn), vitamin A, and folate. Biofortified wheat, rice, maize, and other different crops biofortified with essential micronutrients as well as high levels of provitamin A, amino acids (such as tryptophan and lysine),

Phytoremediation and Biofortification: Strategies for Sustainable Environmental and Health Management.
Nand Kumar Singh, Shadma Afzal, & Tariq Aftab (Eds.)

protein and oils have great potential to benefit the world's poor people, in terms of both economics and health.

12.1 INTRODUCTION

Organic fertilizers (OFs) management in terms of environmental effect and crop yield is important, especially in regions of arid and semiarid. Biofortification is an agricultural process of enhancing the micronutrients, proteins, oils, vitamins, and other content of a food crop through 1 of 3 main non-mutually exclusive agronomic ways: (i) plant breeding; (ii) genetic engineering which includes genetic modification and transgenesis; and (iii) use of enriched fertilizers in soil and leaves or management of agricultural practices. Biofortified crops may be produced using different agricultural practices. Such techniques make it possible to enhancement not only the yield but also the nutritional value of different crops.

Given that more than 3 billion people in the world suffer from micronutrient deficiencies [1], so the need to enrich cereals, legumes, and oilseeds requires that strategies to optimize production and quality of these products be evaluated. Farmer's utilization one of the common fertilization ways including soil use through irrigation, foliar (leaf solubility), and seed mixing. Soil consumption of macro- and micro-nutrients, in addition to the low efficiency of their uptake by the plant also very costly and therefore alternative methods such as foliar spraying can be used. On the other hand, most of the world's soils such as Asia, have a high calcareous nature and pH, low organic matter (OM), and micronutrients solubility in these soils is low [2, 3]. So that foliar is one of the fastest methods to meet the needs. The fertilizer utilization (foliar) is also used in this way to save fertilizer and in addition to the positive economic aspect; the environment is protected from pollution of chemicals, which is very effective in achieving sustainable agriculture. In this connection, Esfandiari et al. [4] reported that selecting fertilizers and combining them with each other and with non-root feeding (sprinkling leaf with fertilizer solutions) had a strong effect on both yield and quality of grain bread wheat. Also, Abdoli et al. [5] reported that the Zn foliar application was more effective in increasing concentrations of Fe, Zn, and Mn as well as content ascorbic acid (AsA) in grain of wheat. They stated that the various Zn treatments ways significantly increased concentration of Zn and decreased content of phytate, as well as decreased phytate/Zn molar ratios in grain of wheat. In other hand, Rajaie & Charkhandeh [6] stated that the simultaneous use of iron (Fe), zinc (Zn), and manganese (Mn) could significantly enhancement the grain yield and its components (such as 1,000 grain weight, number of grains per ear, harvest index, and biological yield), soluble carbohydrates in stem, and grain protein of barley. The highest Fe, Zn, and Mn concentrations in leaf and grain were observed when each one of the nutrients was individually sprayed on aerial parts of plant. In research of Hedayatpour et al. [7], foliar application of Fe and Zn for amount 2 and 4 g L^{-1} at stem elongation + early flowering stages caused an increase of grain Fe, grain Zn, grain protein percent, grain oil percent, and oil yield in canola. Pourjamshid et al. [8] reported that the micronutrients foliar application of Zn, Fe, and Mn increased significantly the measured traits of wheat (such as grain yield and its components as well as protein content in grain) in both terminal drought stress and non-stress conditions. Meanwhile, the use of Zn spray showed the greatest impact in reducing the damage caused by drought stress on these traits.

The application of micronutrient fertilizers is an effective strategy for incrementing the yield and quality properties of different crops. Nowadays, in addition to the quantity of production, its quality and stability are taken into account. In the following, the impacts of the use of macro- and micro-element fertilizers on the quality performance of crops, including elements contained in grains, proteins, and oils, are discussed.

12.2 MACRO- AND MICRO-NUTRIENTS

The micro- and macro-nutrients have a main role in plant nutrition. Zinc and Fe are main micronutrients in the growth of plant and their deficiency in plants, especially in cereals, is a major nutritional problem globally. Iron as a micronutrient is critical for formation of chlorophyll, photosynthesis, and respiration as well as its main in the enzyme systems of plants. Zinc is required as a structural and functional component of many enzymes and proteins as well as increment yield and its components of crops. Manganese is a micronutrient that is necessary for photosynthesis, nitrogen (N) metabolism and to form other compounds required for metabolism of plant [9]. Boron (B) plays a main role in the cell wall structure, cytoplasmic membrane stability, and possibly glucose transport and carbohydrate metabolism [10]. The results of research by various researchers on the impact of micronutrients on the content of nutrients in grain at different crops are presented in Table 12.1. In this connection, Ebrahimi et al. [11] reported that the foliar application nutrients (Zn, Fe, and mixed Zn + Fe) had a considerable enhancement in the concentration of Zn and Fe in grain as well as protein of grain and grain yield of wheat. In general, foliar Zn and Fe application caused an increment Zn and Fe concentration of grain in compared to control (39.85 and 89.43%, respectively). Moreover, the most amounts of carbohydrate percentage, Zn, and Mn in grain of mung bean were achieved from treatments of micronutrients (Zn + Mn) foliar application. Soleimani et al. [12] indicated that the impact of foliar Zn and Mn application on the concentration of Zn in grain of safflower was significant, leading to increasing 55.9 and 59.6%, respectively. Also, Jafar Dokht et al. [13] reported that the combined foliar use of Zn + Mn led to increasing 3.15% of the percentage of N in grain, 39.6% of the content of Zn, and 96.7% of the content of Mn in the grain of mung bean. But these micronutrients use led to decreasing phosphorus (P) uptake in the plant [13]. This may be due to the effect of Mn antagonist on P uptake. Moradi Telavat et al. [14] in research on safflower cultivars (*Carthamus tinctorius* L.) reported that the zinc sulfate ($ZnSO_4$) spraying had a considerable impact on content of Zn and P in seed, oil yield, seed yield and its components. Foliar $ZnSO_4$ in various growth stages increased seed Zn content, however, content of P in seed decreased. Abdoli et al. [5], Zeidan et al. [15] and Javanmard & Asadi Danalo [16] also pointed to the decrement in grain P concentration due to the use of Zn. Excessive consumption of this element (zinc) can reduce the growth of shoots and roots, reduce the absorption of other nutrients such as P and increase the absorption of Fe [17]. Application of Zn in Zn-deficit soils reduces P uptake from the root, thus reducing its concentration in the grain. This reduces the amount of phytate in the grain and thus reduces the ratio of phytate to Zn in grain with the use of Zn [9].

Mostafavi Rad et al. [18] research on some winter rapeseed (*Brassica napus* L.) cultivars (such as Zarfam, Okapi, Modena, and Licord) as affected by micronutrient elements, showed that the Okapi cultivar had the greatest grain yield as affected by combined application of Mn + Zn (4,011 kg ha^{-1}) treatment. In this research, Licord cultivar showed the greatest grain Zn content at combined application of Mn + Fe treatment. The highest grain copper (Cu) content was obtained in Modena cultivar and combined application of Zn + Fe treatment. The highest grain Fe content was shown in Modena cultivar, Fe, and combined application of Zn + Fe treatments. Also, the greatest grain Mn content was obtained in Modena cultivar and combined application of Mn + Zn treatment. Other researchers have suggested an increase content of micronutrients such as Fe in the grain of bread wheat, sorghum, and lentil [19–22], Zn in the grain of sunflower [23], and Fe and Zn in the grain of red bean [24] due to the application of these nutrients. The content of micronutrients in the grain depends on: (i) the amount of uptake of these elements by the root (soil) or leaf (foliar application) during the grain development phase; and (ii) the re-transfer of these elements from the plant tissue to the grain through the phloem [2].

Panjtandoust et al. [25] reported that the kernel oil in both ways of Fe application (soil and foliar applications) decreased notably whereas seed protein and also essential minerals such as active Fe, total Fe, Mg, Ca, K, P, and N of peanut (*Arachis hypogaea* L.) enhanced significantly. In both ways of Fe application, higher doses of Fe applications (20 kg ha^{-1} and 4 g L^{-1} of Fe-EDDHA) had considerable positive impacts on the content of essential elements. Altogether, Fe foliar-applied minerals had less effects compared to Fe soil application. A similar result was obtained in soybean plant, which increased the amount of total Fe (up to 50%) by consuming Fe-EDDHA [26]. On the other hand, Fe foliar application increases the nitrogen (N) and potassium (K) content of bread wheat [16] and peanut [27]. Hossenabadi et al. [28] also in their research on wheat concluded that zinc sulfate ($ZnSO_4$), manganese sulfate ($MnSO_4$) and iron sulfate ($FeSO_4$) increased the amount of K in the grain by 0.68, 0.67, and 0.66% compared to control treatment, respectively.

Khiavi et al. [29] reported that the concomitant application of B + Zn led to increment in their amounts in seed that is important from rapeseed fortification view of point. Therefore, regarding to beneficial impacts of B on grain yield and concomitant application of Zn + B on quality of grain simultaneous spraying of Zn sulfate ($ZnSO_4$) and boric acid (H_3BO_3) with 3 g L^{-1} and 2 g L^{-1}, respectively, in stages of early stemming and flowering may be recommended for the similar regions.

Khoram Ghahfarokhi et al. [30] indicated that application of humic acid (HA) had a considerable impact on the Zn, N, and P content as well as grain yield of safflower (*Carthamus tinctorius* L.). On the other hand, foliar application of compost tea significantly influenced grain yield and content of Zn and Cu in grain of safflower. It seems soil application of HA by amount of 1,500 kg ha^{-1} along with foliar application of compost tea was the best treatment for producing grain yield and greater absorption of nutrition in safflower plant.

12.3 PROTEIN AND AMINO ACIDS

Various studies have been done on the impact of micronutrients on grain protein. For example, Kamaraki & Galavi [31] reported that the maximum grain yield, 1,000 grain weight, and plant height were relevant to Fe treatment and the utmost number of fertile capitols per plant and protein percentage were relevant to Zn treatments and the superlative oil percentage was relevant to joint foliar application of Fe + Zn. They stated that the micronutrients application had positive impacts on qualitative and quantitative traits of safflower (*Carthamus tinctorius* L.). Takkar & Nayar [32] reported that the use of Zn increases the amino acids lysine and histidine in wheat. Application of $ZnSO_4$ caused an enhancement in content of all amino acids (except isoleucine) in grain of wheat. Treatments of Zn application by different doses and different stages increased the total amino acids by 4.55 and 8.18%, respectively compared to no Zn situation (Figure 12.1). Ning et al. [33] reported that the important Zn enrichment in flour paralleled to the increment of amino acid profiles such as lysine, glycine, glutamate, and aspartate, but not to the content of total amino acid in grain of wheat. In plants, cysteine, histidine, aspartate, glutamate, and methionine are considered as substantial Zn ligands in structure of protein [34], which potentially determines the Zn storage in seeds. Afshani et al. [35]; and Habibi et al. [36] reported that the maximum protein (23.6 and 24.62%, respectively) of rapeseed was obtained from Zn fertilizer (compared with B and S). Sulfur (S) is part of the structure of the amino acids' cysteine and methionine and therefore, is part of the structure of the protein. Both of these amino acids (cysteine and methionine) are needed to make other S compounds in coenzymes and plant by-products. Also, Javanmard & Asadi Danalo [16] stated that the combined application of OFs and micronutrients increased N percentage and content of protein in grain. Integrated application of poultry manure + $ZnSO_4$ + $MnSO_4$ + $FeSO_4$ increased content of K in grain, content protein in grain, protein yield, grain yield by 41.86, 6.38, 54.24, and 44.86%, respectively over the no application of OFs and micronutrients (control condition). The highest grain P content was achieved by the $FeSO_4$ application, followed by combined application of poultry manure + $FeSO_4$ in wheat under rainfed conditions.

Increasing the percentage of protein under $MnSO_4$ and $ZnSO_4$ spraying conditions can be attributed to the indirect effect of micronutrients in increasing N uptake, i.e., the plant has more and optimal use of soil N by having micronutrients available, which led to increased protein production. On the other hand, micronutrients such as Mn participate in the structure of some proteins as well as in protein metabolism, and in this way may also increase the amount of grain protein. Zinc is an element that plays an essential role in protein production. Zinc deficiency can be the main cause of decreased RNA-polymerase activity and reduced amino acids and thus reduced protein synthesis [9]. The results of research by various researchers on the impact of micronutrients on grain protein in various crops are presented in Table 12.1. In this connection, Ebrahimi et al. [11] stated the lowest (7.87%) and highest (11.45%) grain protein of wheat obtained from control (no micronutrients application) and mixed consumption of Zn + Fe, respectively. It has been reported that foliar Zn application at a rate of 2.4 kg Zn ha^{-1} in canola increased the grain protein content (GPC) by 6% compared to the control [37]. Other researchers have pointed to the positive effects of the combined use of Fe and Zn on grain protein of bread wheat [11], lentil [22], safflower [31], barley [38], and canola or rapeseed [39, 40]. Pourgholam et al. [41] reported with foliar Fe application in rapeseed that the amount of active Fe in the plant and the activity of NR enzyme, which is essential for the amino acid cycle, increase. As a result, grain protein is increased by using of Fe. Payandeh et al. [40] reported that with increasing micronutrient amounts, grain yield, leaf area index, protein yield, content of protein in grain, and percentage of Fe and Zn enhancement in rapeseed plants. Utmost of these traits were obtained from control irrigation (optimal) and spraying of 6 g L^{-1}, which did not have a considerable difference with the treatment of 4 g L^{-1}. In explaining the reason for the increment of the content protein in grains by the micronutrient's application on Zn and Fe, it should be noted that these nutrients directly contribute to both the process of synthesis of protein and gene expression. The Zn deficiency prevents the activation of several of antioxidant enzymes, which results in oxidative harm to protein molecules, which also decreases function of protein in plant. In this research, spraying of micronutrient decreased the damage caused by water stress to grain yield and also increased the percentage of Fe and Zn grains. On the other hand, Pourjamshid et al. [8] stated that the application of micronutrients, especially Zn, as a foliar application, can decrease the harmful impacts caused by terminal water stress and improve the agronomic characters, physiological traits, and content of protein in grain of bread wheat in the Ramhormoz region of Iran. Joorabi et al. [42] in research on soybean stated that in normal conditions, nano-Zn chelate treatment (containing 20% Zn absorbable ion) has no effect on protein yield, but in water deficit situations, foliar nano-Zn chelate application increased the mentioned trait by 39.3%. The researchers said that protein yield was a function of content of protein and grain yield and those changes in protein yield were primarily related to grain yield [43].

Some researchers believe that the type or quality of stored proteins is far more effective than their amount, but most studies have shown that both the quality and quantity of these proteins are related to the quality of wheat bread. Wheat storage proteins are prolamins that give the dough a special elastic, resilient, and sticky state. These proteins improve the quality properties of the dough to produce the final product of bread flour [44]. In a study, Seyedi et al. [45] reported that the highest amount of protein in grain, percentage of protein, N percentage, and P percentage on the grain of bread wheat were obtained from 100% (N + P) fertilizer application. Kavian Athar & Aboutalebian [46] reported that the phosphate placement at 30 kg ha^{-1} of N incremented grain protein by 20.7%, respectively, and also increased the concentration of P in seed by 34%. Foliar $ZnSO_4$ application increased the concentration of Zn and protein in seed by 13% and 16%, respectively, compared to soil broadcasted application. In general, application of 30 kg ha^{-1} N starter with phosphate placement + foliar $ZnSO_4$ spraying is an appropriate treatment to improve the quality and quantity of common bean (*Phaseolus vulgaris* L.) seeds. Nitrogen (N) and P chemical fertilizers increase the amount of N entering the grain from the vegetative parts compared to carbohydrates and increase the grain N concentration and protein percentage [47].

Barat Zadeh et al. [48] stated that the highest protein percentage was obtained by spraying with 4 L ha^{-1} of nano-potassium fertilizer and 30 mM AsA with 22.02 and 23.57%, respectively and the lowest one was in control. In general, spraying with 4 L ha^{-1} of nano-potassium fertilizer and 30 mM of AsA increases the quantitative and qualitative yields of cowpea.

The amount of sediment SDS has a positive correlation with the quality of the bakery [49]. In the study of Eisvand et al. [20], Shiraz & Marvdasht cultivars of bread wheat responded better to its foliar application in terms of sediment SDS and had more sediment than non-foliar application (control treatment). So, the quality of the bakery increases with the application of Fe foliar application.

12.4 OILS AND FATTY ACIDS (FAS)

Crops of oil could be considered as a valuable resource for providing the plant oil for cooking. Various research has been done on the effect of micronutrients such as Fe and Zn on seed oil. For example, Rezaei Chiyaneh et al. [50] reported that the maximum oil percentage (21.66%) was produced by foliar application of Mn as well as utmost oil yield (124.20 kg ha^{-1}) was produced by concomitant use of Fe + Zn + Mn treatment. So that oil yield and grain yield increased by 44.18 and 31.27% yields more than control in pot marigold (*Calendula officinalis* L.), respectively. Also, Farajzadeh Memari Tabrizi [51] reported that the maximum oil grain and oil yield of amaranth, respectively (33.59%) and (6.083 g $crop^{-1}$) if a foliar application with Zn by

7 per 1,000 concentrations were obtained. She stated that the maximum of the grain yield, oil content, and oil yield was achieved by foliar application 7 per 1,000 this element (Zn) as well as 7 per 1,000 microelements (Fe + Zn + Mn) concentration of protein content increased.

Canola or rapeseed is one of the important oils crops the third oil crop after soybean and oil palm. Omidian et al. [52] reported that foliar use considerably affected grain yield and its components as well as content of oil and protein in seed of rapeseed (*Brassica napus* L.). Foliar $ZnSO_4$ use at elongation of stem and flowering stages increased grain yield by 11% and 17%, respectively. Also, content of oil and protein in grain were increased by the use of $ZnSO_4$. The highest content of protein and oil (27.3 and 35.1%, respectively) of rapeseed were obtained from the use of $ZnSO_4$ at the flowering stage when compared to non-use of micronutrient treatment. Jashni et al. [53] reported that the greatest oil percent (48.47%) was also revealed at Thiobacillus bacteria with integrated application of Zn and Fe (2 cc L^{-1}), while the lowest oil percent (31%) of rapeseed (*Brassica napus* L.) was obtained at control treatment. Kalantar Ahmadi & Shoushi Dezfouli [54] reported that the highest oil content (45.6%) of canola (*Brassica napus* L.) measured in full irrigation and application of Zn (dose of 0.006) and B (dose of 0.006). Hedayatpour et al. [7] in their research on canola stated that the considerable interaction between Zn and Fe foliar use was observed for grain protein percentage, oil yield, and grain yield. Utmost oil yield (2124.5 kg ha^{-1}) and grain protein percentage (37.7%) was obtained from foliar use of Zn-Fe pair combination at 4 g L^{-1}. Increasing Zn from 0 to 4 g L^{-1} considerably increased grain oil percentage (from 38.67 to 43.17%) and concentration of Zn in grain (from 22.07 to 33.28 mg kg^{-1}). Increasing Fe from 0 to 4 g L^{-1} significantly increased grain Fe concentration (from 64.29 to 79.18 mg kg^{-1}). Ultimately, the treatment of the twice foliar application of $ZnSO_4$ and $FeSO_4$ (at first in elongation of stem and then second in early flowering) and at 4 g L^{-1} was suggested to get more quality and quantity of canola (*cv.* Talaye) in Zarghan region of Iran. Also, Habibi et al. [36] reported that the maximum and minimum oil content of rapeseed (*cv.* Hyola 401) obtained from combined application of Zn + B + S (42.58%) and non-application nutrients (38.37%). It has been reported that foliar application of Zn at a rate of 2.4 kg Zn ha^{-1} in canola increased the grain oil content by 9% compared to the control [37].

Researchers reported that rapeseed is a very sensitive plant to Zn deficiency, and the use of Zn fertilizers containing $ZnSO_4$ increased grain yield and seed oil percentage, which is sometimes accompanied by a decrease in

TABLE 12.1 Effect of Different Application of Nutrients on the Changes of Concentration of Macro- and Micro-Nutrients in Grain, Grain Protein Percent, Protein Yield, Grain Oil Percent, and Oil Yield in Different Crops

Method of Application	Nutrient Type	Application of Nutrients at Rates	Time of Application	Increase of Oil Over Untreated Control (%)										Crop type	References
				Grain Nitrogen (N)	Grain Phosphorus (P)	Grain Potassium (K)	Grain Iron (Fe)	Grain Zinc (Zn)	Grain Manganese (Mn)	Grain Protein Percent	Protein Yield	Grain Oil Percent	Oil yield		
Soil	Fe-EDDHA	5 kg ha^{-1}	Sowing	14.4	2.8	0.0	15.4	–	–	6.4	–	–2.2	–	Peanut (variety NC_2)	[25]
Soil	Fe-EDDHA	10 kg ha^{-1}	Sowing	13.3	3.7	18.0	65.4	–	–	10.2	–	–3.1	–	Peanut (variety NC_2)	[25]
Soil	Fe-EDDHA	15 kg ha^{-1}	Sowing	26.0	13.1	19.7	73.1	–	–	13.7	–	–4.3	–	Peanut (variety NC_2)	[25]
Soil	Fe-EDDHA	20 kg ha^{-1}	Sowing	26.0	19.6	21.3	80.8	–	–	20.5	–	–5.6	–	Peanut (variety NC_2)	[25]
Soil	Zn (Zn-EDTA)	1.5 kg ha^{-1}	Before of sowing	–	–	–	–	218.7	–	37.3	52.2	5.3	16.7	Rapeseed	[36]
Soil	B (Borax)	1.5 kg ha^{-1}	Before of sowing	–	–	–	–	145.7	–	8.8	25.7	5.8	22.3	Rapeseed	[36]
Soil	S (Sulfur)	100 kg ha^{-1}	Before of sowing	–	–	–	–	121.6	–	13.1	37.6	7.2	30.5	Rapeseed	[36]
Soil	Poultry manure	2 ton ha^{-1}	Sowing	1.9	12.1	14.0	–	–	–	–0.2	18.7	–	–	Wheat (*cv.* Sardari)	[16]
Foliar	Fe-EDDHA	1 g L^{-1}	Beginning of branching + flowering	–0.6	0.0	0.0	0.0	–	–	5.8	–	–1.6	–	Peanut (variety NC_2)	[25]
Foliar	Fe-EDDHA	2 g L^{-1}	Beginning of branching + flowering	8.8	2.8	19.7	26.9	–	–	8.6	–	–2.1	–	Peanut (variety NC_2)	[25]
Foliar	Fe-EDDHA	3 g L^{-1}	Beginning of branching + flowering	23.2	11.2	21.3	69.2	–	–	12.8	–	–4.3	–	Peanut (variety NC_2)	[25]
Foliar	Fe-EDDHA	4 g L^{-1}	Beginning of branching + flowering	21.7	15.9	21.3	73.1	–	–	15.4	–	–4.9	–	Peanut (variety NC_2)	[25]

TABLE 12.1 *(Continued)*

Method of Application	Nutrient Type	Application of Nutrients at Rates	Time of Application	Increase of Oil Over Untreated Control (%)										Crop type	References
				Grain Nitrogen (N)	Grain Phosphorus (P)	Grain Potassium (K)	Grain Iron (Fe)	Grain Zinc (Zn)	Grain Manganese (Mn)	Grain Protein Percent	Protein Yield	Grain Oil Percent	Oil yield		
Foliar	Zn	3 in 1,000	Stemming + clustering	–	–5.5	0.0	–	44.6	8.2	13.9	32.1	–	–	Mung bean	[13]
Foliar	Mn	3 in 1,000	Stemming + clustering	–	–12.5	0.0	–	2.0	101.5	10.9	36.7	–	–	Mung bean	[13]
Foliar	Zn + Mn	3 in 1,000	Stemming + clustering	–	–10.2	1.3	–	39.7	96.7	18.0	44.7	–	–	Mung bean	[13]
Foliar	Fe ($FeSO_4$)	3 in 1,000	Tillering + flowering	–9.4	69.7	–9.3	–	–	–	–10.8	0	–	–	Wheat (*cv.* Sardari)	[16]
Foliar	Zn ($ZnSO_4$)	1 in 1,000	Tillering + flowering	–12.5	–3.0	7.0	–	–	–	–14.0	–4.2	–	–	Wheat (*cv.* Sardari)	[16]
Foliar	Mn ($MnSO_4$)	3 in 1,000	Tillering + flowering	–4.4	45.5	20.9	–	–	–	–6.3	–0.1	–	–	Wheat (*cv.* Sardari)	[16]
Foliar	Fe	3 in 1,000	Stemming + flowering	–	–	–	–12.3	109.4	–38.7	–	–	5.7	–24.0	Winter rapeseed (*cv.* Zarfam)	[18]
Foliar	Zn	3 in 1,000	Stemming + flowering	–	–	–	–21.1	87.3	–3.7	–	–	–3.2	–20.4	Winter rapeseed (*cv.* Zarfam)	[18]
Foliar	Mn	3 in 1,000	Stemming + flowering	–	–	–	2.3	2.1	43.7	–	–	–3.4	–32.3	Winter rapeseed (*cv.* Zarfam)	[18]
Foliar	Zn ($ZnSO_4$)	3 in 1,000	End of rosette + before of flowering	–	–	–	0.6	36.4	0.2	–	–	1.8	11.6	Safflower	[12]
Foliar	Fe ($FeSO_4$)	3 in 1,000	End of rosette + before of flowering	–	–	–	3.6	–1.4	–2.7	–	–	0.7	5.0	Safflower	[12]

TABLE 12.1 *(Continued)*

Method of Application	Nutrient Type	Application of Nutrients at Rates	Time of Application	Increase of Oil Over Untreated Control (%)										Crop type	References
				Grain Nitrogen (N)	Grain Phosphorus (P)	Grain Potassium (K)	Grain Iron (Fe)	Grain Zinc (Zn)	Grain Manganese (Mn)	Grain Protein Percent	Protein Yield	Grain Oil Percent	Oil yield		
Foliar	Mn ($MnSO_4$)	3 in 1,000	End of rosette + before of flowering	–	–	–	1.6	0.7	44.3	–	–	1.4	8.3	Safflower	[12]
Foliar	Zn	2 g L^{-1}	Stem elongation + early flowering	–	–	–	3.9	30.2	–	13.5	–	4.7	13.4	Canola (*cv.* Talaye)	[7]
Foliar	Zn	4 g L^{-1}	Stem elongation + early flowering	–	–	–	1.5	50.8	–	18.2	–	11.6	34.4	Canola (*cv.* Talaye)	[7]
Foliar	Fe	2 g L^{-1}	Stem elongation + early flowering	–	–	–	9.7	0.6	–	0.0	–	–2.0	1.7	Canola (*cv.* Talaye)	[7]
Foliar	Fe	4 g L^{-1}	Stem elongation + early flowering	–	–	–	23.2	1.0	–	9.5	–	0.4	8.4	Canola (*cv.* Talaye)	[7]
Foliar	Zn ($ZnSO_4$)	5 in 1,000	Stem elongation	–	–	–	–	64.9	–	–	–	13.8	88.5	Safflower	[14]
Foliar	Zn ($ZnSO_4$)	5 in 1,000	Branching	–	–	–	–	41.5	–	–	–	12.5	111.6	Safflower	[14]
Foliar	Zn ($ZnSO_4$)	5 in 1,000	Flowering	–	–	–	–	52.3	–	–	–	10.1	89.4	Safflower	[14]
Foliar	Zn ($ZnSO_4$)	5 in 1,000	Seed filling	–	–	–	–	47.8	–	–	–	11.9	61.7	Safflower	[14]
Foliar	Fe	0.2%	Beginning of stemming + beginning of spiking	–	–	–	187.2	–26.5	–15.6	9.4	–	–	–	Barley	[6]
Foliar	Zn	0.2%	Beginning of stemming + beginning of spiking	–	–	–	–22.6	22.1	–9.8	8.1	–	–	–	Barley	[6]
Foliar	Mn	0.2%	Beginning of stemming + beginning of spiking	–	–	–	–14.3	–20.0	122.1	7.1	–	–	–	Barley	[6]

TABLE 12.1 *(Continued)*

Method of Application	Nutrient Type	Application of Nutrients at Rates	Time of Application	Increase of Oil Over Untreated Control (%)										Crop type	References
				Grain Nitrogen (N)	Grain Phosphorus (P)	Grain Potassium (K)	Grain Iron (Fe)	Grain Zinc (Zn)	Grain Manganese (Mn)	Grain Protein Percent	Protein Yield	Grain Oil Percent	Oil yield		
Foliar	Zn (1.5%), Fe (2.5%), Mn (0.5%)	2 in 1,000	Stemming	–	–	–	4.7	15.1	–	9.5	20.7	–	–	Rapeseed (*cv.* Hayola 401)	[40]
Foliar	Zn (1.5%), Fe (2.5%), Mn (0.5%)	4 in 1,000	Stemming	–	–	–	14.6	33.4	–	18.9	55.9	–	–	Rapeseed (*cv.* Hayola 401)	[40]
Foliar	Zn (1.5%), Fe (2.5%), Mn (0.5%)	6 in 1,000	Stemming	–	–	–	16.6	43.6	–	21.9	63.0	–	–	Rapeseed (*cv.* Hayola 401)	[40]
Foliar	Fe (nano-fertilizer)	2 in 1,000	Flowering	–	–	–	40.0	–	–	12.4	18.0	–	–	Lentil	[22]
Foliar	Fe (nano-fertilizer)	2 in 1,000	Podding	–	–	–	42.9	–	–	14.4	23.4	–	–	Lentil	[22]
Foliar	Fe (nano-fertilizer)	2 in 1,000	Flowering + podding	–	–	–	71.9	–	–	22.5	49.3	–	–	Lentil	[22]
Foliar	Fe (nano-fertilizer)	4 in 1,000	Flowering	–	–	–	64.3	–	–	14.1	28.3	–	–	Lentil	[22]
Foliar	Fe (nano-fertilizer)	4 in 1,000	Podding	–	–	–	58.7	–	–	17.0	33.4	–	–	Lentil	[22]
Foliar	Fe (nano-fertilizer)	4 in 1,000	Flowering + podding	–	–	–	82.0	–	–	28.0	69.9	–	–	Lentil	[22]
Foliar	Zn (Zn-EDTA)	2 in 1,000	Tillering + shoot elongation + grain filling	–	–	–	60.1	39.9	–	36.1	–	–	–	Bread wheat	[11]
Foliar	Fe (Fe-EDDHA)	3 in 1,000	Tillering + shoot elongation + grain filling	–	–	–	89.4	10.5	–	39.4	–	–	–	Bread wheat	[11]

TABLE 12.1 *(Continued)*

Method of Application	Nutrient Type	Application of Nutrients at Rates	Time of Application	Increase of Oil Over Untreated Control (%)										Crop type	References
				Grain Nitrogen (N)	Grain Phosphorus (P)	Grain Potassium (K)	Grain Iron (Fe)	Grain Zinc (Zn)	Grain Manganese (Mn)	Grain Protein Percent	Protein Yield	Grain Oil Percent	Oil yield		
Foliar	Zn (Zn-EDTA) + Fe (Fe-EDDHA)	2 in 1,000 + 3 in 1,000	Tillering + shoot elongation + grain filling	–	–	–	85.5	23.3	–	45.5	–	–	–	Bread wheat	[11]
Foliar	Ascorbic acid	50 ppm	8-leaf Stage + silks appearance stage	–	–	–	–	–	–	–0.5	0.1	–1.3	–0.9	Corn	[74]
Foliar	Ascorbic acid	100 ppm	8-leaf Stage + silks appearance stage	–	–	–	–	–	–	–1.1	5.7	–1.7	5.5	Corn	[74]
Foliar	Zn, Fe, Mn (Micro-Kimia liquid fertilizer)	2 in 1,000	Stemming	–	–	–	–	–	–	15.0	48.3	–	–	Barley	[38]
Foliar	Zn, Fe, Mn (Micro-Kimia liquid fertilizer)	4 in 1,000	Stemming	–	–	–	–	–	–	18.8	73.3	–	–	Barley	[38]
Foliar	Zn ($ZnSO_4$)	3 in 1,000	Before flowering + after pollination	–	–	–	–	–	–	29.7	–	32.0	–	Safflower	[31]
Foliar	Fe ($FeSO_4$)	4 in 1,000	Before flowering + after pollination	–	–	–	–	–	–	22.6	–	40.7	–	Safflower	[31]
Foliar	B (H_3BO_3)	2 in 1,000	Before flowering + after pollination	–	–	–	–	–	–	14.2	–	21.6	–	Safflower	[31]
Foliar	Fe ($FeSO_4$) + Zn ($ZnSO_4$)	4 in 1,000 + 3 in 1,000	Before flowering + after pollination	–	–	–	–	–	–	20.6	–	45.6	–	Safflower	[31]
Foliar	B (H_3BO_3) + Fe ($FeSO_4$)	2 in 1,000 + 4 in 1,000	Before flowering + after pollination	–	–	–	–	–	–	7.1	–	20.3	–	Safflower	[31]
Foliar	B (H_3BO_3) + Zn ($ZnSO_4$)	2 in 1,000 + 3 in 1,000	Before flowering + after pollination	–	–	–	–	–	–	14.8	–	24.9	–	Safflower	[31]

TABLE 12.1 *(Continued)*

Method of Application	Nutrient Type	Application of Nutrients at Rates	Time of Application	Increase of Oil Over Untreated Control (%)										Crop type	References
				Grain Nitrogen (N)	Grain Phosphorus (P)	Grain Potassium (K)	Grain Iron (Fe)	Grain Zinc (Zn)	Grain Manganese (Mn)	Grain Protein Percent	Protein Yield	Grain Oil Percent	Oil yield		
Foliar	Zn ($ZnSO_4$)	5 in 1,000	Stem elongation	–	–	–	–	–	–	7.7	–	1.8	–	Rapeseed	[52]
Foliar	Zn ($ZnSO_4$)	5 in 1,000	Flowering	–	–	–	–	–	–	11.0	–	4.8	–	Rapeseed	[52]
Foliar	Zn (1.5%), Fe (2.5%), Mn (0.5%)	2 in 1,000	Stemming	–	–	–	–	–	–	–	–	11.8	23.2	Canola	[57]
Foliar	Zn (1.5%), Fe (2.5%), Mn (0.5%)	4 in 1,000	Stemming	–	–	–	–	–	–	–	–	17.1	53.5	Canola	[57]
Foliar	Zn (1.5%), Fe (2.5%), Mn (0.5%)	6 in 1,000	Stemming	–	–	–	–	–	–	–	–	19.8	60.1	Canola	[57]

TABLE 12.2 Effect of Different Application of Nutrients on the Changes of Protein and Oil Percent, Protein and Oil Yield, and Fatty Acids in Different Crops

Method of Application	Nutrient Type	Application of Nutrients at Rates	Time of Application	Increase of Oil Over Untreated Control (%)													Crop Type	References
				Grain Protein Percent	Protein Yield	Grain Oil Percent	Oil Yield	Palmitic Acid (16:0)	Stearic Acid (18:0)	Oleic Acid (18:1)	Linoleic Acid (18:2)	Linolenic Acid	α–Linoleic Acid (18:3)	Arachidonic Acid (20:4)	Arachidic Acid (20:0)	Trichosanoic Acid (23:0)		
Soil	Fe–EDDHA	5 kg ha^{-1}	Sowing	6.4	–	–2.2	–	–0.9	–30.8	0	3.3	–	–8.3	10.8	9.0	8.5	Peanut (variety NC_2)	[25]
Soil	Fe–EDDHA	10 kg ha^{-1}	Sowing	10.2	–	–3.1	–	–5.9	8.7	–0.7	–0.3	–	6.9	14.5	19.4	22.2	Peanut (variety NC_2)	[25]
Soil	Fe–EDDHA	15 kg ha^{-1}	Sowing	13.7	–	–4.3	–	–3.9	24.6	–1.1	–2.0	–	4.8	24.1	14.0	17.1	Peanut (variety NC_2)	[25]
Soil	Fe–EDDHA	20 kg ha^{-1}	Sowing	20.5	–	–5.6	–	–5.4	16.8	–1.9	1.7	–	7.6	10.8	12.6	17.1	Peanut (variety NC_2)	[25]
Soil	Zn (Zn–EDTA)	1.5 kg ha^{-1}	Sowing	–	–	5.3	–	3.2	–0.2	9.2	5.1	9.4	–	–	–	–	Rapeseed (*cv.* Hayola 401)	[58]
Soil	B (Borax)	1.5 kg ha^{-1}	Sowing	–	–	5.8	–	0.6	–46.5	13.7	11.1	18.1	–	–	–	–	Rapeseed (*cv.* Hayola 401)	[58]
Soil	S	100 kg ha^{-1}	Sowing	–	–	7.2	–	1.6	–27.7	13.4	10.1	16.7	–	–	–	–	Rapeseed (*cv.* Hayola 401)	[58]
Soil	B + S	1.5 kg ha^{-1} + 100 kg ha^{-1}	Sowing	–	–	9.3	–	7.6	–27.5	23.2	17.6	22.9	–	–	–	–	Rapeseed (*cv.* Hayola 401)	[58]
Soil	B + Zn	1.5 kg ha^{-1} + 1.5 kg ha^{-1}	Sowing	–	–	8.7	–	14.7	–7.7	30.5	28.7	32.9	–	–	–	–	Rapeseed (*cv.* Hayola 401)	[58]
Soil	S + Zn	100 kg ha^{-1} + 1.5 kg ha^{-1}	Sowing	–	–	8.7	–	22.4	–25.1	17.7	11.9	20.8	–	–	–	–	Rapeseed (*cv.* Hayola 401)	[58]
Soil	B + S + Zn	1.5 kg ha^{-1} + 100 kg ha^{-1} + 1.5 kg ha^{-1}	Sowing	–	–	11.0	–	16.4	–26.2	36.4	14.2	36.6	–	–	–	–	Rapeseed (*cv.* Hayola 401)	[58]
Foliar	Fe–EDDHA	1 g L^{-1}	Beginning of branching + flowering	5.8	–	–1.6	–	–2.1	–0.6	–2.9	4.1	–	–8.3	6.0	1.8	–3.4	Peanut (variety NC_2)	[25]
Foliar	Fe–EDDHA	2 g L^{-1}	Beginning of branching + flowering	8.6	–	–2.1	–	–8.1	25.2	–3.0	2.2	–	18.1	18.1	22.5	24.8	Peanut (variety NC_2)	[25]
Foliar	Fe–EDDHA	3 g L^{-1}	Beginning of branching + flowering	12.8	–	–4.3	–	–10.5	9.0	–3.6	1.4	–	45.8	45.8	49.1	55.6	Peanut (variety NC_2)	[25]

TABLE 12.2 *(Continued)*

Method of Application	Nutrient Type	Application of Nutrients at Rates	Time of Application	Increase of Oil Over Untreated Control (%)													Crop Type	References
				Grain Protein Percent	Protein Yield	Grain Oil Percent	Oil Yield	Palmitic Acid (16:0)	Stearic Acid (18:0)	Oleic Acid (18:1)	Linoleic Acid (18:2)	Linolenic Acid	α–Linoleic Acid (18:3)	Arachidonic Acid (20:4)	Arachidic Acid (20:0)	Trichosanoic Acid (23:0)		
Foliar	Fe–EDDHA	4 g L^{-1}	Beginning of branching + flowering	15.4	–	–4.9	–	–11.3	50.5	–5.0	4.0	–	22.9	22.9	24.3	14.5	Peanut (variety NC_2)	[25]
Foliar	Fe	3 in 1,000	Stemming + flowering	–	–	5.7	–24.0	185.7	–	6.4	–11.8	2.4	–	–	–	–	Winter rapeseed (*cv.* Zarfam)	[18]
Foliar	Zn	3 in 1,000	Stemming + flowering	–	–	–3.2	–20.4	186.5	–	3.0	6.4	–2.0	–	–	–	–	Winter rapeseed (*cv.* Zarfam)	[18]
Foliar	Mn	3 in 1,000	Stemming + flowering	–	–	–3.4	–32.3	83.0	–	10.4	3.4	–24.7	–	–	–	–	Winter rapeseed (*cv.* Zarfam)	[18]
Foliar	Kadostim	0.5 in 1,000	Start of flowering + end of flowering + before grain filling	–7.0	352.9	–8.7	170.4	16.3	12.2	22.8	6.2	13.4	–	–	–	–	Soybean	[59]
Foliar	Aminol forte	2 in 1,000	Start of flowering + end of flowering + before grain filling	–31.4	121.5	–13.0	72.4	5.4	–11.9	14.7	2.3	6.1	–	–	–	–	Soybean	[59]
Foliar	Organic manure	10 ton ha^{-1}, was added to the water with the ratio of 0.1	–	–30.5	170.8	–10.9	112.9	18.9	–2.5	15.2	4.3	10.4	–	–	–	–	Soybean	[59]
Foliar	Compost	10 ton ha^{-1}, was added to the water with the ratio of 0.1	–	–14.7	261.1	–8.7	137.1	6.7	–2.2	13.3	4.4	12.0	–	–	–	–	Soybean	[59]
Foliar	Nano–Fe chelate	2 in 1,000	Start of flowering + end of flowering + before grain filling	0.0	461.7	3.5	257.8	18.4	13.3	29.1	12.8	41.9	–	–	–	–	Soybean	[59]
Foliar	Nitrogen fertilizer (Urea)	5 in 1,000	–	–35.2	46.0	–17.4	14.5	–4.0	–0.8	6.2	1.1	–1.1	–	–	–	–	Soybean	[59]

ursic acid, which is a good feature [55]. Shahsavari et al. [56] also reported that the use of Zn improved the metabolism of fatty acids (FAs), especially saturated FAs (palmitic acid, stearic acid, behenic acid) and unsaturated FA (linoleic acid) in rapeseed plant. He stated that the increase in seed oil due to Zn consumption was due to an increase in saturated FAs, which could be due to preventing the oxidation of the above fats. A positive and considerable correlation was reported between Zn consumption and oil yield of rapeseed [7, 36] and safflower [12, 14, 57]. The results of research by various researchers on the effect of micronutrients on the changes of concentration of oil percent, oil yield, and FAs in different crops are presented in Table 12.2.

The quality and composition of seed oil is important for two reasons: (i) its role in human nutrition; and (ii) its direct impact on oil quality during storage. Various research has been done on the effect of micronutrients on oil quality (Table 12.2). For example, the use of 10, 15, and 20 kg ha^{-1} of Fe (form Fe-EDDHA) in soil led to increased stearic acid, α-linoleic acid, arachidonic acid, arachidic acid, and trichosanoic acid but decreased oleic acid, palmitic acid, and linoleic acid in peanut [25]. On the other hand, it has been reported that use of 1.5, 1.5, and 100 kg ha^{-1} as fertilizers of B (borax), Zn (Zn-EDTA), and S in soil led to increased palmitic acid, linoleic acid, oleic acid, and linolenic acid but decreased stearic acid of rapeseed [58]. Kalantar Ahmadi & Shoushi Dezfouli [54] stated that the highest levels of stearic acid (1.68%), palmitic acid (4.0%), oleic acid (56.33%), and linoleic acid (24.48%) in complete-irrigation and foliar application of B, Zn, B + Zn, and B (all of micronutrients with a concentration of 6 per 1,000) were obtained in rapeseed, respectively. Due to the fact that oleic acid is the highest amount of FAs in rapeseed oil, the quality of seed oil can be improved with proper nutritional management. In this case, Tousi et al. [59] reported that the interaction impact between treatments showed that foliar nano-Fe chelate application of at the second harvest (when the more than 95% of pods as yellowing and main stem seed pods had 14–15% moisture) produced the greatest oleic acid (27.1%), oil yield (1,014 kg ha^{-1}), protein yield (1,503 kg ha^{-1}), and grain yield (4,062 kg ha^{-1}) of soybean cultivar Williams. They stated that spray nano-Fe chelate application at second harvest increased the quality and quantity (stearic acid, palmitic acid, linoleic acid, oleic acid, and linolenic acid) of soybean production. The high level of linoleic FA in soybean oil compared to other FAs has increased the quality of its oil. It seems that with spray application of nano-Fe chelate, due to the increase of green growth, improvement of leaf chlorophyll, an increase of capacity, and photosynthetic process more reproductive organs is made and the share of their nutrients and material increases. As a result, while increasing yield components and grain yield, oil, and protein yields are increased [59–61]. Some researchers have reported that the use of Fe reduced the amount of seed oil in canola [7] and peanut [25], while some researchers increased the amount of seed oil by 0.4, 0.7, 3.5, and 40.7 in the plants of canola [7], safflower [12], soybean [59], and safflower [31], respectively. These different results are probably due to the type of Fe fertilizer used, amount of used, type of used and/or the time of use.

About the time of application of the elements on the seed oil, Moradi Telavat et al. [14] reported that the foliar Zn application significantly increased content of oil in grain and grain yield of safflower; however, there was no considerable difference among spray Zn applications in various growth stages. Interaction impact of cultivar and Zn spraying was considerable on content of oil in grain. Greatest content of grain oil in *cv.* Soffeh obtained in Zn foliar in elongation of stem stage and in *cv.* Local Isfahan in stages of flowering and branching. In general, oil yield increased in foliar Zn use compared to non-Zn application (control), except for application of Zn in grain filling stage.

Yadollahi et al. [62] reported that application of OFs increased oil yield, content of seed oil and carbohydrates of safflower (*Carthamus tinctorius* L). Integrated use of HA and compost increased oil yield by 99.7% over the non-application of OFs (control). They stated that the use of compost and HA can be combined to develop cultivation of safflower in Sistan zone of Iran, especially in mild stress situations. On the other hand, Soleimani et al. [12] indicated that the impact of foliar micronutrients application on oil yield and grain yield of safflower (*Carthamus tinctorius* L.) was considerable. Oil yield was considerably higher in the combined spray application of Fe + Zn + Mn (421 kg ha^{-1}) and Zn + Mn (411 kg ha^{-1}) by increasing 16.6 and 14.7% in comparison with control. They concluded that the greatest oil yield and grain yield were obtained from combined foliar applications of Zn + Mn and Fe + Zn + Mn, respectively. The results of other studies have shown that the elements of the Zn and Mn have a greater effect on the amount of oil produced in oilseeds than Fe element [63, 64]. Adequate Zn concentrations increase fat metabolism and increase the amount of fat stored in stored tissues such as grains [65].

The highest oil percentage, oil yield, and grain yield of sunflower (*cv.* Hisun 25) were obtained from the consumption of 600 kg S/ha [66]. Asare & Scarisbrick [67] also pointed to a positive and considerable correlation among S consumption with oil percentage and oil yield of rape. Ravi et al.

[68] stated that the impact of using S with Fe and Zn on yield, oil content and protein percentage of safflower was significant. This indicates the positive effects of the above elements on the percentage of oil and the percentage of grain protein. In return, boron (B) use caused by considerable plant B content, but had no considerable on grain oil and N, and leaf N content [69]. While the positive impact of foliar spraying of micronutrients such as Zn and B on seed oil of safflower and canola has been reported [10, 31, 54].

12.5 VITAMINS

Vitamin C is also known as ascorbic acid (AsA). AsA is a small water-soluble molecule with high antioxidant properties [70]. External application of AsA increases resistance to environmental stresses (such as deficit and drought stress) and reduces the harmful effects of oxidative stresses [71]. Application of $ZnSO_4$ at various growth stages caused an increase in the AsA in wheat grain (Figure 12.2). Esfandiari et al. [4] reported that the different Zn treatments increased the content of AsA in wheat compared control. So that foliar Zn use at booting + soft dough stages and stage of flowering resulted in 15- and 16-times higher amounts of AsA, respectively than non-use Zn (control). Towo et al. [72] reported that the AsA in maize, red sorghum, wheat, and finger millet fortified with nutrients in particular micronutrients. Bharti et al. [73] reported that the content of AsA increased gradually with increasing values of Zn (soil + foliar) in both the crop seasons of wheat. On the other hand, Towhidi Moghaddam [74] reported that the AsA foliar application increased protein and oil percentage as well as protein and oil yield of corn in normal irrigation and irrigation withholding. So, the AsA use can scavenge reactive oxygen species (ROS) and decrease accumulation of proline and antioxidant enzyme activity, which in turn, leads to increased quality and grain yield with improving growth situations.

In connection with the effect of boron (B) on vitamin C, Tariq et al. [75] reported that the B (0.5 kg ha^{-1}) applied as a spray had considerably increased tuber volume, tuber per plant, and plant height as well as enhanced the quality in terms of vitamin C amount, content of B and starch of potato compared to other use ways (fertigation and soil application). Maximum vitamin C content (17.5 mg/100 g) was obtained at B spray by dose 0.5 kg ha^{-1} which was statistically at par to fertigation by dose 1 kg ha^{-1}. They stated that concentration of B in tubers was strongly correlated with starch and vitamin C contents. Besides the potato; B also improved the quality of other crops like Hajduk et al. [76] reported maximum content of AsA in tomato (by 32.57 mg/100 g) with the spray use of 1.25 g L^{-1} B. The results of the above research indicate the positive effect of B on the improvement of vitamin C.

12.6 CONCLUSION

Using micronutrients such as zinc (Zn) and iron (Fe) can play a main role in improving seed yield, content of protein in seed and enrichment of Zn and Fe in different crops. Spraying of Zn in various growth stages increased content of Zn in grain, however, content of grain phosphorus (P) decreased as well as decreased grain phytate/Zn molar ratios. Use of micronutrients such as Zn caused an increase in amino acids content in grain, grain protein percent, protein yield, and content of AsA in grain. Also, consumption of micronutrients such as Zn, boron (B) and sulfur (S) in soil led to increased grain oil percent, and oil yield as well as oleic acid, palmitic acid, linoleic acid, and linolenic acid but decreased stearic acid of oil plants. In generally, a foliar application of Zn plus Fe is recommended for simultaneous Fe and Zn biofortification of crops cultivated in soils within lower these micronutrients availability.

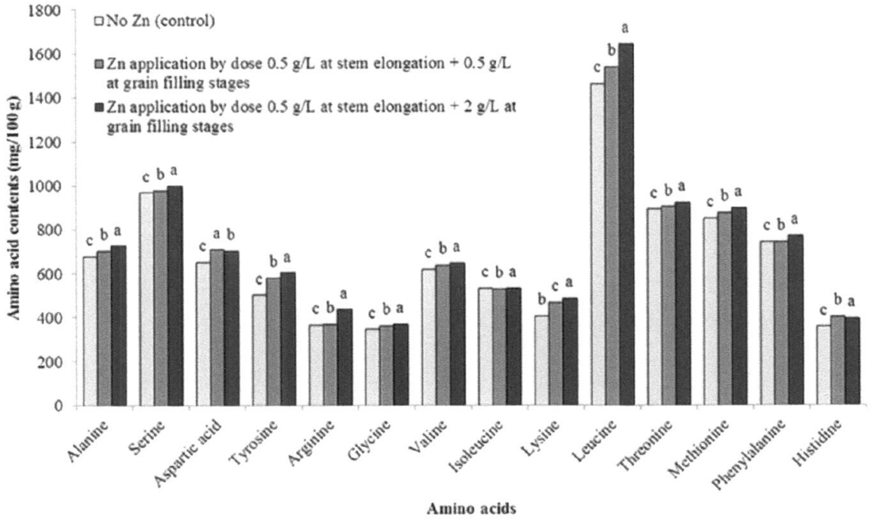

FIGURE 12.1 The effect of foliar zinc application on contents of amino acids in grain of wheat.

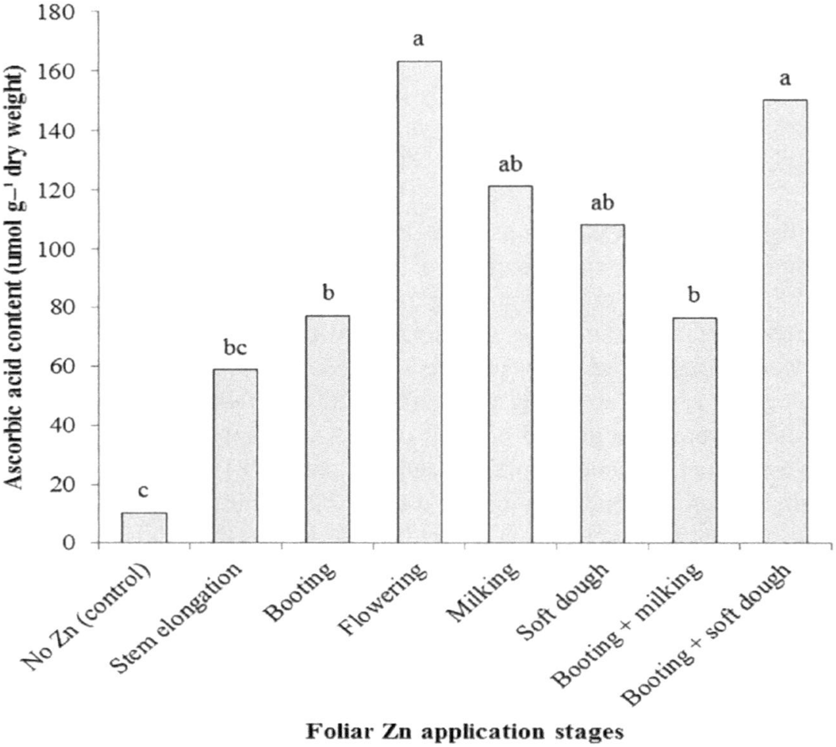

FIGURE 12.2 Effect of foliar zinc application at various growth stages on ascorbic acid content in grain of wheat.

KEYWORDS

- **agriculture**
- **biofortification**
- **crops**
- **essential oils**
- **malnutrition**
- **minerals**
- **vitamins**

REFERENCES

1. WHO/WFP/UNICEF, (2007). Preventing and controlling micronutrient deficiencies in populations affected by an emergency: multiple vitamin and mineral supplements for pregnant and lactating women, and for children aged 6 to 59 months. Retrieved from: https://www.who.int/publications/m/item/WHO-WFP-UNICEF-statement-micronutrients-deficiencies-emergency (accessed on 7 April 2006).
2. Alloway, B. J., (2008). *Zinc in Soils and Crop Nutrition* (2th edn., p. 136). Brussels: International Zinc Association (IZA).
3. Dare, G. B., Rasouli, S. M. H., & Khodaverdilu, H., (2012). Adsorption and retention of Mn, Zn and Cu in some acidic and non-acidic soils. *Journal of Soil Management and Sustainable Production, 2,* 63–77.
4. Esfandiari, E., Abdoli, M., Mousavi, S. B., & Sadeghzadeh, B., (2016). Impact of foliar zinc application on agronomic traits and grain quality parameters of wheat grown in zinc deficient soil. *Indian Journal of Plant Physiology, 21*(3), 263–270. doi: 10.1007/s40502-016-0225-4.
5. Abdoli, M., Esfandiari, E., Sadeghzadeh, B., & Mousavi, S. B., (2016). Zinc application methods affect agronomy traits and grain micronutrients in bread and durum wheat under zinc-deficient calcareous soil. *Yuzuncu Yil University Journal of Agricultural Sciences, 26*(2), 202–214.
6. Rajaie, M., & Charkhandeh, M., (2020). Improvement of the yield and grain protein in barley (*Hordeum valgare* L.) by iron, manganese and zinc foliar spray. *Iranian Journal of Agronomy and Plant Breeding, 16*(1), 1–14.
7. Hedayatpour, R., Movahhedi, D. M., Khademhamzeh, H. R., & Morshedi, S. M., (2014). Effect of foliar application of zinc and iron on yield quantity and quality of canola (*Brassica napus cv.* Talaye) in Zarghan region, Fars. *Journal of Oil Plants Production, 1*(1), 33–42.
8. Pourjamshid, A., Ghaysari, M., Sharifi, N. A., & Salemi, F., (2020). Effect of micronutrients foliar application on some physiological and agronomic traits of bread wheat (*Triticum aestivum* L.) under complete irrigation and terminal drought stress condition. *Environmental Stresses in Crop Sciences., 13*(2), 401–412. doi: http://dx.doi.org/10.22077/escs.2020.2202.1550.
9. Marschner, H., (1995). *Mineral Nutrition of Higher Plants* (p. 889). Academic press. New York, USA.
10. Yang, M., Shi, L., Xu, F. S., & Wang, Y. H., (2009). Effect of boron on dynamic change of seed yield and quality formation in developing seed of *Brassica napus*. *Journal of Plant Nutrition, 32,* 785–797.
11. Ebrahimi, Z., Biabani, A., Mohammadi, R., Sabouri, H., & Rahemi, K. A., (2021). Effect of wheat enrichment by foliar application of zinc and iron on quantitative and qualitative traits at different phenological stages. *Journal of Crop Breeding, 38,* 138–148.
12. Soleimani, R., Nourgholipour, F., & Moshiri, F., (2017). Effect of foliar application of Zn, Fe and Mn on seed yield and micronutrient contents of safflower (*Carthamus tinctorius* L.). *Iranian Journal of Crop Sciences., 19*(1), 1–12.
13. Jafar, D. R., Mosavi, N. S. M., Mehraban, A., & Basiri, M., (2015). Effect of water stress and foliar micronutrient application on physiological characteristics and nutrient uptake in mung bean. *Electronic Journal of Crop Production, 8*(1), 121–141.

14. Moradi, T. M. R., Roshan, F., & Siadat, S. A., (2015). Effect of foliar application of zinc sulfate on minerals content, seed and oil yields of two safflower cultivars (*Carthamus tinctorius* L.). *Iranian Journal of Crop Sciences, 17*(2), 153–164.
15. Zeidan, M. S., Mohamed, M. F., & Hamouda, H. A., (2010). Effect of foliar fertilization of Fe, Mn and Zn on wheat yield and quality in low sandy soils fertility. *World Journal of Agricultural Sciences, 6*(6), 696–699.
16. Javanmard, A., & Asadi, D. A., (2017). Effect of poultry manure and micronutrient foliar application on some wheat quantity and quality characteristics under rainfed conditions. *Journal of Crop Science Research in Arid Regions, 1*(1), 13–26. doi: 10.22034/csrar.01.01.02.
17. Rosen, J. A., (1977). Pike, C. S., Golden, M. L., Zinc, iron, and chlorophyll metabolism in zinc-toxic corn. *Plant Physiology, 59*(6), 1085–1087.
18. Mostafavi, R. M., Jadidi, E., & Babaei, T., (2014). Effect of micronutrient elements on seed yield, qualitative traits and oil in winter rapeseed varieties. *Journal of Crops Improvement, 16*(3), 627–639.
19. Khan, H., Hassan, Z. U., & Maitlo, A. A., (2006). Yield and micronutrients content of bread wheat (*Triticum aestivum* L.) under a multi-nutrient fertilizer Hal-tonic. *International Journal of Agriculture and Biology, 8,* 366–370.
20. Eisvand, H. R., Esmaeili, A. R., & Mohammadi, M., (2014). Effects of iron oxide nanoparticles on some quantity, quality and physiological characteristics in wheat (*Triticum aestivum* L.) at Khoramabad climate. *Iranian Journal of Field Crop Science, 45*(2), 287–298. doi: 10.22059/ijfcs.2014.51907.
21. Jokar, L., & Ronaghi, A., (2015). Effect of foliar application of different Fe levels and sources on growth and concentration of some nutrients in sorghum. *Journal of Science and Technology of Greenhouse Culture, 6*(2), 163–174.
22. Mehraban, A., (2017). Effect of foliar application of iron on yield, yield component, and grain protein of lentil crop. *Journal of Plant Environmental Physiology, 45,* 27–37.
23. Torabian, S., Zahedi, M., & Khoshgoftar, M. A. H., (2016). Effects of foliar spray of two kinds of zinc oxide on the growth and ion concentration of sunflower cultivars under salt stress. *Journal of Plant Nutrition, 39,* 172–180.
24. Saeedi, A. R. A., & Yadavi, A. R., (2016). The effect of iron and zinc irrigation and irrigation levels on quantitative and qualitative characteristics of red bean. *Journal of Iranian Cereals Research, 6*(1), 54–65.
25. Panjtandoust, M., Soroosh, Z. A., & Ghanati, F., (2010). Effect of iron soil and spray applied on some qualify characteristics of peanut (*Arachis hypogaea* L.) plants in a calcareous soil. *Journal of Plant Biology, 2*(5), 37–50.
26. Schenkeveld, W. D. C., Dijcker, R., Reichwein, A. M., Temminghoff, E. J. M., & Riemsdijk, W. H., (2008). The effectiveness of soil-applied FeEDDHA treatments in preventing iron chlorosis in soybean as a function of the o,o-FeEDDHA content. *Plant and Soil, 303,* 161–176.
27. Patel, M. S., Sutar, D. M., & Kanizaria, M. V., (1993). Effect of foliar application of iron and sulfur in curing chlorosis in groundnut. *Journal of the Indian Society of Soil Science., 41,* 103–105.
28. Hossenabadi, A., Galavi, A., & Heydari, M., (2006). Study of the effects of foliar application of Fe, Zn and Mn on wheat *cv.* hormone quantity and quality properties in Sistan condition. *New Findings in Agriculture, 1*(2), 103–110.

29. Khiavi, M., Khorshidi, M. B., Ismaeili, M., Azarabadi, S., Faramarzi, A., & Emaratpardaz, J., (2010). Effect of foliar application of boron and zinc on yield and some qualitative characteristics of two rapeseed (*Brassica napus* L.) cultivars. *Journal of Water and Soil Science, 20*(3), 31–45.
30. Khoram, G. A., Rahimi, A., Torabi, B., & Maddah, H. S., (2015). Effect of humic acid application and foliar spraying of compost tea and vermiwash on nutrient absorption and chlorophyll content of safflower (*Carthamus tinctorius* L.). *Journal of Oil Plants Production, 2*(1), 71–84.
31. Kamaraki, H., & Galavi, M., (2012). Evaluation of foliar Fe, Zn and B micronutrients application on quantitative and qualitative traits of safflower (*Carthamus tinctorius* L.). *Journal of Agroecology, 4*(3), 201–206.
32. Takkar, P. N., & Nayar, V. K., (1990). Reports of wheat grain grown on manganese deficient soil on method and rate of manganese sulphate application. *Fertilizer News, 36,* 55–57.
33. Ning, P., Fei, P., Wu, T., Li, Y., Qu, C., Li, Y., Shi, J., & Tian, X., (2021). Combined foliar application of zinc sulphate and selenite affects the magnitude of selenium biofortification in wheat (*Triticum aestivum* L.). *Food and Energy Security, 1–12*, e342. doi: 10.1002/fes3.342.
34. Broadley, M., Brown, P., Cakmak, I., Rengel, Z., & Zhao, F., (2012). Function of nutrients: Micronutrients. In: Marschner, P., (ed.), *Marschner's Mineral Nutrition of Higher Plants* (3rd edn., pp. 191–248). Academic Press.
35. Afshani, S., Amirnia, R., & Hadi, H., (2015). The effect of iron and zinc spray application on yield and yield components of autumn canola under low irrigation conditions. *Iranian Journal of Agricultural Research., 13*(1), 52–43.
36. Habibi, M., Majidian, M., Shoja, T., & Rabiee, M., (2016). Effects of S, B and Zn, on seed yield, nutrient concentration and seed quality of rapeseed (*Brassica napus* L.). *Journal of Oil Plants Production., 2*(2), 1–12.
37. Morshedi, A., & Naghibi, H., (2004). Effects of foliar application of Cu and Zn on yield and quality of canola grain (*Brassica napus* L.). *Journal of Agricultural Sciences and Natural Resources., 11*(3), 15–22.
38. Yousefipor, M., Lack, Sh., & Payandeh, Kh., (2019). Evaluation of the combined effect of biological and chemical phosphorous fertilizers and micronutrient on seed and protein yield of barley (*Hordeum vulgare* L.). *Journal of Crop Ecophysiology., 13*(1), 103–120.
39. Hedayatpur, R., Movahedi, D. M., Khadem, H. H. R., & Morshedi, S. M., (2014). Effect of foliar application of zinc and iron on yield quantity and quality of canola (*Brassica napus cv.* Talaye), in Zarghan region, Fars. *Journal of Oil Plant Production., 1*(1), 43–33.
40. Payandeh, Kh., Mojaddam, M., & Derogar, N., (2020). Study of quality and yield of rapeseed, Hayola 401 cultivar, with applying iron, zinc and manganese compound fertilizer under irrigation cut stress. *Environmental Stresses in Crop Sciences., 13*(1), 10–119. doi: 10.22077/escs.2019.1800.1414.
41. Pourgholam, M., Nemati, N., & Oveysi, M., (2013). Effect of zinc and iron under the influence of drought on prolin, protein and nitrogen leaf of rapeseed (*Brassica napus*). *Annals of Biological Research., 4*(7), 200–203.
42. Joorabi, S., Eisvand, H. R., Ismaili, A., & Nasrolahi, A. H., (2020). Effects of Zn nano-chelate foliar application on some physiological traits and grain yield of soybean under water deficit stress. *Journal of Plant Process and Function, 35,* 73–86.

43. Ghotavi, H., Moafporian, Gh., & Bahrani, A., (2011). Effect of zinc sulfate spraying and irrigation intervals on yield, yield components and corn protein content. *Journal of Plant Ecophysiology., 4*(1), 80–92.
44. Payne, P. I., Holt, L. M., Jackson, E. A., & Law, C. N., (1984). Wheat storage proteins: Their genetics and their potential for manipulation by plant breeding. *Philosophical Transactions of the Royal Society of Landon: Series B, Biological Sciences, 304,* 359–371.
45. Seyedi, M., Mojaddam, M., Babaei, N. T., & Derogar, N., (2018). Study of the chemicals and biological interaction effects on quantitative and qualitative characteristics of some bread wheat cultivars in Shoushtar climatic. *Journal of Plant Production Science, 8*(1), 1–11.
46. Kavian, A. N., & Aboutalebian, M. A., (2020). Quantitative and qualitative reaction of common bean (*Phaseolus vulgaris* L.) to phosphorous and zinc sulfate application method under different rates of nitrogen starter fertilizer. *Iranian Journal of Field Crop Science, 51*(3), 17–32. doi: 10.22059/ijfcs.2019.249549.654430.
47. Yasari, E., & Patwardhan, A. M., (2007). Effects of azotobacter and *Azospirillium* inoculations and chemical fertilizers on growth and productivity of canola. *Asian Journal of Plant Sciences, 6*(1), 77–82.
48. Barat, Z. S., Saki, N. T., & Babaei, N. T., (2019). Effect of potassium nano-chelate and ascorbic acid on grain yield and some qualitative characteristics of cowpea (*Vigna unguiculata* L. Kamran cultivar). *Journal of Plant Production, 9*(2), 149–159.
49. Eisvand, H. R., (2002). *Effect of Drought Stress and Timing of Nitrogen Application on Growth, Yield, Baking Quality and the Pattern of Storage Proteins Expression in Wheat Grain (Triticum aestivum var Mahdavi).* MSc. Thesis, University of Tehran, Iran.
50. Rezaei, C. E., Zehtab, S. S., Pirzad, A., & Rahimi, A., (2015). Effect of foliar application of iron, zinc and manganese micronutrients on yield and yield components and seed oil of pot marigold (*Calendula officinalis* L.). *Journal of Horticultural Science, 29*(1), 95–102.
51. Farajzadeh, M. T. E., (2018). Nutritional assessment of micronutrients on quantitative and qualitative characteristics of amaranth seed production (Koniz). *Iranian Journal of Seed Science and Research, 5*(3), 63–75. doi: 10.22124/jms.2018.2935.
52. Omidian, A., Siadat, S. A., Naseri, R., & Moradi, M., (2012). Effect of foliar application of zinc sulphate on grain yield, oil and protein content in four rapeseed (*Brassica napus* L.) cultivars. *Iranian Journal of Crop Sciences, 14*(1), 16–28.
53. Jashni, R., Fateh, E., & Aynehband, A., (2017). Effect of thiobacillus and nitrocara biological fertilizers and foliar application of zinc and iron on some qualitative characteristic and remobilization of rapeseed (*Brassica napus* L.). *The Plant Production (Scientific Journal of Agriculture), 40*(1), 1–14.
54. Kalantar, A. S. A., & Shoushi, D. A. A., (2019). Effects of foliar application of micronutrients on seed yield and oil quality of canola (*Brassica napus* L. *cv.* Hyola 401) under drought stress conditions. *Iranian Journal of Crop Sciences, 21*(3), 237–253. doi: 10.29252/abj.21.3.237.
55. Grant, C. A., & Baileg, L. D., (2000). Fertility management in canola production. *Canada Journal of Plant Science, 73,* 65–670.
56. Shahsavari, N., Jais, H. M., & Shirani-Rad, A. H., (2014). Responses of canola oil quality characteristics and fatty acid composition to zeolite and zinc fertilization under drought stress. *International Journal of Agricultural Science, 4*(1), 49–59.

57. Payandeh, Kh., Mojadam, M., & Derogar, N., (2018). Application of micronutrient elements on quantitative and qualitative yield of rapeseed under drought tension conditions. *Crop Physiology Journal, 38,* 23–37.
58. Habibi, M., Majidian, M., & Rabiei, M., (2014). Effect of boron, zinc and sulfur elements on grain yield and fatty acid composition of rapeseed. *Journal of Crops Improvement, 16*(1), 69–84.
59. Tousi, P., Tajbakhsh, M., & Esfahani, M., (2014). Effect of spray application of nano-Fe chelate, amino acid compounds and magnetic water on protein content and fatty acids composition of oil of soybean (*Glycine max*) in different harvest time. *Iranian Journal of Crop Sciences, 16*(2), 125–136.
60. Malakouti, M. J., Khogar, Z., & Khademi, Z., (2004). *New Method OF Feeding* (p. 868). Sana Publications. Tehran.
61. Monica, R. C., & Cremonini, R., (2009). Nanoparticles and higher plants. *Caryologia, 62,* 161–165.
62. Yadollahi, P., Asgharipour, M. R., Kheiri, N., & Ghaderi, A., (2015). Effects of drought stress and different types of organic fertilizers on the yield and yield components of safflower (*Carthamus tinctorius* L.). *Journal of Oil Plants Production, 1*(2), 27–40.
63. Kanwal, N., Hanif, M. A., Khan, M. M., Ansari, T. M., & Khalil-Ur-Rehman, R., (2016). Effect of micronutrients on vegetative growth and essential oil contents of *Ocimum sanctum*. *Journal of Essential Oil Bearing Plants, 19,* 980–988.
64. Rezaeieh, K. A. P., Gurbuz, B., & Eivazi, A., (2016). Effects of different zinc levels on vegetative growth and essential oil contents of some Iranian and Turkish cumin (*Cumin cyminum* L.) genotypes. *Journal of Essential Oil Bearing Plants, 19,* 1181–1191.
65. Hanif, M. A., Nawaz, H., Ayub, M. A., Tabassum, N., Kanwal, N., Rashid, N., Saleem, M., & Ahmad, M., (2017). Evaluation of the effects of zinc on the chemical composition and biological activity of basil essential oil by using Raman spectroscopy. *Industrial Crops and Products, 96,* 91–101.
66. Nakhzari, M. A., Ghaffari, S., Rahemi, K. A., & Salahi, F. M., (2016). The effect of different levels of sulfur and thiobacillus fertilizer on yield, yield components and oil percentage of sunflower in Gonbad area. *Journal of Oil Plants Production, 2*(2), 25–34.
67. Asare, E., & Scarisbrick, H., (1995). Rate of nitrogen and sulfur fertilizers on yield, yield components and seed quality of oilseed rape (*Brassica napus* L.). *Field Crops Research, 44*(1), 41–46.
68. Ravi, S., Channal, H. T., Hebsur, N. S., Patil, B. N., & Dharmatti, P. R., (2008). Effect of sulfur, zinc and iron nutrition on growth, yield, nutrient uptake and quality of safflower (*Carthamus tinctorius* L.). *Karnataka Journal Agriculture Science, 32,* 382–385.
69. Moradi, T. M. R., Kazemi, Z., & Siadat, S. A., (2016). Effect of boron application on some of morphological characteristics and grain yield, oil and nitrogen of spring canola (*Brassica napus* L.) in different sowing dates. *Journal of Oil Plants Production, 3*(1), 55–69.
70. Noctor, G., & Foyer, C. H., (1998). Ascorbate and glutathione: Keeping active oxygen under control. *Annual Review of Plant Physiology and Plant Molecular Biology, 49,* 249–279.
71. Shalata, A., & Neumann, P. M., (2001). Exogenous ascorbic acid (vitamin C) increases resistance to salt stress and reduces lipid peroxidation. *Journal of Experimental Botany, 52,* 2207–2211.

72. Towo, E., Mgoba, C., Ndossi, G. D., & Kimboka, S., (2006). Effect of phytate and iron-binding phenolics on the content and availability of iron and zinc in micronutrients fortified cereal flours. *African Journal of Food, Agriculture, Nutrition and Development, 6,* 1–14.
73. Bharti, K., Pandey, N., Shankhdhar, D., Srivastava, P. C., & Shankhdhar, S. C., (2013). Improving nutritional quality of wheat through soil and foliar zinc application. *Plant, Soil and Environment, 59*(8), 348–352.
74. Towhidi, M. H. R., (2017). Effect of foliar application of ascorbic acid on quantitative and qualitative traits as well as some biochemical changes in leaves of grain corn (*Zea maize* L.) under water deficit stress. *Iranian Journal of Field Crop Science, 48*(2), 365–375. doi: 10.22059/ijfcs.2017.140380.654018.
75. Tariq, M., Ahmad, B., Adnan, M., Mian, I. A., Khan, S., Fahad, S., Saleem, M. H., et al., (2022). Improving boron use efficiency via different application techniques for optimum production of good quality potato (*Solanum tuberosum* L.) in alkaline soil. *PLoS One, 17*(1), 1–13. e0259403. doi: 10.1371/journal.pone.0259403.
76. Hajduk, E., Gąsior, J., Właśniewski, S., Nazarkiewicz, M., & Kaniuczak, J., (2017). Influence of liming and mineral fertilization on the yield and boron content of potato tubers (*Solanum tuberosum* L.) and green mass of fodder sunflower (*Helianthus annuus* L.) cultivated in loess soil. *Journal of Elementology, 22*(2), 411–426.

CHAPTER 13

Biofortification of Iron for Nutrition and Simultaneous Bioremediation of Iron-Contaminated Soil

SAPNA CHANDWANI and NATARAJAN AMARESAN

C. G. Bhakta Institute of Biotechnology, Uka Tarsadia University, Maliba Campus, Bardoli, Surat, Gujarat, India

ABSTRACT

Iron is a micronutrient which is considered essential for plants to carry out physiological processes such as photosynthesis, respiration, oxygen transport, oxygen metabolism, DNA and RNA synthesis, cellular proliferation, electron transfer, etc. Although iron is present in bulk amounts on the surface of the earth, but still, it is not available in soil for plants. Iron solubility is very low under aerobic conditions because of the prevalence of ferric ions (Fe^{3+}), which are generally found as oxihydroxide polymers, particularly in calcareous soils, affecting the iron supply for plant uptake. Increased amount of ferrous ions (Fe^{2+}) is generated in flooded soils, i.e., in anaerobic and acidic soils, by the reduction of iron oxides (Fe^{3+}) leading to Fe toxicity due to iron uptake in excess amount. This iron toxicity affects the plant growth and agricultural productivity hampering the potential of plants in terms of plant uptake and resulting in slow growth rate and low biomass. Therefore, an efficient strategy for iron uptake by plants is the secretion of microbial phytosiderophores. Recently, plant-based techniques such as phytoremediation/bioremediation were considered as environmentally friendly and low-cost technologies to clean iron-polluted soils via plant breeding, plant volatizing, and plant extraction. Biofortification and bioremediation are closely connected similar to the two sides of a coin. Therefore, this chapter

Phytoremediation and Biofortification: Strategies for Sustainable Environmental and Health Management. Nand Kumar Singh, Shadma Afzal, & Tariq Aftab (Eds.)

deals with the transgenic technology that increases the iron accumulation in plants or staple crops, thereby using it for both biofortification and simultaneous bioremediation of iron.

13.1 INTRODUCTION

Iron (Fe) found to be the fourth most abundant element in the earth's crust, which is not readily available in soil. During aerobic conditions, Fe binds to soil particles thereby forming insoluble complexes at neutral or alkaline pH [1]. Monotonous plant-based diets consumption with little meat provides low bioavailability of Fe because of the inhibition of Fe absorption in the presence of polyphenols and phytates [2, 3]. Fe absorption is less than 10% in plant-based diets resulting in the prevalence of Fe deficiency worldwide [3]. Fe deficiency affects the health and development of human beings and hampers their socio-economic development. It has been found that around 2 billion of the world's population are Fe deficient [4], 21% of female teenagers between the age group 12–18 years, and 18% of women between age group of 16–64 years in the UK are Fe deficient [5]. Around 2% of children age group 1–2 years, 9–11% of non-pregnant women age group 16–49 is Fe deficient, with 2-fold higher frequency in less educated, poor minority populations [6]. According to WHO, in developing countries, 39% of children aged 5 years, 42% of women, 48% of 5–14 years age group children and 52% of pregnant women are anemic [4]. As a result, a major threat towards the health and development of human populations around the world is the deficiency of Fe and to combat with Fe deficiency, diversification or dietary modification, crops biofortification, food fortification or Fe supplementation are adopted.

Iron generally exists in two forms – ferrous (Fe^{+2}) and ferric (Fe^{+3}) forms and plays an important role in the physiological and metabolic functions of plants [1]. Fe is a micronutrient that is required in critical amounts and if present in deficient or excess amounts, growth, and development of plants may be severely affected [7]. In some regions, Fe toxicity is less severe than Fe deficiency; thus, many crop plants are not able to tolerate excess of iron, especially in acidic soils. For example, in Asia, South America and humid tropical regions of Africa, where soils are acidified or often flooded, therefore plant growth and development is hampered by excess Fe [8, 9]. Setter et al. [10] reported that Fe toxicity affects wheat shoot tissues and found a strong negative relationship between dry weight (DW) of shoot and

Fe concentration in wheat plants [8, 10]. Root and shoot elongation of wheat are also inhibited by the excess Fe, and a linear relationship has been found between wheat shoot elongation and Fe exposure [11]. Furthermore, wheat grain yield is also adversely affected by excess Fe [12]. Wheat is considered as the major staple food crops of India and Fe is absorbed via its roots by the strategy II through the uptake and release of Fe-chelator mugineic acid phytosiderophores from the rhizosphere. Bacteria take up both associative organic Fe and dissociative inorganic Fe more efficiently via combining with siderophores through different ATP binding cassette transporters. It has been revealed that when Fe is deficient, plant growth-promoting bacteria (PGPB) provides a basis for a microbiological strategy for Fe biofortification via producing different metabolites and regulating transporters. On the other hand, PGPB with siderophore production ability helps to reduce Fe toxicity in case of iron abundancy through negative regulation of Fe uptake and strong bioleaching capacity [13, 14]. Therefore, this chapter mainly focuses on the recent knowledge of Fe deficiency and Fe toxicity and PGPB role in the biofortification of Fe for mineral nutrition and simultaneous bioremediation of Fe toxicity in soil.

13.2 BIOFORTIFICATION OF FE BY SIDEROPHORE PRODUCING PGPB

Currently, many PGPB have been reported to improve nutrient absorption in the plants, as these bacteria secrete different types of signals triggering the phytosiderophore secretion and plants hormones to regulate the uptake of Fe [15]. These bacteria produce siderophores, volatile compounds, and organic acids, and help in biofilm formation [16, 17]. These molecules improve the bioavailability of Fe in plants by solubilizing and mobilizing Fe in the soil. In addition, it is possible that bacteria producing siderophores may activate genes that act to play a role in Fe deficiency in plants. For example, Verma et al. [18]; and Frietas et al. [19] revealed that *Paenibacillus* spp. and *Bacillus* spp. plays an essential role in coping with shortage of iron and fortifying Fe to plants [18, 19]. Further, it has been reported that *Paenibacillus polymyxa* BFKC01 activates divalent metal transporter *irt1* and membrane bound Fe chelate reductase *fro2* and related genes, thereby increasing the release of phenolic compounds in plants [20]. Similarly, PGP *Bacillus altitudinis* increased Fe content of pigeonpeas and chickpeas by 12–18% [21] whereas, *Bacillus pichinotyi* inoculation increased the Fe content of the grains up to 70% [22].

Kumar et al. [23] reported that *Enterobacter* sp. consortium, *Bacillus megaterium,* and *Arthrobacter chlorophenolicus* increased Fe content of grains by 42% under field conditions and by 49% under pot conditions [23]. In another set of study, Kumar et al. [11] showed increased 16–18% Fe content in grains under pot and field condition via co-inoculation of *Microbacterium arborescens*, *Serratia marcescens,* and *Enterobacter* sp. [11]. Sharma et al. [24] reported that PGPB-producing siderophores showed promising results when Fe fortified bacterial siderophore was applied in wheat plant, significant increase in grain yield by 45% was noticed and concentration of Fe in grains by 200% was recorded [24]. Similarly, Chandwani et al. [1] revealed through their research findings that inoculation of Fe-tolerant siderophore producing bacteria enhanced nitrogen, phosphorus, and Fe uptake in the rice plant [1] whereas Rasouli-Sadaghiani et al. [25] reported that only siderophore producing bacteria *Pseudomonas putida* improved Fe availability in wheat contrasting to *Pseudomonas fluorescens* and *Pseudomonas aeruginosa* [25].

13.2.1 MECHANISM OF FE TRANSPORT ACROSS CELL MEMBRANE VIA SIDEROPHORE PRODUCING MICROORGANISMS (CONSIDERING ESCHERICHIA COLI AS A MODEL ORGANISM)

Escherichia coli, a Gram-negative bacterium, has three iron stress-induced systems, a high affinity for transport of Fe and a low affinity for inorganic iron. Iron transport is very well characterized in the *E. coli* characteristics, thereby allowing it to serve as a role model organism. The primary siderophore, enterobactin (a catecholate siderophore) and aerobactin (a hydroxamate siderophore) in some strains are two different types of siderophores which are synthesized by *E. coli.* Two separate systems are involved in the transport of these siderophores: one for enterobactin and the other for some hydroxamate and aerobactin siderophores, such as coprogen, certain ferrichromes and ferrioxamines, and rhodotorulic acid production via other bacteria and fungi [26, 27].

Bagg & Neilands [28] through genetic studies reported that several gene products in *E. coli* which are negatively regulated via Fe-binding protein known as *Fur* involved in ferric uptake regulation are required for siderophore transport system [28]. Around 16 genes are required for enterobactin synthesis and transport systems, seven are required for siderophore production, eight genes that are present on the outer envelope, cytoplasmic membrane and periplasm for ferrienterobactin transport proteins and one gene for a protein in the cell cytoplasm that removes iron from the siderophore. *E. coli* require genes for a periplasmic protein FhuD and two cytoplasmic membrane proteins FhuB and FhuC, and individual genes involved in receptor proteins for each ferric hydroxamate type of siderophore transport. All such genes are present within the bacterial chromosome with the exception of the genes required for transport and synthesis of aerobactin, that are to be carried on pColV plasmids. Three gene products ExbB, ExbD, and TonB (not iron regulated) are needed for ferric-hydroxamate, ferrienterobactin, vitamin B12 transport systems and ferric dicitrate where the protein TonB helps in energy transfer from cytoplasm to the outer membrane and proteins ExbB and ExbD are involved in physically stabilizing TonB and are located in the cytoplasmic membrane [29]. After transporting to the periplasm, the cytoplasmic membrane proteins set for subsequent transport of ferrienterobactin involves FepB, FepC, FepD, and FepG, for ferric-dicitrate siderophores FecB, FecC, FecD, and FecE are involved and for ferric-hydroxamate siderophore transport into the cytoplasm FhuB, FhuC, and FhuD are involved [30].

Siderophore transporter or receptor systems are not well characterized in other bacteria and fungi compared to *E. coli* due to unavailability of mutant strains required to identify genes that code for siderophore transport proteins [26]. *E. coli* has highly specific receptors whereas other microorganisms have receptors for various siderophore types comprising different degrees of specificity. Therefore, for making use of siderophores produced by bacteria and fungi, microbes employ two different strategies: Firstly, a combination of highly specific receptors is employed; secondly, only one or a few broad specificity receptors are required. From an ecological perspective, the benefits of using such strategies were not clearly examined, but it was hypothesized that for recognizing specific chelates, highly specific receptors were designed for siderophore binding with lower K_m and increased efficiency to scavenge Fe-siderophores and also at very low concentrations. This system is useful for a particular type of siderophore. Recently, there has been evidence that some microorganisms have receptor proteins which rely on the basis of adjacent functional groups and structural features of the metal center for recognizing siderophores. For ferrated siderophores, the metal center has three bidentate ligands with octahedral coordination around the Fe^{3+} atom and two enantiomeric forms for imparting a twist whether left-or right-handed to the coordinating ligands while hydroxamate siderophores have cis and trans configurations [26].

13.3 IRON UPTAKE MECHANISMS FOR SIDEROPHORES

1. **Removal of iron in the cytoplasm through siderophore transporter and via chelate destruction or deferrated chelate recycling:** Microorganisms use three different mechanisms involving siderophore-mediated transport throughout their cell membranes. One of a review article by Scarborough [31] stated that a membrane-spanning proteins are involved in these mechanisms, which may bind to the substrate undergoing a conformational change, thereby facilitating transportation into the cell [31]. However, the location of iron release differs from the other mechanisms. For example, a transporter or permease is involved in one mechanism for the formation of intact ferric-siderophore complexes. Iron gets cleaved from the siderophore after transport with the help of cytoplasmic reductase, reducing the chelated Fe^{3+} and dissociating Fe^{2+} from unstable ferrous complex. Then the desferrisiderophore is resecreted back to the environment (e.g., aerobactin: recycled) or destroyed (e.g., enterobactin: non-recycled). Nonreductive removal of iron is also possible via hydrolytic destruction of the chelate [27].
2. **Removal of iron at the cell surface with direct shuttle mechanism without concomitant siderophore transport:** The second type of mechanism called direct shuttle mechanism, also referred to as taxi by others, where the ferric type siderophore binds with the cell surface receptor, thereby cleaving iron and transporting it simultaneously without concomitant desferrisiderophore transport [32]. It involves a reductive process in which Fe^{2+} is transferred to carrier protein and in contrast involves a non-reductive system where direct exchange of ligand of Fe^{3+} to carrier occurs. This is how Fe is removed from the cell surface via a direct shuttle mechanism [32].
3. **Removal of iron at a site distant from the ion channel with indirect shuttle mechanism:** The third mechanism is called the indirect shuttle mechanism (extracellular dissociation), where the reductive Fe is removed from a site distant to the carrier protein. The mechanism may be not much efficient, as Fe transport is not directly linked to Fe^{2+} release. The dissociative Fe^{2+} may reoxidize after reduction and precipitated on the cell surface or diffuse to the carrier protein or ion channel for removal of inorganic iron [33]. Based on the current knowledge and understanding of highly specific transporter/receptor

> systems, ferric siderophore transporter mechanism was taken into account for high-affinity microbial accretion of iron. For siderophore binding, recognition, and transport, the simple transporter employs a single protein, although the transporters can employ multiple protein systems, as in the case of ferric-siderophore transporter systems in the membranes and *E. coli* periplasm. In the case of *N. crassa,* a multicomponent system was proposed, i.e., a fungus has a common transporter and two separate receptors for ferric-coprogen and ferric-ferrichrome [33].

Very few microorganisms make use of direct shuttle systems compared to siderophore transporters. It is very difficult to differentiate such systems from recycling siderophore transport systems, as none of them accumulate siderophore by the cell. In some microorganisms, for making use of siderophores which they do not produce, direct shuttle mechanism has been reported. A review article by Muller & Raymond [34], stated Ferrioxamine B (FOB) utilization by *Bacillus, Neurospora, and Salmonella* [34]. Based on such examples, a direct shuttle mechanism has been emerged as a reductase complex or as a specific receptor or reductase complex, making use of various types of siderophore production by other microbes. Iron removal at the surface of the cell using a chelate with a protein recognizing general properties at the metal center and avoiding complications involved in transporting diverse, relatively large intact ferric siderophores. Thus, the primary mechanism that makes use of a production and transportation of a siderophore via a microorganism is the direct shuttle with a high specificity mechanism. For example, *Ustilago sphaerogena* and ferrichrome A and *R. pilmanae* and rhodotorulic acid [32, 35].

13.4 BIOFORTIFICATION OF IRON BY AGRONOMIC PRACTICES

Biofortification via agronomic practices is the process of application of micronutrient fertilizers to plant leaves (foliar) or soil to enhance the micronutrient concentration of edible parts of food crops [36]. Agronomic biofortification is one of the most appropriate approaches to attenuate micronutrient deficiencies in the human population and the sustainable production of food and feed [37]. Micronutrient deficiency sometimes called hidden hunger, is mainly associated with poor diet. Agronomic biofortification thus helps to increase crop yield as well as nutrient content of staple crops, and thus can alleviate micronutrient deficiency in plants and humans who can consume them [36]. In this era of green revolution, staple crops such as wheat, maize, rice, millet, sweet potato, legumes, and sorghum were primarily focused on meeting the requirements of rapidly growing population [38].

Agronomic biofortification temporarily increases the nutritional value of crops and thus fulfills human nutritional requirements [39]. Agronomic biofortification of Zn, Se, and Fe is the most widely performed. There are a number of techniques by which the biofortification is achieved. Foliar and soil applications are mostly carried out in the case of biofortification. Foliar application of micronutrient-rich fertilizers stimulates more efficient uptake of nutrients and proficient allocation in edible parts of the plant rather than soil application in the case of green leafy vegetables. Foliar application requires a large surface area for better absorption of nutrients in sufficient amount [40]. However, a limitation of foliar application is that immediate rainfall can easily wash off applied fertilizers. Such techniques are proficient for micronutrients that can be absorbed by plants, such as Zn, Fe, and Se, but cannot be used for micronutrients that are synthesized by plants [41].

The effectiveness of biofortification is measured by a variety of variables, such as the state of micronutrients and their availability for plant absorption, fertilization techniques, nature of soil, cropping systems, and soil alteration [42]. Fe is also a micronutrient, and many plants suffer from Fe deficiency because it is usually present in an insoluble form in soil. Soil application of Fe in soil is in insoluble forms and hence Fe is precipitated, which cannot be absorbed by plants [36]. An experiment conducted under greenhouse conditions with soil application of Fe and Zn on wheat, Zn and Fe concentration did not improve significantly [43]. The most efficient agronomic practice for enrichment of Fe is foliar application of Fe or litter fertilization [36]. Foliar applications increased the Fe concentration in wheat and rice [44]. There are many cases in which no response was recorded in the foliar application of inorganic or chelated Fe fertilizers [45].

Macronutrients play important roles in plants. In several research studies, a positive interaction has been reported between NPK fertilizers (nitrogen, phosphorus, and potassium fertilizers) and Fe concentration in grains [46–49]. The presence of nitrogen increased the Fe content by 15% in rice, and potassium increased the Fe content in rice [50]. The reason for the increased Fe content is that nitrogen and potassium are associated with root development, re-localization, and shoot transport, which elicit iron uptake in rice grains [36, 45, 47, 51]. Phosphorus can reduce toxic effect in plants and increase Fe and Zn uptake because of its dilution effect [36].

Among the various approaches used to enhance the iron concentration in rice, foliar fertilization is a rapid and efficient method. In general, Fe compounds solubility is low, but reducing conditions and low pH facilitate the conversion of Fe^{3+} to plant available Fe^{2+} [52]. Several researchers have verified that it is more profitable to use various cost-effective approaches instead of single-costlier approach to cap deficiency of nutrients [53]. To solve a problem of deficiency of micronutrients in the human population, plant researchers are focusing on the strategies of fertilizers application or plant breeding strategies to increase the bioavailability and concentration of mineral in edible portions of crop plants [54–57].

13.5 IRON TOXICITY IMPACTS IN PLANTS AND ITS BIOREMEDIATION STRATEGIES

Iron is an important micronutrient for the plant cycle as it regulates various physiological processes, such as oxygen transport, respiration, photosynthesis, cellular proliferation, electron transfer, oxidative metabolism, and DNA and RNA synthesis [1, 58]. However, if present in excess amounts, it significantly hampers growth and development of the plants, photosynthetic efficiency, carbon metabolism, respiration, and enzyme activity that cause oxidative stress via generation of free-catalyzed reactive oxygen species (ROS) [59, 60]. Fe is generally present in ferric oxides form (Fe^{3+}) in soils which is an insoluble form, whereas high amount of ferrous ions (Fe^{2+}) is generated by the reducing ferric oxides during anaerobic conditions [61]. Due to the Fe interflow from upland slopes, the soluble iron content in the soil increases, approximately 1,500 ppm of Fe was found to expose sites [62]. In addition, industrial discharge from various industries also affects the Fe content of the soil [63]. Poor water management also deteriorates the properties of the soil and promotes Fe^{2+} uptake in soil, thereby resulting in iron toxicity as Fe^{2+} is either absorbed by plants or is stored in leaves. Various anthropogenic activities are involved for release of Fe in the environment, such as the textile industry, tanneries, steel industry, paint, pigment manufacturing, laundry bluing, sludge disposal in water treatment plants, blueprint papers, etc. [64, 65]. Fe also imposes toxicity on important biomolecules, inducing changes in the structure of the photosynthetic apparatus and leading to ferroptotic cell death resulting in carbon metabolism retardation [60]. To improve yield as well as quality of crops, some agricultural practices such as the use of microbes, organic amendments, bioremediation, and phytoremediation, and several biological strategies can provide fruitful results for purifying Fe-contaminated soil and for enhancing quality and production of crops in soils having Fe contamination [28].

13.6 MICROBES-ASSISTED REMEDIATION USING SIDEROPHORE PRODUCING ABILITY

Microorganisms with high efficiency in changing the redox state, having the ability to precipitate or sequester heavy metals can be used for Fe remediation (Figure 13.1). Moreover, various fungal species can remediate Fe, such as *Agraicus bitorguis, A. bisporus*, *Pleurotus floridianus* [66, 67], *Mucor*, *Aspergillus* spp. [68], *Rhizopus,* and *Penicillium* [68]. Oladipo et al. [69] reported a higher ability of Fe bioremediation of extracted gold mine strains of fungus, like *Trichoderma ghanense, Rhizopus microspores,* and *Fomitopsis meliae* [69]. Botebol et al. [70] reported the ability of different algal species to uptake Fe [70], and the study by Sutak et al. [71] reported the *Emiliania huxleyi* role in Fe uptake and concluded that it prefers Fe^{2+} over Fe^{3+} [71]. To eliminate Fe toxicity [72], three different species of soil microalgae (*Chamydiminas* spp. MM7, *Chlorococcum* spp. MM11, and *Chlorella* spp. MM3) were used and revealed *Chlorella* to be more tolerant than other two species [72]. Initially, PGPRs were reported to promote growth and development in plants; however, PGPRs were also reported to ameliorate the phytotoxicity of heavy metals [73]. He et al. [13] reported associative organic and dissociative inorganic uptake of Fe by PGPB using siderophores having different ATP-binding cassette transporters. Further, his study also revealed the Fe toxicity reduction in wheat plants by using PGPB bioleaching ability, which helps in downregulation of Fe-uptake transporters [13]. To alleviate the adverse effects of Fe toxicity in rice plants, *Paenibacillus* spp. FeS53 has also been previously reported [74]. Due to low environmental effects and high colonization rate, endophytic bacteria are more potent in remediating Fe toxicity [13].

He et al. [13] reported that endophytic *Bacillus altituinis* WR 10 had the potential of improving Fe tolerance and absorption via upregulation of ferritin-encoding genes; this strain was found to have a high content of siderophores and IAA production, as well as capable of producing ACC deaminase and phytase activity, indicating its potent role in alleviating Fe toxicity [13]. Luna et al. [75] revealed 95% Fe removal efficiency by *Candida sphaerica* due to its biosurfactant producing ability [75]. Zhou et al. [76] reported the

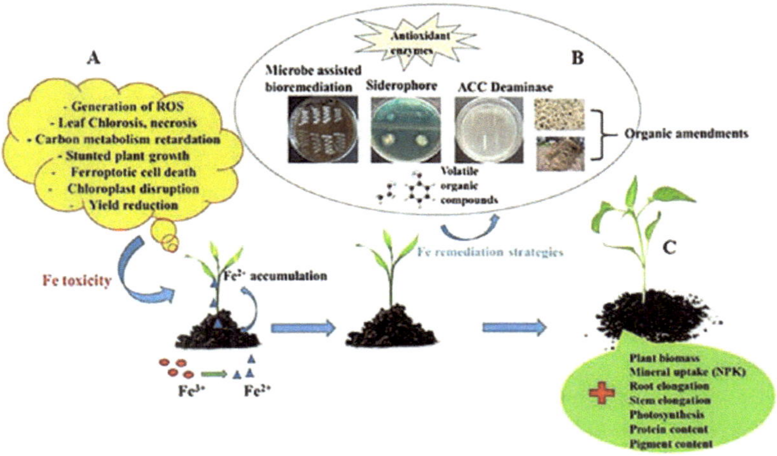

FIGURE 13.1 Overall mechanism of Fe based toxicity and Fe remediation in plants. (A) Effect of Fe toxicity includes generation of ROS, leaf chlorosis and necrosis, carbon metabolism retardation, stunted plant growth, ferroptotic cell death, chloroplast destruction, yield reduction, etc. (B) Depiction of Fe remediation through different techniques, including microbe assisted remediation, use of volatile organic compounds, use of organic amendments, use of antioxidant machinery, use of siderophore and ACC deaminase producing PGPB, etc. (C) Implications of Fe remediation techniques resulting in increased plant biomass, mineral uptake, protein content, photosynthetic rate, pigment content, root, and shoot elongation, etc. [denotes increase].

efficiency of both endophytes and rhizobacteria of *Bacillus* spp. in protecting wheat plant against Fe toxicity [76]. Chandwani et al. [1] reported the role of *Bacillus tequilensis, B. weidmanii, B. cereus, B. paramycoides*, and *Pseudomonas aeruginosa* in mitigating Fe stress from highly polluted Fe-stressed sites, as these microbes were found to be Fe-tolerant and have the ability to produce siderophore, IAA, and ACC deaminase [1].

13.7 FE REMEDIATION VIA USING ORGANIC AMENDMENTS

A decline in the redox potential associated with soil flooding is the major cause of Fe toxicity. Generally, when high amount of reducible Fe binds to organic matter (OM), Fe toxicity occurs rapidly, thereby increasing the symptoms of iron toxicity. The most important parameter that affects the biological and physicochemical properties of soil is the OM. Phytoavailability and metal transport are controlled by adsorption reactions in aquatic and terrestrial environments. Soil reaction (pH), OM, specific surface area, mineral nutrients in soil, cation exchange capacity, OM, clay content and background electrolytes are some of the factors affecting adsorption reactions [77]. The mobility of metals, especially Fe, is greatly affected by nature and amount of OM, as metals form complexes and chelates of varying stability with OM, ensuring plant growth enhancement in metal-contaminated soil via organic amendments [78]. Generally, to improve physicochemical properties, fertility, and to immobilize metals such as Fe in soils, metal-contaminated soils are amended with OM [79]. The stability of metals is altered by the conversion of OM into immobile forms in soils that bind to OM, carbonates, or oxides [79].

To immobilize Fe in soils, a variety of organic materials such as wood ash, sawdust, manure (farmyard, poultry, and pig manure), composts from different origins, sewage sludge, wood chips, biosolids, and bark chips are used [80] (Figure 13.1). The metal availability in soils is greatly influenced by the amount and source of OM and the time of application [81]. Even the influence of OM (whether fresh or decomposed OM) may vary in terms of the interaction of OM with metals; fresh OM bound with metals and are released during OM decomposition, and thus, as a result, mobility of metals increases due to less OM in soils [82]. Oxidation, hydrolysis, or depolymerization results in the transformation of OM, releasing dissolved OM and increasing metal availability [83]. Decomposition of OM increases retention of metal in soil due to conversion of low molecular weight (LMW) organic compounds to high molecular weight compounds during decomposition like amino acids, humic substances (HS), and organic acids, thereby forming complexes with metals that could alter metal availability [84]. An economic approach for remediating Fe-contaminated soil is the use of organic amendments to ensure better soil fertility and quality and a safe environment through C sequestration; however, this approach is still not yet familiar on a large scale. To adopt this technology commercially, scientists still need to focus on key challenges in this approach.

In one of the research studies conducted by Baruah et al. [85], during early stages of inundation, 60% to 80% of decomposed OM supplied in rice soils was mineralized through microbes using Fe^{3+} as an electron acceptor [85]. Bashir et al. [86] reported formation of complexes of OM with Fe^{2+} and that Fe^{2+} is present in the form of chelates [86]. In acid sulfate soils and alluvial clay, easily decomposed OM application such as the use of green manure should be avoided [87]. Moreover, the efflux of organic exudates is regulated by the microbial action of the rhizoflora, which, in turn is regulated by the root membrane. Therefore, root permeability is increased by potassium deficiency, accelerating the microbial action of Fe-reducing bacteria in the rhizosphere as well as LMW metabolites exudation [87]. To overcome Fe toxicity to aroeira (*Schinus therebinthi folius*), Dobbss et al. [88] reported the use of vermicompost HS as these substances mimic as a biostimulant, blocking excess of Fe outside the root system and increases CAT, POD, and APX activity [88].

13.8 IRON REMEDIATION USING ANTIOXIDANT MACHINERY

Under normal as well as stress conditions, antioxidant substances are considered as a strong prerequisite for the survival of plants. Several research studies have explored the role of antioxidants in protecting plants from Fe damage [89]. Antioxidants, such as ascorbate peroxidase (APX), superoxide dismutase (SOD), dehydroascorbate reductase (DHAR), ascorbate (AsA), and glutathione (GSH) reductase protect against tissue damage caused by iron toxicity and act against ROS for prevention of damage caused due to generation of ROS [90, 91] (Figure 13.1). During Fe toxicity, the antioxidant defense machinery is activated for maintaining Fe homeostasis at the cellular level [92]. Verma & Pandey [93] reported an increase in proline accumulation due to reduced degradation or higher synthesis in green grams as scavenging agents of O_2^- and OH^+ and proline accumulated more in roots than in shoot [93]. Mehraban et al. [94] reported lipid peroxidation increase, H_2O_2, peroxidase (POD) and a decrease in catalase (CAT) activity during Fe stress conditions. These authors reported that during Fe toxicity, free radical metabolism shifts towards the production of ROS like H_2O_2 due to the reduced activity of CAT and reported lower CAT activity and high phenolic content in rice [94]. Majerus et al. [95] revealed that after three days stress exposure, the glutathione (GSSH) oxidized content was not affected in roots and sheaths but inside leaf lamina concomitantly increased with a reduction of GR inhibition. They found that the earlier concentration of ascorbate (AsA) was in a small amount, but in aerial parts, it was significantly higher and 10-fold high in lamina compared to sheaths. They also found that in Fe-stressed plants, the concentration of AsA culminated after 24 h in laminae and in roots after 48 h. On the other hand, AsA concentration increased in sheaths and decreased in laminae after 72 h of applied stress, and this decrease was noted due to dehydroascorbate (DHA) and monodehydroascorbate (MDHA) synthesis [95]. Stein et al. [61] reported higher SOD activity under Fe stress in rice [61] and also reported in maize (*Zea mays*) [97]. Moreover, it has also been reported that under higher Fe stress, APX and POX activities increase in different type of plants such as rice [61], waterweed (*Elodea nuttallii*), maize, potato (*Solanum tuberosum*), Brazilian pepper tree (*Schinus terebinthifolius*) [64, 97, 99, 100]. Wu et al. [101] reported that during Fe stress, L-glactono-1-4-lactone and AsA application exogenously decreased lipid peroxidation [101]. The most important secondary metabolites that act as antioxidants are phenolics and polyphenols. A study reported that in case of Fe toxicity, polyphenol oxidase content remained same, while the phenolic content significantly increased in the shoots of rice plants under stress conditions [94]. Under high Fe stress conditions, improved concentrations of phenols, flavonoids, and carotenoids have also been reported [90].

13.9 PHYTOREMEDIATION OF FE TOXICITY

Phytoremediation of Fe is a bioremediation process that directly uses green plants for stabilization, removal, transfer, and degradation of excess Fe (contaminants) in the soil. Obnoxious free-floating macrophyte behavior during Fe phytoremediation has been reported in many studies [102]. The use of several microphytes such as *Pistia stratiotes, Salvinia culcullata, Eichhomia crassipes*, and *Lemna minor* has also been reported for Fe phytoremediation [103]. The phytostabilization impact of Alexandrian laurel (*Calophyllum inophyllum*) reported to have 100% survival in plants with no toxicity [104]. For enhancing Fe phytoremediation efficiency, 24 willow (*Salix* spp.) clones were experimented by Yang et al. [105] whereas for remediating Fe from landfill leachates, a role of kenaf (*Hibiscus cannabinus* L.) was reported and 0.06–0.58 mg of Fe per plant DW was estimated [106]. High Fe phytoremediation efficiency of sunflower was reported by Chauhan & Mathur [107], and phytoremediation of mixed metal solutions including Fe by *P. cummunis*, *Cyperus esculentus*, and *Typha augustifolia* has also been reported [107, 108]. It was revealed that *Cyperus esculentus* and *Typha augustifolia* are root accumulators and *P. cummunis* is a shoot accumulator for Fe, and the efficient uptake of Fe >1,000 μg/g was reported in all three plants [107, 108]. Phytoremediation techniques can be categorized as in subsections.

13.9.1 PHYTOEXTRACTION

The ability of plants, i.e., absorption of organic and inorganic pollutants with the help of roots and stems is chosen for the contaminants removal in the soil during the removal or harvesting of the plant [109]. This technique is considered to be valid for the recovery of contaminated areas, but the application of such technology requires more time than other techniques, therefore it is very difficult to apply such techniques to heavily polluted areas [110]. Fe-phytoextraction ability of *Psoralea pinnata* was observed up to three months, and it was found that *P. pinnata* was able to accumulate 55% of Fe [111]. The plants used in this method accumulated contaminant elements 100 times more compared to other plants [110]. *Euphorbiacea*, *Brassicacea*, *Lamiaceae*, *Scrophulariaceae*, and *Asteraceae* have been shown to accumulate heavy metals [110]. Moreover, by using Fe-tolerant bacteria, the phytoextraction ability of rapeseed oil (*Brassica napus*) was explored by Jinal et al. [14] and found that soil inoculants enhanced 57.91–128.31% increased Fe uptake [14].

13.9.2 PHYTOSTABILIZATION

Phytostabilization is a technique which converts toxic substances into less toxic substances using the plants' ability to store and immobilize heavy metals. As the toxicity of Fe is reduced by binding with biomolecules such as phytochelatins (PCs) and metallothioneins (MTs) through sedimentation, sorption, complexation, and reduction of metal valences [112]. This technique has a great advantage in protecting groundwater from heavy metal contaminants leaching deeper into the soil layers [113]. Thus, compared to other remediation practices that are very costly, phytostabilization offers a feasible alternative for restoring a functional ecosystem in a more substantial manner [113]. Phytostabilization of Fe using the macrophyte *Vossia cuspidata* and seasonal variations effect on the phytostabilization ability were reported by Galal et al. [114]. A study reported the ability of *Leucaena leucocephala* for absorption, translocation, and tolerance of Fe and its phytoremediation potential and photosynthetic mechanism with different concentrations of Fe in soil (0, 100, 200, 300, 400, and 500 mg Fe/dm^3 [115]. It has been reported that after cultivation of *Leucaena leucocephala,* Fe availability in the soil decreased, indicating that the Fe absorption capacity of the plant aggregated mainly in the roots, and only a small percentage of Fe was translocated to the shoots [115]. This confers that the species was found to be tolerant to the several doses of Fe analyzed and phytostabilize soil Fe, although plant growth is influenced by different concentrations of Fe present in soil and Fe accumulation is less. However, the high amount of Fe accumulation per biomass at 200 mg Fe/dm^3 was 9.48 mg because it is mainly a phytostabilization of Fe [115].

13.9.3 PHYTOSTIMULATION

It is a technique for stimulating compounds released by the roots to enhance the activity of the microbes, thereby boosting the phytoremediation process [116]. One of the studies revealed that inoculation of three arbuscular mycorrhizal fungi (AMF) (*Aculospora, Scutellospora*, and *Glomus*) and four plant growth-promoting rhizobacteria (PGPR) belonging to genera *Pseudomonas, Azotobacter, Paenibacillus,* and *Streptomyces* enhanced the uptake of iron in *Pennisetum glaucum* and *Sorghum bicolor* [117]. Further, it has been reported that presence of ethylene in low amount stimulate root elongation, whereas inhibits DNA synthesis and cell division when present in high amount. However, Chandwani & Amaresan [118] revealed that ethylene concentration in plants can be regulated or reduced using the enzyme ACC deaminase (1-aminocyclopropane-1-carboxylase deaminase), thereby reducing abiotic stress (environment stressors) in plants via balancing ethylene production [118, 122]. Plants with PGPR linked to the roots of the plant using exudates release this enzyme as a carbon and energy source for degrading Fe contaminants [119] (Figure 13.1).

13.9.4 PHYTOVOLATILIZATION

In this process soil contaminants are removed with the help of plants and are released into the atmosphere after readily changing into vapor via the leaves or foliage system [120]. During this process, contaminants are converted into volatile forms because of the combined effects of the metabolic potential of plants and microbes present in the rhizosphere [119]. The advantage of using this approach compared to other phytoremediation strategies is that phytovolatilization removes heavy metal (metalloid) contaminants and disperses them into gaseous compounds without the need of plant disposal or harvesting [121]. However, a high risk is associated with its application in the field as it is a remedial strategy where the pollutants are not removed completely; in fact, the pollutants are transferred from the soil to the atmosphere, and hence, the ambient air gets contaminated via toxic volatile compounds or via precipitation; they may be redeposited into the soil [121].

13.9.5 PHYTOFILTRATION

Phytofiltration is widely used for sterilizing polluted water or underground wastes by using plant roots, shoots, or seedlings [121, 122]. This process via roots is termed as rhizofiltration, via shoots caulofiltration and via seedlings termed as blastofiltration. Heavy metals are either adsorbed by roots of the plant or onto the root surface during rhizofiltration. Rhizosphere pH may change due to root exudates, resulting in the heavy metals precipitation on plant roots, thereby reducing the heavy metals movement to underground water [123].

For the process of rhizofiltration, the plants to be grown hydroponically to develop a large root system in clean water to assimilate the plants. After establishment of plants, clean water is then replaced with polluted water and transferred to the contaminated site for heavy metals remediation [124]. Plants are harvested and disposed of once the roots became saturated and ideally for rhizofiltration; plants (both terrestrial and aquatic) with high biomass production, tolerance to heavy metals like Fe, and having a dense root system are preferred [124]. During rhizofiltration, terrestrial plants sunflower (*H. annuus*) and Indian mustard (*B. juncea*) have been reported to show good capacity for heavy metal accumulation. Whereas aquatic species such as duckweed, poplar, azolla, hyacinth, and cattail have been used for the remediation of wetland water because of their high tolerance, fast growth, high accumulation of heavy metals, and high biomass production [98, 125]. Mustapha et al. [96] revealed the ability of *Typha latifolia, Cyperus alternifolius*, and *Cynodon dactylon* to translocate Fe from wastewater and revealed that a high amount of Fe accumulation was found in roots, followed by leaves and stems [96].

13.10 CONCLUSION AND FUTURE PROSPECTS

Iron malnutrition in humans and Fe toxicity are both major challenges that require the efforts of researchers from multiple disciplines, such as biotechnology, plant biology, plant breeding, environmental sciences, and plant nutrition. For the biofortification of Fe, the major targets are to enhance the trace element availability in the rhizosphere, from where it is translocated from roots to shoots and then redistributed to the grain tissues. Siderophore, volatile compounds, and organic acids play a significant role. Fe remediation methods include microbial-assisted remediation, phytoremediation, siderophore-producing PGPB inoculation, use of antioxidants, and use of nutrients and organic amendments. In case of soil having contaminants, phytostabilization can be the suitable option and for phytoextraction, metal/metalloid hyperaccumulation traits combined with high biomass production are critical. Both Fe remediation and fortification can be achieved via gaining knowledge of the mechanisms relying on ion transport, Fe acquisition, and homeostasis in plants. Moreover, there is an uttermost need to apply these technologies under real conditions.

KEYWORDS

- **antioxidants**
- **biofortification**
- **bioremediation**
- **chelation**
- **iron**
- **micronutrient**
- **siderophore**

REFERENCES

1. Chandwani, S., Chavan, S. M., Paul, D., & Amaresan, N., (2022). Bacterial inoculations mitigate different forms of iron (Fe) stress and enhance nutrient uptake in rice seedlings (*Oryza sativa* L.). *Environmental Technology & Innovation*, 102326.
2. Zimmermann, M. B., Chaouki, N., & Hurrell, R. F., (2005). Iron deficiency due to consumption of a habitual diet low in bioavailable iron: A longitudinal cohort study in Moroccan children. *The American Journal of Clinical Nutrition, 81*(1), 115–121.
3. Hurrell, R., (2002). How to ensure adequate iron absorption from iron-fortified food. *Nutrition Reviews, 60*, S7–S15.
4. WHO, U., (2001). UNU iron deficiency anemia. Assessment. *Prevention and Control. A Guide for Program* (pp. 1–114). *Managers.* https://www.scirp.org/(S(czeh2tfqyw2orz553k1w0r45))/reference/referencespapers.aspx?referenceid=1029425 (accessed on 22 June 2023).
5. Heath, A. L. M., & Fairweather-Tait, S. J., (2002). Clinical implications of changes in the modern diet: Iron intake, absorption and status. *Best Practice & Research Clinical Haematology, 15*(2), 225–241.
6. Scholl, T. O., (2005). Iron status during pregnancy: Setting the stage for mother and infant. *The American Journal of Clinical Nutrition, 81*(5), 1218S–1222S.

7. Yaseen, M., Abbas, T., Aziz, M. Z., Wakeel, A., Yasmeen, H., Ahmed, W., & Naveed, M., (2018). Microbial assisted foliar feeding of micronutrients enhance growth, yield and biofortification of wheat. *Int. J. Agric. Biol, 20*, 353–360.
8. Khabaz-Saberi, H., Rengel, Z., Wilson, R., & Setter, T. L., (2010). Variation for tolerance to high concentration of ferrous iron (Fe^{2+}) in Australian hexaploid wheat. *Euphytica, 172*(2), 275–283.
9. Guo, J. H., Liu, X. J., Zhang, Y., Shen, J. L., Han, W. X., Zhang, W. F., & Zhang, F. S., (2010). Significant acidification in major Chinese croplands. *Science, 327*(5968), 1008–1010.
10. Setter, T. L., Waters, I., Sharma, S. K., Singh, K. N., Kulshreshtha, N., Yaduvanshi, N. P. S., & Cakir, M., (2009). Review of wheat improvement for waterlogging tolerance in Australia and India: The importance of anaerobiosis and element toxicities associated with different soils. *Annals of Botany, 103*(2), 221–235.
11. Kumar, A., Maurya, B. R., Raghuwanshi, R., Meena, V. S., & Islam, M. T., (2017). Co-inoculation with *Enterobacter* and rhizobacteria on yield and nutrient uptake by wheat (*Triticum aestivum* L.) in the alluvial soil under indo-gangetic plain of India. *Journal of Plant Growth Regulation, 36*(3), 608–617.
12. Khabaz-Saberi, H., Barker, S. J., & Rengel, Z., (2012). Tolerance to ion toxicities enhances wheat (*Triticum aestivum* L.) grain yield in waterlogged acidic soils. *Plant and Soil, 354*(1), 371–381.
13. He, L., Yue, Z., Chen, C., Li, C., Li, J., & Sun, Z., (2020). Enhancing iron uptake and alleviating iron toxicity in wheat by plant growth-promoting bacteria: Theories and practices. *Int. J. Agric. Biol., 23*(1), 190–196.
14. Jinal, H. N., Gopi, K., Prittesh, P., Kartik, V. P., & Amaresan, N., (2019). Phytoextraction of iron from contaminated soils by inoculation of iron-tolerant plant growth-promoting bacteria in *Brassica juncea* L. Czern. *Environmental Science and Pollution Research, 26*(32), 32815–32823.
15. Roesti, D., Gaur, R., Johri, B. N., Imfeld, G., Sharma, S., Kawaljeet, K., & Aragno, M., (2006). Plant growth stage, fertilizer management and bio-inoculation of arbuscular mycorrhizal fungi and plant growth promoting rhizobacteria affect the rhizobacterial community structure in rain-fed wheat fields. *Soil Biology and Biochemistry, 38*(5), 1111–1120.
16. Sah, S., Singh, N., & Singh, R., (2017). Iron acquisition in maize (*Zea mays* L.) using *Pseudomonas* siderophore. *3 Biotech, 7*(2), 1–7.
17. Liu, D., Yang, Q., Ge, K., Hu, X., Qi, G., Du, B., & Ding, Y., (2017). Promotion of iron nutrition and growth on peanut by *Paenibacillus illinoisensis* and *Bacillus* sp. strains in calcareous soil. *Brazilian Journal of Microbiology, 48*, 656–670.
18. Verma, P., Yadav, A. N., Khannam, K. S., Kumar, S., Saxena, A. K., & Suman, A., (2016). Molecular diversity and multifarious plant growth promoting attributes of bacilli associated with wheat (*Triticum aestivum* L.) rhizosphere from six diverse agro-ecological zones of India. *Journal of Basic Microbiology, 56*(1), 44–58.
19. Freitas, M. A., Medeiros, F. H., Carvalho, S. P., Guilherme, L. R., Teixeira, W. D., Zhang, H., & Paré, P. W., (2015). Augmenting iron accumulation in cassava by the beneficial soil bacterium *Bacillus subtilis* (GBO3). *Frontiers in Plant Science, 6*, 596.
20. Zhou, C., Guo, J., Zhu, L., Xiao, X., Xie, Y., Zhu, J., & Wang, J., (2016). *Paenibacillus polymyxa* BFKC01 enhances plant iron absorption via improved root systems and

activated iron acquisition mechanisms. *Plant Physiology and Biochemistry, 105*, 162–173.

21. Gopalakrishnan, S., Vadlamudi, S., Samineni, S., & Sameer, K. C. V., (2016). Plant growth-promotion and biofortification of chickpea and pigeonpea through inoculation of biocontrol potential bacteria, isolated from organic soils. *Springerplus, 5*(1), 1–11.
22. Yasin, M., El-Mehdawi, A. F., Anwar, A., Pilon-Smits, E. A., & Faisal, M., (2015). Microbial-enhanced selenium and iron biofortification of wheat (*Triticum aestivum* L.)-applications in phytoremediation and biofortification. *International Journal of Phytoremediation, 17*(4), 341–347.
23. Kumar, A., Maurya, B. R., & Raghuwanshi, R., (2014). Isolation and characterization of PGPR and their effect on growth, yield and nutrient content in wheat (*Triticum aestivum* L.). *Biocatalysis and Agricultural Biotechnology, 3*(4), 121–128.
24. Sharma, S., Chandra, S., Kumar, A., Bindraban, P., Saxena, A. K., Pande, V., & Pandey, R., (2019). Foliar application of iron fortified bacteriosiderophore improves growth and grain Fe concentration in wheat and soybean. *Indian Journal of Microbiology, 59*(3), 344–350.
25. Rasouli-Sadaghiani, M., Malakouti, M. J., Khavazi, K., & Miransari, M., (2014). Siderophore efficacy of fluorescent pseudomonades affecting labeled iron (59 Fe) uptake by wheat (*Triticumae stivum* L.) genotypes differing in Fe efficiency. In: *Use of Microbes for the Alleviation of Soil Stresses* (pp. 121–132). Springer, New York, NY.
26. Crowley, D. E., Wang, Y. C., Reid, C. P. P., & Szaniszlo, P. J., (1991). Mechanisms of iron acquisition from siderophores by microorganisms and plants. In: *Iron Nutrition and Interactions in Plants* (pp. 213–232). Springer, Dordrecht.
27. Payne, S. M., & Neilands, I. B., (1988). Iron and virulence in the family *Enterobacteriaceae. CRC Critical Reviews in Microbiology, 16*(2), 81–111.
28. Bagg, A., & Neilands, J. B., (1987). Ferric uptake regulation protein acts as a repressor, employing iron (II) as a cofactor to bind the operator of an iron transport operon in *Escherichia coli. Biochemistry, 26*(17), 5471–5477.
29. Fischer, E., Günter, K., & Braun, V., (1989). Involvement of ExbB and TonB in transport across the outer membrane of *Escherichia coli*: Phenotypic complementation of exb mutants by overexpressed tonB and physical stabilization of TonB by ExbB. *Journal of Bacteriology, 171*(9), 5127–5134.
30. Burkhardt, R., & Braun, V., (1987). Nucleotide sequence of the fhuC and fhuD genes involved in iron (III) hydroxamate transport: Domains in FhuC homologous to ATP-binding proteins. *Molecular and General Genetics MGG, 209*(1), 49–55.
31. Scarborough, G. A., (1985). Binding energy, conformational change, and the mechanism of transmembrane solute movements. *Microbiological Reviews, 49*(3), 214–231.
32. Carrano, C. J., & Raymond, K. N., (1978). Coordination chemistry of microbial iron transport compounds. 10. Characterization of the complexes of rhodotorulic acid, a dihydroxamate siderophore. *Journal of the American Chemical Society, 100*(17), 5371–5374.
33. Huschka, H., Naegeli, H. U., Leuenberger-Ryf, H., Keller-Schierlein, W., & Winkelmann, G., (1985). Evidence for a common siderophore transport system but different siderophore receptors in *Neurospora crassa. Journal of Bacteriology, 162*(2), 715–721.
34. Müller, G., & Raymond, K. N., (1984). Specificity and mechanism of ferrioxamine-mediated iron transport in *Streptomyces pilosus. Journal of Bacteriology, 160*(1), 304–312.

35. Ecker, D. J., & Emery, T., (1983). Iron uptake from ferrichrome A and iron citrate in *Ustilago sphaerogena. Journal of Bacteriology, 155*(2), 616–622.
36. De Valença, A. W., Bake, A., Brouwer, I. D., & Giller, K. E., (2017). Agronomic biofortification of crops to fight hidden hunger in sub-Saharan Africa. *Global Food Security, 12*, 8–14.
37. Qureshi, M. T., Ahmad, M. F., Iqbal, N., Waheed, H., Hussain, S., Brestic, M., & Noorka, I. R., (2021). Agronomic bio-fortification of iron, zinc and selenium enhance growth, quality and uptake of different sorghum accessions. *Plant, Soil and Environment, 67*(10), 549–557.
38. Jan, B., Bhat, T. A., Sheikh, T. A., Wani, O. A., Bhat, M. A., Nazir, A., & Rashid, A., (2020). Agronomic bio-fortification of rice and maize with iron and zinc: A review. *Int. Res. J. Pure Appl. Chem., 21*, 28–37.
39. Cakmak, I., & Kutman, U. Á., (2018). Agronomic biofortification of cereals with zinc: A review. *European Journal of Soil Science, 69*(1), 172–180.
40. Lawson, P. G., Daum, D., Czauderna, R., Meuser, H., & Härtling, J. W., (2015). Soil versus foliar iodine fertilization as a biofortification strategy for field-grown vegetables. *Frontiers in Plant Science,* 450.
41. Lyons, G., (2018). Biofortification of cereals with foliar selenium and iodine could reduce hypothyroidism. *Frontiers in Plant Science, 9*, 730.
42. Bruulsema, T. W., Heffer, P., Welch, R. M., Cakmak, I., & Moran, K., (2012). Fertilizing crops to improve human health: A scientific review. *Better Crops, 2*, 96.
43. Melash, A. A., & Mengistu, D. K., (2020). Improving grain micronutrient content of durum wheat (*Triticum turgidum* var. Durum) through agronomic biofortification to alleviate the hidden hunger. *Advances in Agriculture, 2020.*
44. Shahzad, Z., Rouached, H., & Rakha, A., (2014). Combating mineral malnutrition through iron and zinc biofortification of cereals. *Comprehensive Reviews in Food Science and Food Safety, 13*(3), 329–346.
45. García-Bañuelos, M. L., Sida-Arreola, J. P., & Sánchez, E., (2014). Biofortification-promising approach to increasing the content of iron and zinc in staple food crops. *Journal of Elementology, 19*(3).
46. Sperotto, R. A., Ricachenevsky, F. K., De Abreu, W. V., & Fett, J. P., (2012). Iron biofortification in rice: It's a long way to the top. *Plant Science, 190*, 24–39.
47. Prasad, R., Shivay, Y. S., & Kumar, D., (2014). Agronomic biofortification of cereal grains with iron and zinc. In: *Advances in Agronomy* (Vol. 125, pp. 55–91). Academic Press.
48. Velu, G., Ortiz-Monasterio, I., Cakmak, I., Hao, Y., & Singh, R. Á., (2014). Biofortification strategies to increase grain zinc and iron concentrations in wheat. *Journal of Cereal Science, 59*(3), 365–372.
49. Bindraban, P. S., Dimkpa, C., Nagarajan, L., Roy, A., & Rabbinge, R., (2015). Revisiting fertilizers and fertilization strategies for improved nutrient uptake by plants. *Biology and Fertility of Soils, 51*(8), 897–911.
50. Gregorio, G. B., Senadhira, D., Htut, H., & Graham, R. D., (2000). Breeding for trace mineral density in rice. *Food and Nutrition Bulletin, 21*(4), 382–386.
51. Petry, N., Boy, E., Wirth, J. P., & Hurrell, R. F., (2015). The potential of the common bean (Phaseolus vulgaris) as a vehicle for iron biofortification. *Nutrients, 7*(2), 1144–1173.

52. Fang, Y., Wang, L., Xin, Z., Zhao, L., An, X., & Hu, Q., (2008). Effect of foliar application of zinc, selenium, and iron fertilizers on nutrients concentration and yield of rice grain in China. *Journal of Agricultural and Food Chemistry, 56*(6), 2079–2084.
53. Bohra, A., Jha, U. C., & Kumar, S., (2016). Enriching nutrient density in staple crops using modern "-omics" tools. In: *Biofortification of Food Crops* (pp. 85–103). Springer, New Delhi.
54. White, P. J., & Broadley, M. R., (2009). Biofortification of crops with seven mineral elements often lacking in human diets–iron, zinc, copper, calcium, magnesium, selenium and iodine. *New Phytologist, 182*(1), 49–84.
55. Çakmak, İ., Torun, A. Y. F. E. R., Millet, E., Feldman, M., Fahima, T., Korol, A., & Özkan, H., (2004). Triticum dicoccoides: An important genetic resource for increasing zinc and iron concentration in modern cultivated wheat. *Soil Science and Plant Nutrition, 50*(7), 1047–1054.
56. Graham, R. D., Welch, R. M., Saunders, D. A., Ortiz-Monasterio, I., Bouis, H. E., Bonierbale, M., & Twomlow, S., (2007). Nutritious subsistence food systems. *Advances in Agronomy, 92*, 1–74.
57. Pfeiffer, W. H., & McClafferty, B., (2007). HarvestPlus: Breeding crops for better nutrition. *Crop Science, 47*, S-88.
58. Kroh, G. E., & Pilon, M., (2020). Regulation of iron homeostasis and use in chloroplasts. *International Journal of Molecular Sciences, 21*(9), 3395.
59. Onyango, D. A., Entila, F., Egdane, J., Pacleb, M., Katimbang, M. L., Dida, M. M., & Drame, K. N., (2020). Mechanistic understanding of iron toxicity tolerance in contrasting rice varieties from Africa: 2. Root oxidation ability and oxidative stress control. *Functional Plant Biology, 47*(2), 145–155.
60. Zahra, N., Hafeez, M. B., Shaukat, K., Wahid, A., & Hasanuzzaman, M., (2021). Fe toxicity in plants: Impacts and remediation. *Physiologia Plantarum, 173*(1), 201–222.
61. Stein, R. J., Duarte, G. L., Spohr, M. G., Lopes, S. I. G., & Fett, J. P., (2009). Distinct physiological responses of two rice cultivars subjected to iron toxicity under field conditions. *Annals of Applied Biology, 154*(2), 269–277.
62. Yoshida, S., (1981). *Fundamentals of Rice Crop Science.* Los Banos: IRRI.
63. Deka, J., & Sarma, H. P., (2012). Heavy metal contamination in soil in an industrial zone and its relation with some soil properties. *Archives of Applied Science Research, 4*(2), 831–836.
64. Xing, W., Li, D., & Liu, G., (2010). Antioxidative responses of *Elodea nuttallii* (Planch.) H. St. John to short-term iron exposure. *Plant Physiology and Biochemistry, 48*(10, 11), 873–878.
65. Audebert, A., & Fofana, M., (2009). Rice yield gap due to iron toxicity in West Africa. *Journal of Agronomy and Crop Science, 195*(1), 66–76.
66. Nagy, B., Măicăneanu, A., Indolean, C., Mânzatu, C., Silaghi-Dumitrescu, L., & Majdik, C., (2014). Comparative study of Cd (II) biosorption on cultivated *Agaricus bisporus* and wild *Lactarius piperatus* based biocomposites. Linear and nonlinear equilibrium modelling and kinetics. *Journal of the Taiwan Institute of Chemical Engineers, 45*(3), 921–929.
67. Prasad, Y. L., & Sachin, D. R., (2013). Biosorption of Cu, Zn, Fe, Cd, Pb and Ni by non-treated biomass of some edible mushrooms. *Asian Journal of Experimental Biological Sciences, 4*(2), 190–195.

68. Fulekar, M. H., Sharma, J., & Tendulkar, A., (2012). Bioremediation of heavy metals using biostimulation in laboratory bioreactor. *Environmental Monitoring and Assessment, 184*(12), 7299–7307.
69. Oladipo, O. G., Awotoye, O. O., Olayinka, A., Bezuidenhout, C. C., & Maboeta, M. S., (2018). Heavy metal tolerance traits of filamentous fungi isolated from gold and gemstone mining sites. *Brazilian Journal of Microbiology, 49*, 29–37.
70. Botebol, H., Sutak, R., Scheiber, I. F., Blaiseau, P. L., Bouget, F. Y., Camadro, J. M., & Lesuisse, E., (2014). Different iron sources to study the physiology and biochemistry of iron metabolism in marine micro-algae. *Biometals, 27*(1), 75–88.
71. Sutak, R., Botebol, H., Blaiseau, P. L., Léger, T., Bouget, F. Y., Camadro, J. M., & Lesuisse, E., (2012). A comparative study of iron uptake mechanisms in marine microalgae: Iron binding at the cell surface is a critical step. *Plant Physiology, 160*(4), 2271–2284.
72. Subramaniyam, V., Subashchandrabose, S. R., Thavamani, P., Chen, Z., Krishnamurti, G. S. R., Naidu, R., & Megharaj, M., (2016). Toxicity and bioaccumulation of iron in soil microalgae. *Journal of Applied Phycology, 28*(5), 2767–2776.
73. Zhuang, X., Chen, J., Shim, H., & Bai, Z., (2007). New advances in plant growth-promoting rhizobacteria for bioremediation. *Environment International, 33*(3), 406–413.
74. Souza, R. D., Ambrosini, A., & Passaglia, L. M., (2015). Plant growth-promoting bacteria as inoculants in agricultural soils. *Genetics and Molecular Biology, 38*, 401–419.
75. Luna, J. M., Rufino, R. D., & Sarubbo, L. A., (2016). Biosurfactant from *Candida sphaerica* UCP0995 exhibiting heavy metal remediation properties. *Process Safety and Environmental Protection, 102*, 558–566.
76. Zhou, G. W., Yang, X. R., Su, J. Q., Zheng, B. X., & Zhu, Y. G., (2018). B*acillus ferrooxidans* sp. Nov., an iron (II)-oxidizing bacterium isolated from paddy soil. *Journal of Microbiology, 56*(7), 472–477.
77. Seregin, I., & Kozhevnikova, A. D., (2006). Physiological role of nickel and its toxic effects on higher plants. *Russian Journal of Plant Physiology, 53*(2), 257–277.
78. Chamon, A. S., Gerzabek, M. H., Mondol, M. N., Ullah, S. M., Rahman, M., & Blum, W. E. H., (2005). Influence of cereal varieties and site conditions on heavy metal accumulations in cereal crops on polluted soils of Bangladesh. *Communications in Soil Science and Plant Analysis, 36*(7, 8), 889–906.
79. Walker, D. J., Clemente, R., & Bernal, M. P., (2004). Contrasting effects of manure and compost on soil pH, heavy metal availability and growth of *Chenopodium album* L. in a soil contaminated by pyritic mine waste. *Chemosphere, 57*(3), 215–224.
80. Sabir, M., Hanafi, M. M., Aziz, T., Ahmad, H. R., Zia-Ur-Rehman, M., Saifullah, G. M., & Hakeem, K. R., (2013). Comparative effect of activated carbon, pressmud and poultry manure on immobilization and concentration of metals in maize (*Zea mays*) grown on contaminated soil. *Int. J. Agric. Biol., 15*, 559–564.
81. Arnesen, A. K. M., & Singh, B. R., (1998). Plant uptake and DTPA-extractability of Cd, Cu, Ni and Zn in a Norwegian alum shale soil as affected by previous addition of dairy and pig manures and peat. *Canadian Journal of Soil Science, 78*(3), 531–539.
82. Singh, B. R., & Oste, L., (2001). In: Situ immobilization of metals in contaminated or naturally metal-rich soils. *Environmental Reviews, 9*(2), 81–97.
83. Martínez, C. E., Jacobson, A. R., & McBride, M. B., (2003). Aging and temperature effects on DOC and elemental release from a metal contaminated soil. *Environmental Pollution, 122*(1), 135–143.

84. Misra, S. G., & Pande, P., (1974). Effect of organic matter on availability of nickel. *Plant and Soil, 40*(3), 679–684.
85. Baruah, K. K., Das, S., & Das, K., (2007). Physiological disorder of rice associated with high levels of iron in growth medium. *Journal of Plant Nutrition, 30*(11), 1871–1883.
86. Bashir, K., Ishimaru, Y., & Nishizawa, N. K., (2010). Iron uptake and loading into rice grains. *Rice, 3*(2), 122–130.
87. Baruah, K. K., & Bharali, A., (2015). *Physiological Basis of Iron Toxicity and Its Management in Crops: Recent Advances in Crop Physiology. 2,* 203–224.
88. Dobbss, L. B., Dos, S. T. C., Pittarello, M., De Souza, S. B., Ramos, A. C., & Busato, J. G., (2018). Alleviation of iron toxicity in *Schinus terebinthifolius* Raddi (*Anacardiaceae*) by humic substances. *Environmental Science and Pollution Research, 25*(10), 9416–9425.
89. Saikia, T., & Baruah, K. K., (2012). Iron toxicity tolerance in rice (Oryza sativa) and its association with anti-oxidative enzyme activity. *Journal of Crop Science, 3*(3), 90.
90. Ahammed, G. J., Wu, M., Wang, Y., Yan, Y., Mao, Q., Ren, J., & Chen, S., (2020). Melatonin alleviates iron stress by improving iron homeostasis, antioxidant defense and secondary metabolism in cucumber. *Scientia Horticulturae, 265,* 109205.
91. Stein, R. J., Lopes, S. I. G., & Fett, J. P., (2014). Iron toxicity in field-cultivated rice: Contrasting tolerance mechanisms in distinct cultivars. *Theoretical and Experimental Plant Physiology, 26*(2), 135–146.
92. Pinto, S. D. S., Souza, A. E. D., Oliva, M. A., & Pereira, E. G., (2016). Oxidative damage and photosynthetic impairment in tropical rice cultivars upon exposure to excess iron. *Scientia Agricola, 73,* 217–226.
93. Verma, L., & Pandey, N., (2017). The effect of iron toxicity on seed germination and early seedling growth of green gram (*Vigna radiata* L. Wilczek). In: *Int.* (Vol. 6, No. 8, pp. 1427–1430).
94. Mehraban, P., Zadeh, A. A., & Sadeghipour, H. R., (2008). Iron toxicity in rice (*Oryza sativa* L.), under different potassium nutrition. *Asian J. Plant Sci., 7*(3), 251–259.
95. Majerus, V., Bertin, P., & Lutts, S., (2007). Effects of iron toxicity on osmotic potential, osmolytes and polyamines concentrations in the African rice (*Oryza glaberrima* Steud.). *Plant Science, 173*(2), 96–105.
96. Mustapha, H. I., Van, B. J. J. A., & Lens, P. N. L., (2018). Fate of heavy metals in vertical subsurface flow constructed wetlands treating secondary treated petroleum refinery wastewater in Kaduna, Nigeria. *International Journal of Phytoremediation, 20*(1), 44–53.
97. Kumar, P., Tewari, R. K., & Sharma, P. N., (2008). Modulation of copper toxicity-induced oxidative damage by excess supply of iron in maize plants. *Plant Cell Reports, 27*(2), 399–409.
98. Dhanwal, P., Kumar, A., Dudeja, S., Chhokar, V., & Beniwal, V., (2017). Recent advances in phytoremediation technology. *Advances in Environmental Biotechnology,* 227–241.
99. Chatterjee, C., Gopal, R., & Dube, B. K., (2006). Impact of iron stress on biomass, yield, metabolism and quality of potato (*Solanum tuberosum* L.). *Scientia Horticulturae, 108*(1), 1–6.
100. Kuki, K. N., Oliva, M. A., Pereira, E. G., Costa, A. C., & Cambraia, J., (2008). Effects of simulated deposition of acid mist and iron ore particulate matter on photosynthesis and the generation of oxidative stress in *Schinus terebinthifolius* Radii and *Sophora tomentosa* L. *Science of the Total Environment, 403*(1–3), 207–214.

101. Wu, L. B., Ueda, Y., Lai, S. K., & Frei, M., (2017). Shoot tolerance mechanisms to iron toxicity in rice (*Oryza sativa* L.). *Plant, Cell & Environment, 40*(4), 570–584.
102. Jayaweera, M. W., Kasturiarachchi, J. C., Kularatne, R. K., & Wijeyekoon, S. L., (2008). Contribution of water hyacinth (*Eichhornia crassipes* (Mart.) Solms) grown under different nutrient conditions to Fe-removal mechanisms in constructed wetlands. *Journal of Environmental Management, 87*(3), 450–460.
103. Singh, M. M., & Rai, P. K., (2016). A microcosm investigation of Fe (iron) removal using macrophytes of Ramsar lake: A phytoremediation approach. *International Journal of Phytoremediation, 18*(12), 1231–1236.
104. Chaturvedi, N., Dhal, N. K., & Reddy, P. S. R., (2012). Phytostabilization of iron ore tailings through *Calophyllum inophyllum* L. *International Journal of Phytoremediation, 14*(10), 996–1009.
105. Yang, W., Zhao, F., Ding, Z., Wang, Y., Zhang, X., Zhu, Z., & Yang, X., (2018). Variation of tolerance and accumulation to excess iron in 24 willow clones: Implications for phytoextraction. *International Journal of Phytoremediation, 20*(13), 1284–1291.
106. Meera, M., & Agamuthu, P., (2012). Phytoextraction of As and Fe using *Hibiscus cannabinus* L. from soil polluted with landfill leachate. *International Journal of Phytoremediation, 14*(2), 186–199.
107. Chauhan, P., & Mathur, J., (2020). Phytoremediation efficiency of *Helianthus annuus* L. for reclamation of heavy metals-contaminated industrial soil. *Environmental Science and Pollution Research, 27*(24), 29954–29966.
108. Chandra, R., & Yadav, S., (2011). Phytoremediation of Cd, Cr, Cu, Mn, Fe, Ni, Pb and Zn from aqueous solution using phragmites cummunis, *Typha angustifolia* and *Cyperus esculentus*. *International Journal of Phytoremediation, 13*(6), 580–591.
109. Suman, J., Uhlik, O., Viktorova, J., & Macek, T., (2018). Phytoextraction of heavy metals: A promising tool for clean-up of polluted environment? *Frontiers in Plant Science*, 1476.
110. Adiloğlu, S., (2017). Heavy metal removal with phytoremediation. *Advances in Bioremediation and Phytoremediation*, 115–126.
111. Ochonogor, R. O., & Atagana, H. I., (2014). Phytoremediation of heavy metal contaminated soil by *Psoralea pinnata*. *International Journal of Environmental Science and Development, 5*(5), 440.
112. Cheraghi, M., Lorestani, B., Khorasani, N., Yousefi, N., & Karami, M., (2011). Findings on the phytoextraction and phytostabilization of soils contaminated with heavy metals. *Biological Trace Element Research, 144*(1), 1133–1141.
113. Shackira, A. M., & Puthur, J. T., (2019). Phytostabilization of heavy metals: Understanding of principles and practices. In: *Plant-Metal Interactions* (pp. 263–282). Springer, Cham.
114. Galal, T. M., Gharib, F. A., Ghazi, S. M., & Mansour, K. H., (2017). Phytostabilization of heavy metals by the emergent macrophyte *Vossia cuspidata* (Roxb.) Griff.: A phytoremediation approach. *International Journal of Phytoremediation, 19*(11), 992–999.
115. Bomfim, N. C. P., Aguilar, J. V., De Paiva, W. D. S., De Souza, L. A., Justino, G. C., Faria, G. A., & Camargos, L. S., (2021). Iron phytostabilization by *Leucaena leucocephala*. *South African Journal of Botany, 138*, 318–327.

116. Jia, H., Wang, H., Lu, H., Jiang, S., Dai, M., Liu, J., & Yan, C., (2016). Rhizodegradation potential and tolerance of *Avicennia* marina (Forsk.) Vierh in phenanthrene and pyrene contaminated sediments. *Marine Pollution Bulletin, 110*(1), 112–118.
117. Mishra, V., Gupta, A., Kaur, P., Singh, S., Singh, N., Gehlot, P., & Singh, J., (2016). Synergistic effects of arbuscular mycorrhizal fungi and plant growth promoting rhizobacteria in bioremediation of iron contaminated soils. *International Journal of Phytoremediation, 18*(7), 697–703.
118. Chandwani, S., & Amaresan, N., (2022). Role of ACC deaminase producing bacteria for abiotic stress management and sustainable agriculture production. *Environmental Science and Pollution Research*, 1–17.
119. Whitacre, D. M., (ed.). (2008). *Reviews of Environmental Contamination and Toxicology* (Vol. 202). New York: Springer.
120. Hasegawa, H., Rahman, I. M. M., & Rahman, M. A., (2016). *Environmental Remediation Technologies for Metal-Contaminated Soils*. Minato, Japan: Springer Japan.
121. Yan, A., Wang, Y., Tan, S. N., Mohd, Y. M. L., Ghosh, S., & Chen, Z., (2020). Phytoremediation: A promising approach for revegetation of heavy metal-polluted land. *Frontiers in Plant Science, 11*, 359.
122. Gaiero, J. R., McCall, C. A., Thompson, K. A., Day, N. J., Best, A. S., & Dunfield, K. E., (2013). Inside the root microbiome: Bacterial root endophytes and plant growth promotion. *American Journal of Botany, 100*(9), 1738–1750.
123. Javed, M. T., Tanwir, K., Akram, M. S., Shahid, M., Niazi, N. K., & Lindberg, S., (2019). Phytoremediation of cadmium-polluted water/sediment by aquatic macrophytes: Role of plant-induced pH changes. In: *Cadmium Toxicity and Tolerance in Plants* (pp. 495–529). Academic Press.
124. Wuana, R. A., & Okieimen, F. E., (2011). Heavy metals in contaminated soils: A review of sources, chemistry, risks and best available strategies for remediation. *International Scholarly Research Notices, 2011*.
125. Hooda, V., (2007). Phytoremediation of toxic metals from soil and waste water. *Journal of Environmental Biology, 28*(2), 367.

CHAPTER 14

CRISPR/Cas9-Mediated Genome Engineering for Production of Biofortified Crops

MONALISHA MISHRA, SHADMA AFZAL, and NAND K. SINGH

Department of Biotechnology, Motilal Nehru National Institute of Technology, Allahabad, Prayagraj, Uttar Pradesh, India

ABSTRACT

Genome editing (GE) is a mode of gene modification to develop elite crop varieties expressing desirable traits. There are different GE technologies which could be distinguished based upon the nuclease enzyme being used, namely, meganucleases, zinc finger nuclease (ZFNs), transcription activator like effector nuclease (TALEN), CRISPR/Cas9 technology, CRISPR/Cpf1 system and base editors. Micronutrient deficiency or hidden hunger is the insufficiency of one or multiple micronutrients such as Fe, folate, Zn, vitA, vitB12, and vitD, in the body. Hidden hunger is seen in people who can only afford, for example cheap rice but not fruits, vegetables, etc., which although makes one feel full but lacks essential nutrients. Micronutrient deficiency causes weakness, mental, and physical retardation in children, impaired immune system and can even lead to death. Biofortification is an attempt to develop nutritionally-enriched staple food crops that even lower income communities can include in their diet. GE approach is currently in trend to develop novel crop varieties in far less time as compared to traditional breeding programs. Researchers have developed food crops enhanced with carbohydrate, vitamins, minerals, proteins, and quality fats, along with those efforts have been put to reduce are phytic acid (PA), Cd, etc., like anti-nutrients from the plant. Limitations associated with the release of crops in

Phytoremediation and Biofortification: Strategies for Sustainable Environmental and Health Management. Nand Kumar Singh, Shadma Afzal, & Tariq Aftab (Eds.)

the market, developed by gene editing technology are the biosafety concerns related to off-target mutation and the undesirable effects generated due to unwanted genetic changes in plants and the social acceptance.

14.1 INTRODUCTION

A healthy nutritious food is essential for overall well-being of a human being. It aids in the early recovery from an illness, fight infections, enhances metabolism and keeps oneself energetic. Micronutrient deficiencies or hidden hunger affects billions of individuals globally. Hidden hunger can have serious health consequences for instance poor mental and physical development, and it can be even life threatening. Serious physical and cognitive consequences are caused to children, due to poor nutrition, especially within the first 1,000 days of life of a newborn, which is calculated after the child attains the age of two years. Hidden hunger not just affects human health, but socioeconomic development especially of low- and middle-income countries can also get adversely impacted. Although population from developing countries constitute major proportion of undernourished people but unexpectedly research data has shown, Fe (iron) and I (iodine) deficiency was prevalent in developed countries. The global food security has constraints to feed the ever-increasing world population and climate change also effects the crop produce. A report by United Nations (UN) suggests that by 2030 the global population will be across 8.5 billion and consequently food demand will increase by 11.6 billion tons. This also leads to a conclusion that by 2030 the number of underfed populations would exceed 840 million.

Micronutrient malnutrition could be fought back by delivering adequate nutrition in the daily diet. Advanced breeding programs can help increase essential nutrients in the food crops, such an approach is called biofortification. Although biofortification being an advanced breeding program, requires huge capital investment but over longer period it is cost-effective. Using biofortified staple crops does not require any extra or unusual knowledge in relation to its preparation for the end users.

Biofortification of crops can be achieved by one of the two ways; conventional breeding or gene modification of the plant. Genome modification of crops can again be done in two ways either by introducing a foreign gene with desirable traits into the crop (transgenic crop) or by editing the endogenous genes of the crop and bringing about the required changes in the plant.

The latter approach is called GE or simply gene editing. Traditional breeding methods include crossing elite crop varieties to bring stable desired traits in the crops. Conventional breeding requires a greater number of generations to develop a desired crop variety and hence is more time consuming and laborious when compared to genome modification method of biofortication of crops. But in case of genome modification of the crop extensive research to target the specific gene is required.

This chapter gives a brief insight into what is GE, its types, micronutrient deficiency, how essential nutrients can be enhanced in the diet, clustered regularly interspaced short palindromic repeats/CRISPR associated 9 (CRISPR/Cas9) mediated biofortification of crops and enhancement of various nutritional elements such as carbohydrates, proteins, vitamins, quality fat, pigments, and lowering or elimination of anti-nutrients such as PA, Cd, commercialization of crops raised by CRISPR/Cas technology and ELSI associated with the CRISPR/Cas9 edited crops.

14.2 GENOME EDITING (GE)

GE, also popularly known as gene-editing, is an exemplary approach developed by scientists to tailor the genome or a piece of gene in a most desirable way. Researchers have used this technology in science and research, bringing out prudent products for human welfare. This technology came into picture in late 1900s where a group of scientists performed targeted mutagenesis in the genome of yeast and later mice. Meganucleases, I-SceI derived from yeast, is a pioneering gene editing tool. I-SceI meganuclease is a rare-cutter endonuclease and it recognizes an 18 bp sequence. It was originally identified as the one responsible for homing of introns in mitochondria of yeast [1]. This methodology utilizes addition, removal or alteration of nucleotides at a very precise locations in the whole genome of organisms under study. It can modify genotypic as well as phenotypic characteristics of bacteria, animals, and plants [2]. The general strategy to perform GE is to employ an "engineered nuclease" which comprises of a nuclease domain that cleaves the DNA and a DNA-targeting domain, designed to guide the nucleases to the target nucleotide sequence, once the DNA is cut in the targeted site, the cell will initiate its endogenous DNA double strand break repair machinery [3].

Following are the objectives of genome editing:

- Stable integration and maintenance of genes in the cells;

- Prevent damage to neighboring genes and ill-effects in its expression;
- Integrate large DNA segments;
- Remove undesirable genes or dominant mutations.

The conventional gene addition confronted the above-mentioned problems. GE has wider applicability in the field of health and medicine such as in the prevention and treatment of human genetic diseases which can be due to one gene defect for instance cystic fibrosis, hemophilia, and sickle cell disease or more complicated ailments, such as human immunodeficiency virus (HIV) infection, cardiac disorder, mental disorder, and cancer [4]. This technology is applied on cells and animal models like mice and zebrafish, to understand diseases. It has been used in the field of agriculture to improve crop yield, to enhance nutritional value, and to resist biotic and abiotic stresses. Different GE technologies (Table 14.1) are described on the basis of the type of nuclease being used, namely:

- Meganuclease;
- Zinc finger nuclease;
- TALEN;
- CRISPR-Cas9 technology;
- CRISPR-Cpf1 system;
- Base editors.

14.3 APPROACHES

14.3.1 MEGANUCLEASES

Meganucleases are one-of-a-kind nucleases (endodeoxyribonucleases) which can cause double-stranded break (DSB) at specific target site. Homing endonucleases (HE) stimulate the lateral mobility of genetic elements in an organism, hence the name. The recognition site sequence of a meganuclease can be 14–40 bp long [5]. As these recognition sequences are quite long hence their presence in the genome is rare or occurs once in given genome, which helps in cutting of DNA at specific sites. These can be used to treat viral infections as they can recognize long target sites precisely in the viral genome [6, 7] and hence find potential use in anti-viral therapy by designing meganucleases to target specific sites. These nucleases activate DNA repair pathways and targeted mutation after repeated cleavage of the DNA. HEs are bonded to proteins and large variants like DmoCre and E-Drel are created, providing nucleotide site-specific cleavage [8].

The mechanism of action of meganucleases include two simple steps:

- The recognition of a target specific cleavage site; and
- Splicing out the DNA and causing a cut [5].

The HE from the LAGLIDADG family are most frequently used in genome engineering [9]. Some of the meganucleases widely used in the genetic engineering are I-SceI derived from *Saccharomyces cerevisiae*, I-CreI derived from the chloroplasts of the green algae *Chlamydomonas reinhardtii* and I-DmoI derived from the archaebacterium *Desulfurococcus mobilis*.

14.3.2 ZINC FINGER NUCLEASE (ZFNS)

Zinc finger nucleases (ZFNs) are artificially synthesized nucleases. This technology makes use of the cleaving properties of restriction enzymes and the specific binding feature of protein domain. These nucleases enzymes are designed such that the restriction endonucleases are bonded with zinc-finger-binding domain protein. The zinc-finger protein domain helps in identifying the target site and binds to it, this target site is then cut at a particular codon by FokI restriction endonucleases [10]. The zinc finger binding domain protein are normally expressed in the organisms and acts as transcription factor by binding to specific DNA sequence, more precisely on the promoter of a gene and regulating the gene expression. For example, in humans EGR1 gene codes a zinc finger transcription factor. These zinc finger proteins could be reprogramed and modeled according to ones need [11]. Zinc finger proteins are made up of two to four domains, with each domain possessing an α-helix which fits into the major groove of the DNA double helix where specific amino acids bind selectively to a 3 base pair sequence. The most commonly used zinc fingers have three 'fingers' (DNA-binding domains), hence one can design a unique sequence of 9-base pair length [12]. The technology uses exclusive FokI restriction enzyme, isolated by Hiroyuki and Susumu [89], have a non-specific DNA-cleaving domain and can work separately from the enzyme's DNA-binding domain. ZFNs could be synthesized by fusing two zinc finger proteins, each with three alpha-helix domains (fingers) for a total of 18 base pair specificity, with each having one FokI endonuclease in between them. ZFNs can be homodimeric or heterodimeric based on the FokI nucleases DNA cleavage-domain. Scientists have constructed the FokI cleavage domain on each protein slightly different and hence cleave DNA as *heterodimers. The off-target effects can also be reduced using this novel technology*. CompoZr custom ZFNs, a patented product of Sigma-Aldrich can recognize 24–36 bases and is commercially available for use. One can design their own ZFNs by the help of Zinc Finger Consortium, an open-source protocols for synthesizing ZFNs. ZFNs have shown promising results in the treatment of HIV infection although it's still under clinical trials. The virus uses CCR5 co-receptor of CD4+ to invade the T-cells and cause infection, Samson 1996b discovered that a 32 bp frameshift deletion in CCR5 allele can completely disable gene product and now the allele is called CCR5Δ32 allele. Therefore, several groups of scientists were able to demonstrate that CCR5 could be deleted from the genome using ZFNs and can help resist the infection.

14.3.3 TRANSCRIPTION ACTIVATOR-LIKE EFFECTOR NUCLEASES

Transcription activator-like effector nucleases (TALENs) are revolutionary nucleases which help in gene editing and introduce double stranded break (DSB) at targeted site. TALENs constitute a non-specific FokI nuclease domain similar to ZFNs, fused to a tailored DNA-binding domain [12]. Transcription activator-like effectors (TALEs) give rise to the DNA-binding domain which are highly conserved repeats. TALENs are proteins secreted by the Xanthomonas genus of bacteria to modify genes transcription in plants. Xanthomonas are gram-negative bacteria which cause infection in various plant species by utilizing the TALE nucleases. The first TAL effector protein to be isolated was AvrBs3, from a pathogen of pepper called *Xanthomonas campestris* [13]. AvrBs3 TALE protein activates Bs3 gene which activate plant immune response. The DNA binding domain sequence comprises a series of 33–35 amino acids repeats, which essentially differs by two positions [14, 15]. A single repeat of amino acid series contacts with each base pair in the target DNA sequence, the specificity is determined by the two variant amino acids of the repeat. Hence, there is protein-DNA interaction between the nuclease and the target DNA. Each TALE protein recognizes one specific nucleotide and in a similar way an array of TALEs can be designed to bind to a longer sequence of nucleotides [16]. TALENs are very specific as one TALE protein binds to only one nucleotide and the binding specificity of the neighboring TALEs is not affected [17]. Designing of TALENS is quite easier than ZFNs. TALENs also use *FokI* endonuclease just like ZFNs and dimerization reaction takes place prior to cleavage process [11]. Here the two different TALENs bind proximally to the two complementary strands to make a cut in the target DNA [18].

Researchers have used this technology to alter endogenous genes in *Saccharomyces cerevisiae, Drosophila melanogaster*, roundworm, zebrafish, frog, rat, cattle like cow, rice, potato, crickets, silkworm, and human somatic and pluripotent stem cells [19–28]. One of the variants of TALEN, Platinum TALENs, is most recently used for gene editing. Platinum TALENs are highly active TALENs with variable TALE repeats harboring non-repeat-variable di-residue (non-RVD) variations developed by Sakuma et al. [29] who introduced targeted mutagenesis in cultured human cell lines and rats. Platinum TALENs have also been used in introducing mutations at specific sites in frogs, fungi, plant tissues and potato. TALEN mediated somatic gene therapy was successfully performed on a year-old patient named Layla of United Kingdom, suffering from leukemia in the year 2015. Similarly, this technology was used in improving rice seed storability by mutating LOX3 gene (lipoxygenase) which catalyzes the dioxygenation reaction for the formation hydroperoxide from polyunsaturated fatty acids (FAs) [30].

14.3.4 CRISPR/CAS9 TECHNOLOGY

CRISPR/Cas9 is an ingenious genome-editing tool. As we humans have an adaptive immune system similarly in prokaryotes too adaptive immune response could be seen against viral infection. This secondary immune response in prokaryotes is provided by the CRISPR/Cas9 system [31]. This feature of prokaryotes is now exploited in genetic engineering. Mechanism of Cas9 action on the target DNA (as shown in Figure 14.1).
Following are important components of CRISPR/Cas9 system:

- Cas9 nuclease.
- Two types of RNA: a single guide RNA (sgRNA) which help locate the target sequence by base complementarity (Watson-Crick base pairing) and trans-activating crRNA (tracrRNA).
- A protospacer adjacent motif (PAM), a DNA motif. For each Cas9 protein there is a specific PAM sequence and has a G rich consensus sequence 5′-NGG-′3 where, N is any nucleotide; G is guanine.

In the CRISPR/Cas9 system there is RNA–DNA interaction between the target DNA and recognition module.

Cas9 endonuclease contains two domains:

1. **HNH Domain:** It cuts the complementary strand of CRISPR RNA or crRNA; and

2. **RucV-like Domain:** It cleaves the opposite strand of the double-stranded DNA.

The sgRNA is a 100 nucleotide long artificially designed RNA having a 20 nucleotide (at the 5′ end) long guide sequence, it along with a PAM sequence (protospacer adjacent motif) helps in the recognition of the specific nucleic acid sequence. The loop structure at the 3′-end of sgRNA and the guide sequence anchors the target DNA sequence forming a complex with Cas9 nuclease. Cas9 later cuts the dsDNA forming a double-strand break (DSB) at desired site [32]. This DSB stimulates the cell's natural DNA repair mechanism. The two mechanisms to repair DSB are nonhomologous end-joining (NHEJ) and homology-directed repair (HDR) [33]. To create gene knockout (KO) CRISPR/Cas9 generally initiates NHEJ repair mechanism after a DSB is made and creates mismatches and gene insertion/deletions (indel) whereas for introducing foreign DNA knock-ins, an oligo-template is required which initiates HDR DNA repair mechanism [31]. Cas9 endonuclease can be obtained from various bacterial species, such as *Streptococcus thermophilus, Streptococcus pyogenes, Brevibacillus laterosporus, Staphylococcus aureus*, etc. Cas9 nuclease obtained from the bacteria *Streptococcus pyogenes* is the favorite choice of scientists.

This technology is getting wider recognition due to its simplicity, efficiency, and versatility. The genome of various organisms including humans, animals, and plants have been efficiently edited using CRISPR/Cas9 technology (Figure 14.2).

14.3.5 CRISPR/CPF1

CRISPR/Cpf1 system is an improvised and more advanced form of CRISPR/Cas9 GE system. Cpf1 here stands for **C**RISPR from *Prevotella* and *Francisella*1 and is also known as Cas12a. It is a dual nuclease, as it functions both as an endoribonuclease (processing of crRNA) and an endodeoxyribonuclease (cleaves specific DNA and cause double-stranded breaks). Cpf1 along with its CRISPR RNA (crRNA) is smaller in size in contrast to Cas9 protein [34]. The tracrRNA is also not required to process Cpf1-associated CRISPR repeats into mature crRNAs and therefore a crRNA of 40–45 nucleotide long consisting of the repeats and the spacer sequence is appropriate for gene editing as compared to 100 nucleotides long sgRNA used in CRISPR/Cas9 system [35]. Cpf1 has T-rich PAM consensus sequence, i.e., 5′-TTTN-3 or 5′-TTN-3′ PAM present at the 5′ end of the target sequence.

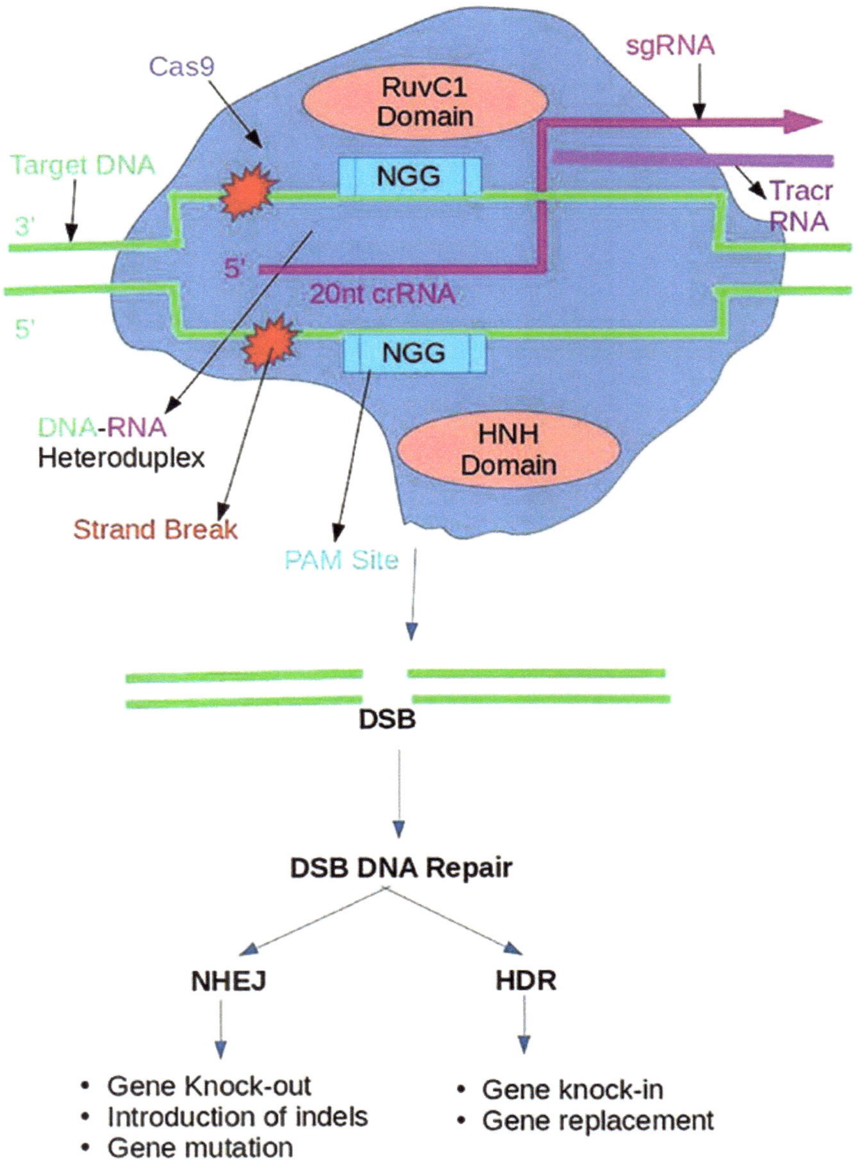

FIGURE 14.1 Schematic illustration of CRISPR/Cas9 protein and its mechanism of action.

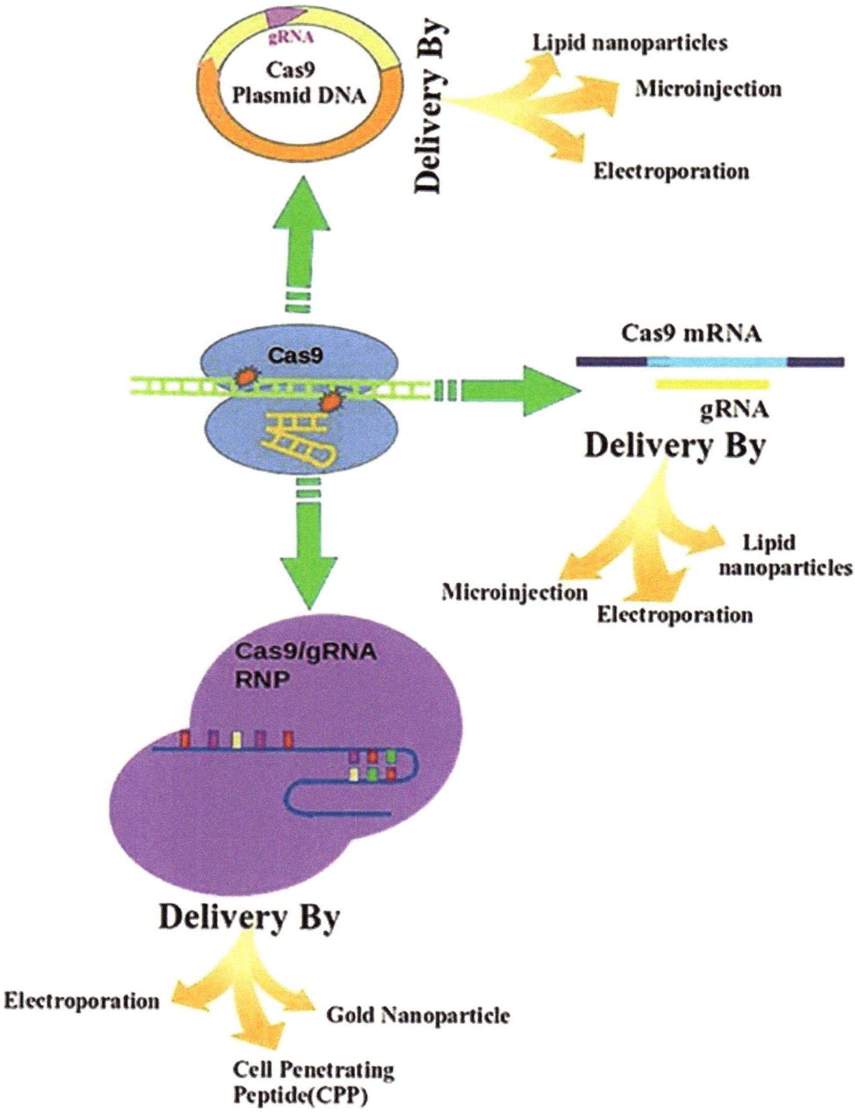

FIGURE 14.2 Three common ways to perform gene-editing by using CRISPR/Cas9 and the delivery method.

The nuclease domain cleaves the target DNA sequence in the downstream of the PAM sequence generating 5-bp overhangs as sticky ends [35]. The Cpf1 processes the pre-crRNA to crRNA acting as an RNase also it has nuclease activity which cleaves the double stranded DNA. Cpf1 can generate, a single promoter regulated multiple crRNAs, making it simpler technology than CRISPR/Cas9. Multiplexed GE can be successfully performed by Cpf1, as multiple loci in the genome can be targeted using a single crRNA array transcript. The off-target activity with Cpf1 cleavage is also comparatively lower than Cas9. The CRISPR/Cpf1 system causes mutation in the genome by deletion or insertion of gene, base editing, and locus tagging. This technology has been efficiently used for gene editing of both monocot and dicot plants such as in tobacco, rice, soybean, wheat, etc.

14.3.6 BASE EDITORS

CRISPR/Cas9-based base editor or simply known as base editors is a robust genome-editing technology. Base editing causes direct and irreversible conversion of one target base into another without cleaving DNA and hence no DSB are generated. Also, it doesn't require a donor template [36]. Base editors show lesser off-target mutations compared to that of CRISPR/Cas9 system, adding to its preciseness. The base editor consists of a catalytically inactive Cas9 domain (dCas9 nuclease), and a cytosine deaminase domain fused to it that converts G-C base pairs to A-T base pairs [37]. This tool by conferring precise gain-of-function or loss-of-function of genes has been efficiently used for gene editing of many cereal crops including wheat, rice, and maize [38]. Yan et al. [39] developed a fluorescence-tracking adenine base editor using the Cas9n-guided TadA: TadA7.10 heterodimer and could efficiently introduce an A to G base conversion in rice.

14.4 MICRONUTRIENT DEFICIENCY

Micronutrient deficiency also referred to as 'hidden hunger' is a chronic hunger; its effect is seen in the long run. Hidden hunger can be described as the deficiency of one or more micronutrients such as Fe (iron), folic acid, Zinc, vitamins like A, B12 and D in an individual. Those who can only afford, for example cheap rice but not fruits, vegetables, etc., which although makes one feel full but lacks essential nutrients, generally suffer from malnutrition. Lack of vital micronutrients cause weakness, mental,

TABLE 14.1 Comparison between Different Gene-Editing Technologies [4, 84]

Property	Meganuclease	ZFN	TALEN	CRISPR/Cas9	CRISPR/Cpf1	Base Editor
Derived from	Microbial mobile genetic elements.	Gene expression regulators of eukaryotes.	*Xanthomonas* bacteria	Archaeal and bacterial adaptive immune system.	Archaeal and bacterial adaptive immune system.	Improvised form of CRISPR/Cas9.
Recognition sequence (bp)	12–40	9–18	24–49	17–20	17–20	17–20
Nuclease	Homing endonuclease	Fok1	Fok1	Cas9	Cpf1	dCas9
Target recognition domain	LAGLIDADG protein domain.	Zinc finger	TALEs	gRNA	gRNA	gRNA
Interaction with the target	Protein-DNA	Protein-DNA	Protein-DNA	Protein-RNA	Protein-RNA	Protein-RNA
Complexity	High	Moderate	Moderate	Low	Low	Moderate
Efficiency	Low	Low	Moderate	Moderate	Moderate	High
Off-target mutations	Moderate	Moderate	Moderate	Low	Low	Very low

Sources: Ref. [4, 84]

and physical retardation in children, poor immune system and can be even fatal. As an instance, iodine deficiency, generally seen in pregnant women, increases the risk of infant mortality, miscarriage, poor brain development of the fetus and still-birth [40]. Over 2 billion people around the globe suffer from micronutrient deficiency of which one-third resides in India. According to a UNICEF report released in October 2019, in India more than 80% adolescent suffer from micro-nutrient deficiency. In INDIA, The National Nutrition Monitoring Board (NNMB) is the body which takes care of the related matters.

The hidden hunger can be cured currently by two approaches: Fortification of food and biofortification of crops. The two approaches are described in detail in subsections.

14.4.1 FORTIFICATION OF FOOD

Micronutrient fortification refers to deliberately adding essential vitamins and minerals in a food product to increase its nutritional value. It is a widely used approach to add micronutrients in the processed food, for example the addition of iodine in salt. Fortification of cheaper food products like flour and salt with Fe and iodine has saved millions of lives. The Universal Salt Iodization program, for fortifying salt, implemented worldwide has successfully resolved iodine deficiency. A teaspoon of iodine, consumed in little amount on a daily basis over a lifetime, is sufficient to prevent deficiency. Some of the other fortified folic acid (since 1998) foods are corn meal, flour, bread, pasta, and rice, fortified grains and grain products with thiamine, or vitamin B-1; riboflavin, or vitamin B-2; niacin, or vitamin B-3; and Fe are available in the market; vitamin D fortified milk (also contains vitamin A), orange juice, soy milk and breakfast cereals; margarine and other butter substitutes fortified with sterols which help in lowering cholesterol; infant formulas fortified with Fe, vitamin K and omega-3 FAs. According to FAO, cereals and cereal-based products, dairy products, fats, and oils, beverages such as tea, and infant formulas are the most commonly fortified food products [40].

This approach can utilize single micronutrient fortification, dual micronutrient fortification and multiple micronutrient fortification, of various micronutrients such as Zn, Fe, Se, I, vitamin A, vitamin B complexes, vitamin C and vitamin E.

Many reports have suggested that fortification of food with I, folic acid, vitamin A and Fe have helped combat serious disease pertaining to nutritional deficiency (Table 14.2) [41]. This is a proven, cheaper and safe strategy for nutritional enhancement and diet improvement whereby micronutrient deficiency could be prevented. A report from the Environmental Working Group (EWG) suggested that these fortified foods can cause nutrient overdoses. Bioavailability of certain nutrients in the whole food would also be changed when compared to the original one. For example, skim milk whose fat has been removed, and then fortified with vitamin A and vitamin D but Vitamins A and D are fat-soluble vitamins, hence someone consuming skim milk would not absorb much of these vitamins as the one drinking whole milk.

14.4.2 BIOFORTIFICATION OF CROPS

Biofortification is a novel approach to bring up staple food crops having enriched nutritional content that even lower income communities can afford. It was primarily developed to address vitamin A, Fe, and Zn deficiencies that commonly occur in low-income countries. Biofortification of crops can be achieved by two ways:

i. Adding mineral fertilizers: In some countries Se-containing fertilizers are added in the soil to improve Se content of wheat; and
ii. Plant breeding: micronutrient-enhanced crop varieties created by the conventional breeding method or genetic modification by using one of the genome-editing tools.

Biofortification of crops is attainable and more economical means of enhancing micronutrients in the food crops and hence a modus operandi to curb the hidden hunger [42]. The main aim of biofortification is to reduce morbidity rates caused by micronutrient deficiency, to increase food security and productivity in developing countries.

14.5 BIOFORTIFICATION OF FOOD CROPS USING CRISPR/CAS9 TECHNOLOGY

The clustered regularly interspaced short palindromic repeats or CRISPR/Cas9 is a growing technology and is being popularly adopted in the plant breeding program worldwide because of its versatility. This methodology has been efficiently used to develop quality crops. Its preciseness in targeting

TABLE 14.2 Nutritional Deficiencies in Humans, the Associated Diseases, Symptoms, and Standard Treatment [42]

Nutrient Deficient	Disease	Symptoms	Treatment
Iron (Fe)	Anemia	Anemia (microcytic, hypochromic); weakened cognitive function, poor body temperature regulation, impaired immune function.	Fe supplements/Fe pills/oral Fe.
Iodine(I)	Mumps, goiter	Stunted growth and neurodevelopmental deficits.	Radioactive iodine treatment. Potassium iodine solution or small dosage of Lugol's iodine.
Vitamin A	Xerophthalmia	Night blindness, xerosis, keratomalacia, hyperkeratotic skin lesions, and Bitot's spots.	Eating vitamin, A-enriched foods. Such as fortified milk, liver, beef, oily fish, chicken, eggs, sweet potatoes, carrots, and green leafy vegetables.
Vitamin C	Scurvy	Inflammation of the gums, tiredness, and impaired connective tissues.	Vitamin C supplements and eating vitamin enriched food like citrus fruits, potatoes, and strawberries.
Vitamin B3	Pellagra	Rough appearance to the skin, pigmented rash over skin when exposed to the sun, tongue bright red in color, tiredness, constipation, vomiting, and diarrhea.	Nicotinamide doses of 250 to 500 mg/day orally.
Vitamin B1	Beri-Beri	• Loss of hunger, tiredness, tingling arms and legs. • Nerve damage, delirium. • Blurry vision, nausea, and vomiting.	Vitamin B1 supplements and B1-enriched foods, such as wholegrain cereals, beans, and beef.
Vitamin D	Osteomalacia (adults), rickets (children)	No symptoms until severe deficiency and can lead to thin, brittle or misshapen bones.	Intake of oral ergocalciferol for eight weeks.
Calcium	Osteoporosis	Muscle cramps and aches tingling in the arms, legs, and feet as well as discomfort while walking or moving.	Calcium supplements.
Protein	Kwashiorkor, Marasmus	Stunted growth in children, swelling, fatty liver, skin degeneration and increased severity of infections.	Increasing intake of protein-containing foods such as nuts like almonds, pistachios, cashews, seeds such as flax seeds and legumes, like beans and pulses.
Selenium (Se)	Keshan disease	Infertility in both males and females, frail muscle, hair loss, mental fog, and impaired immune system.	Increased dietary intake of Se and Se supplementation 55–60 mg/day (composed of sodium selenite or l-selenomethionine).

Source: Ref. [42]

the desirous gene has sped up the breeding programs many folds also it has greatly saved labor and other costs. In recent years, many papers have been available showing the effectiveness of this genome-editing tool in improving crop quality. The CRISPR/Cas9 technology is straightforward, easy to use GE tool. The nucleotide complementarity of the guide RNA (gRNA) to a particular nucleotide sequence governs the specificity of editing also it does not require any complex protein engineering. Researchers have used this tool to study gene function.

14.5.1 IMPROVEMENT OF CARBOHYDRATE CONTENT

Sun and colleagues [43] successfully employed CRISPR/Cas9 technology and developed high amylose rice by varieties by targeting genes coding for two rice starch-branching enzymes (SBE) OsSBEI and OsSBEIIb. The study showed sbeII mutants have amylase content (AC) 25% and resistant starch (RS) 9.8%, significantly more than the wild type of grain, whereas there were no variations seen in sbeI mutants. This leads to a conclusion that the starch quality and quantity could be enhanced by activity of SBEII enzymes, and it is also important for defining the fine structure of starch. A study by Chen et al. [44] suggests the risk of hypertension, diabetes, and colon cancers like serious diseases could be reduced by including food rich in high AC and RS. Hence developing rice varieties high in amylose content can help improve human health. Similarly, Tuncel and colleagues [45] performed CRISPR/Cas9 mediated gene knock-out of starch branching enzymes SBEI and SBEII in tetraploid potatoes. Mutant cells when observed microscopically contained many granules of < 4 μm and granules with multiple hila. Wang et al. [46] in their study applied this technology to introduce mutation in the two starch biosynthetic pathway genes GBSSI (encoding granule-bound starch synthase I), and SBEII (encoding starch branching enzyme II). They found that in the allopolyploid sweet potato, the IbGBSSI-knockout reduced, while the IbSBEII-knockout increased, the amylose percentage and hence it could be concluded that this technology is efficient in improvement of root crops as well. Li et al. [47] used this technique to increase RS content in common wheat (*Triticum aestivum* L.), as modern grain varieties have low RS contents. High-amylose wheat varieties were developed by CRISPR/Cas9 site directed mutagenesis on TaSBEIIa gene in a modern winter wheat cv Zhengmai and a spring wheat cv Bobwhite. Further analysis of triple-null lines' flours revealed significantly increased amylose, RS, protein, and soluble pentosan contents which benefits human health. Maize crop with high amylopectin was developed by Corteva Agriscience in 2016 [48] (earlier DuPont Pioneer) by knocking out the maize waxy gene Wx1 creating a 30 bp deletion (a frameshift mutation) at the exon/intron boundary, using CRISPR/Cas9 technology. The gene encodes the granule-bound starch synthase (GBSS) enzyme responsible for making amylose. In the absence of this enzyme amylase was not synthesized in the endosperm, leading to a high amylopectin (waxy) maize variety with improved digestibility and finds the potential use in bio-industrial applications. Similarly, researchers Andersson et al. [49, 50] at the Swedish Agricultural University produced waxy potatoes by targeting the same waxy gene which helped improve cultivars. Zhang et al. [51] edited wheat waxy genes that encode enzyme important for amylose biosynthesis in wheat endosperm, so as to improve wheat flour quality. They could successfully develop glutinous wheat having 3.6% amylose content as compared to wild varieties possessing 17.2% amylose. To increase the sweetness of maize kernels Dong et al. [52] targeted genes SHRUNKEN2 and WAXY which control the seed phenotype. Mutation in gene SH2 caused accumulation of sugar instead of starch which enhanced the sweetness of kernels on the other hand mutation in WX gene improved the amylose/amylopectin ratio. Gao et al. [53] used CRISPR/Cas9 to recover the elite in-bred maize lines by targeting WAXY gene. Qi et al. [54] targeted the WX locus in the ZC01-DTMwx by CRISPR/Cas9 to produce waxy maize varieties with enhanced amylopectin content.

14.5.2 ENHANCEMENT OF MINERAL CONTENT

Rice being staple food for the majority of population worldwide hence a lot of effort is being put to introduce essential nutrition in the crop. In one such attempt Che and colleagues [55] introduced site directed mutation in OsVIT2 gene using CRISPR/Cas9 approach of gene editing. Knocking-down OsVIT2, which codes for vacuolar iron transporter (VIT), helped in enhancement of grain Fe content. The study revealed, mutation of OsVIT2 enriched Fe distribution to embryo and endosperm of the rice grains, and ultimately in the polished grain Fe content was increased without impacting the rice yield. Se one of the important micronutrients having antioxidant properties, was enriched in rice mutant lines. Sun et al. [56] in their work showed, that gene OsASTOL1 which encodes a chloroplast-localized O-acetylserine (thiol) lyase (OAS-TL) enzyme it has S189N point mutation. This mutation led to the gain-of-function arsenite tolerant 1 (astol1) rice mutant varieties having followed remarkable properties: enhanced uptake of sulfur and Se; tolerance to arsenic toxicity; and decrease in accumulation of arsenic in rice grains. Achary et al. [57] targeted the GRAIN WIDTH and WEIGHT2 (GW2) gene in Indica (var. MTU1010) through CRISPR/Cas9 gene editing technology. In cereal crops GW2 gene codes for an enzyme RING-type E3 ubiquitin ligase responsible for grain weight. They developed GW2 knockout (KO) mutants, the loss of function of OsGW2 resulted in improved accumulation of essential mineral nutrients Fe, Zn, K, P, and Ca in the rice endosperm.

14.5.3 REDUCTION IN PHYTIC ACID (PA) CONTENT TO ENHANCE BIOAVAILABILITY OF NUTRIENTS

Phytic acid (PA) or phytate or inositol hexakisphosphate/IP6 is present in cereal grains, is an important antinutrient, responsible for reduction of the bioavailability of minerals and proteins in human body. PA acts as chelating agent of minerals and proteins and further makes complexes with them this leads to reduction in absorption of essential minerals and proteins, hence its presence in food can lead to malnutrition. Also due to the lack of PA degrading enzymes humans cannot metabolize it. Inositol pentakisphosphate 2-kinase 1 (IPK1) is an important enzyme for PA biosynthesis [58]. Khan et al. [59] targeted this enzyme in rice crop and developed new mutant varieties with reduced PA content. In a similar way Sashidhar et al. [60] reduced PA content by 35% in canola grade rapeseed by knocking-out three BnITPK genes involved in biosynthesis of PA. Saira et al. [61] mutated the gene TaIPK1 (inositol pentakisphosphate 2-kinase 1) by CRISPR/Cas9 to reduce PA content and improve Fe and Zn accumulation in wheat crop.

14.5.4 IMPROVEMENT OF PIGMENTS IN CROPS

Pigments like anthocyanin and carotenoids give plants and their products brilliant color, have beneficial effects on humans as well for instance carotenoids like beta carotene are provitamin A carotenoids, lycopene are non-provitamin A carotenoid with antioxidant properties and could combat certain types of cancers similarly anthocyanins have ability to scavenge oxidizing agents, can act as anti-inflammatory agents, can treat certain viral infections, and has shown anti-cancer properties too. The genes for these pigments are of great interest to researchers as it will serve the human community for treatment of various diseases. The first attempt to increase β-carotene levels by using CRISPR/Cas9 approach was made by Yang et al. [62] in rice endosperm. The genes linked to carotenoid metabolism (OsCYP97A4, OsDSM2, OsCCD4a, OsCCD4b, and OsCCD7) were knocked-out, with an intent to increase the amount of beta-carotene accumulation in rice endosperm but there was no considerable increment in the amount of carotene. Similarly, Li et al. [63] disrupted SGR (stay green protein) gene by CRISPR/Cas9 method. In tomato, SGR protein when mutated a significant surge of lycopene and β-carotene content in mature fruits was observed. An attempt was made by Akira et al. [64] so as to increase β-carotene levels in rice plant by directed gene modification of Osor gene using CRISPR-Cas9 mediated GE tool and resulted in accumulation of 2.7 μg/g β-carotene of fresh weight. Kitaake, a rice cultivar was investigated by Dong and coworkers [65] for targeted gene insertion of 5.2 kb marker-free DNA, a carotenoid biosynthesis cassette, at two genomic safe harbors which consisted of CrtI and maize PSY genes, the two genes are regulated by an endosperm-specific promoter. The gene-edited lines could accumulate 7.9 μg/g dry weight (DW) β-carotene in the endosperm, which lies in the range of the amounts produced in the Golden Rice 2 lines (7–31 μg/g) [66]. Kaur et al. [67] used 'block' strategy and through CRISPR/Cas9-mediated deletion of the lycopene cyclases (LCY) gene produced β-carotene-augmented mature banana fruit pulp named as Golden banana.

14.5.5 IMPROVEMENT OF PROTEIN CONTENT

Researchers in a few years have tried their hands on improving protein content in food crops using CRISPR/Cas9 technology and the results were prompting. Satoko and co-workers [68] could efficiently enhance ɣ-aminobutyric acid (GABA) quantity in *Solanum lycopersicum* by site-directed mutation in SlGAD2 and SlGAD3 genes that regulate enzymatic function. They incorporated a stop codon sequence upstream to the autoinhibitory domain of SlGAD2 and SlGAD3 and enhanced GABA accumulation by 7 to 15-fold in tomato fruit. Tomato genome contains five GAD genes SlGAD1, SlGAD2, SlGAD3, SlGAD4, and SlGAD5), of which SlGAD2 and SlGAD3 genes are important genes required for the development of tomato fruit. Sánchez-León et al. [69] developed a gluten protein free wheat line with reduced immunoreactivity for gluten intolerant consumers such as in Celiac disease. They performed directed mutagenesis and targeted a conserved region adjacent to the 33-mer coding sequence for the alpha-gliadin gene, the main immunodominant peptide in celiac patients and the mutant lines revealed 85% reduced immunoreactivity. Zhang et al. [70] analyzed GW2 gene of *Triticum aestivum* L. hexaploid common wheat. GW2 gene is an important grain weight determinant in cereal crops having three homoeologs (TaGW2-A1, -B1, and -D1). The study revealed that grain protein content (GPC) is regulated by TaGW2 homoeologs. Mutation in either two or all the three homologs helped enhance flour protein content and gluten strength. GABA is a non-proteinaceous amino acid and is synthesized from amino acid glutamic acid catalyzed by the enzyme glutamate decarboxylase (GAD). GABA is an important nutrient to humans as it reduces the risk of hypertension and diabetes and helps in growth and development and stress tolerance in plants. Akama et al. [71] developed GABA-fortified rice lines using CRISPR/Cas9 by knocking-out the GAD3 gene. There was a seven-fold increase in GABA, 2 to 5-fold increases in the levels of amino acids ala, asp, met, phe, leu, and ile, 20% increase in seed protein content levels as well as small increase in the levels of serine and glutamate amino acids whereas the levels of Asn and tryp amino acids reduced to 30% of the wild type lines. Similarly, Camerlengo et al. [72] edited α-amylase/trypsin inhibitor genes in durum wheat to reduce allergen proteins in wheat. Baysal and co-workers [73] observed that knocking out the rice SBEIIb gene, a glucose branching enzyme led to cumulation of amino acids alanine, aspartic acid, and proline content to > 100 μg/g. It also caused the accumulation of several other amino acids, including glycine, lysine, serine, and valine that were not seen in wild type endosperm. Achary et al. [57] reported CRISPR/Cas9 GE of GRAIN WIDTH and WEIGHT2 (OsGW2) gene of Indica (var. MTU1010). GRAIN WIDTH and WEIGHT2 (OsGW2) gene translates into a RING-type E3 ubiquitin ligase. Mutation of the gene resulted in enhanced GPC with accumulation of a thick aleurone layer in the endosperm.

14.5.6 IMPROVEMENT OF OIL CONTENT

Scientists are working to improve monounsaturated fatty acids (MUFA) content, like oleic acid in oils which can provide cardiovascular benefits. One such attempt was made by Jiang and colleagues [74] in Camelina sativa, an oil seed. The FA desaturase enzyme has a vital role to play in the biosynthesis of linoleic acid as it catalyzes the conversion of oleic acid (monounsaturated omega 9 fatty acid) to linoleic acid (polyunsaturated omega 6 fatty acid). They used CRISPR/Cas9 to target the fatty acid desaturase (FAD2) gene with the goal to improve seed oil composition and could successfully obtain Camelina seeds having more than 50% oleic acid content as compared to 16% in wild type seed. This increase was because of significant decrease in the less desirable polyunsaturated FAs, linoleic acid whose amount decreased from ~16% to <4% and α-linolenic acid which decreased from ~35% to <10%. Okuzaki et al. [75] worked on improving quality of *Brassica napus* or rapeseed, an important oil crop. A FA desaturase 2 encoding enzyme FAD2 gene, which catalyzes the desaturation reaction of oleic FA, in Westar cultivar of *Brassica napus* was targeted by the CRISPR/Cas9 system and an increase in oleic acid content compared to wild-type seeds was observed. Similarly, Phat et al. [76] reported that disruption of omega-6 desaturase gene GmFAD2 of *Glycine max* (Soyabean) caused an increase in oleic acid accumulation. Zhai et al. [77] reported that the yellow-seeded desirable phenotype of Brassica crops could be retrieved by site-specific mutagenesis of BnTT8 gene which caused complete blocking of the proanthocyanidin (PA)-specific build-up in the seed coat and amount of seed oil, protein content and altered FA composition were also altered in the mutant seeds. Ma et al. [78] targeted GmFATB1 via CRISPR/Cas9, encoding Acyl-acyl carrier protein thioesterases (FATs) and releases free FAs and acyl-ACP. This mutation decreased significant levels of palmitic acid and stearic acid (saturated fatty acids) contents in soybean leaf whereas unsaturated FA content remains unaffected.

14.5.7 ELIMINATING ANTI-NUTRIENTS

Quite a few substances have undesirous effects on the dietary value of crops, such as heavy metal accumulation as observed in rice crop. It was reported in some studies that the accumulation of Cadmium metal in indica rice cultivars is more when compared to japonica varieties [79]. Zhang et al. [80] used CRISPR/Cas9 editing system for developing improved indica rice cultivars which showed decreased levels of Cadmium build-up in the grain, by knocking out the metal transporter gene OsNramp5. The results revealed cadmium concentration in improved mutant indica rice lines was consistently less than 0.05 mg/kg, as opposed to the high cadmium concentrations from 0.33 to 2.90 mg/kg in grains of wild-type cultivar. Oxalic acid is also an anti-nutrient. Gago and co-workers [81] mutated tomato LEAFYCOTYLEDON1-LIKE4 (L1L4, NF-YB6) genes, principal regulator of biosynthesis for seed storage proteins and FAs, to reduce oxalic acid. L1L4 mutants revealed significant variation in moisture content, titratable acidity and overall metabolite profiles including oxalic and citric acid, fructose, β-carotene, total polyphenols and antioxidants.

14.5.8 IMPROVEMENT OF VITAMINS

Vitamins are an important component of our daily dietary needs. It helps in proper growth and development of human mind and body. CRISPR/Cas9 has also set its foot in enhancing vitamins in the food crops. Zhang et al. [82] successfully increased ascorbate content by ~150% along with increased oxidative stress tolerance, by GE of endogenous upstream open reading frames (uORFs) of LsGGP2, a gene which encodes a key enzyme in vitamin C biosynthesis in lettuce plants. Another group of researchers used a similar approach to design a gRNA targeting the upstream open reading frame (uORF) of SlGGP1 (encoding a vitamin C-biosynthetic enzyme) in accession LA1589 and led to increased foliar ascorbic acid (AsA) content in wild tomatoes [63].

14.6 COMMERCIALIZATION OF CRISPR/CAS9 EDITED FOOD CROPS

The first CRISPR-edited food crop to be commercialized is Sanatech Seed's Sicilian Rouge tomato which is a GABA enriched tomato. It is a product of a Tokyo based firm Sanatech Seeds and is being sold in the Japanese market since September 2021. The CRISPR-edited tomatoes were approved to release in the market in December 2020. The company claims Sicilian Rouge tomato has four to five times more GABA amount than the wild type of fruit. The studies show including GABA in diet helps normalize blood pressure. There is a list of other crops which are undergoing field trails and expected to release in market in coming years. Ground Cherry tomato (*Physalis pruinosa*) is ready to be released in the market although the timeline is yet to be ascertained. The crop has improved productivity along with fruit size and fruit number. Lemmon et al. [83] are working on this crop and is being developed in United States of America. Wild tomatoes (*Solanum pimpinellifolium*) with three times bigger fruit size, 10 times increased yield and 500% enhanced lycopene content are being developed by Zsögön et al. [84] in European Union (EU) and are in trial stage soon to be released in market for consumers. Researchers have been working on improving the shelf life of white button mushrooms and also preventing them from browning quickly [85]. The product is ready to be released in markets and is under scrutiny of USFDA. Similarly waxy corn with reduced amylose and enhanced amylopectin starch has been approved by USFDA for commercialization. Gluten free wheat for gluten intolerant people or those suffering from Celiac disease is being developed by Sánchez-Léon et al. [86] and is in trial stage in EU. Recently in Europe, the UK government granted permission for trial of asparagine free wheat lines generated by UK research institute Rothamsted Research. As suggested by few studies when bread is baked, asparagine can convert into acrylamide, a potential cancer-causing chemical.

14.7 ETHICAL LEGAL AND SOCIAL ISSUES (ELSI) OF CRISPR/CAS9 EDITED CROPS

GE is a revolutionary approach to develop novel crop varieties in far less time as compared to traditional breeding programs. But there are a few drawbacks associated with the methodology which bars it from being completely accepted by the consumers and the concerned regulatory bodies [87]. The social and biosafety concerns are related to off-target mutation and the undesirable effects generated due to unwanted genetic changes in plants. It is of concern that CRISPR-Cas9 can degenerate into filler DNA and might get inserted into non-specific locations in the genome while the DNA repair is taking place [33]. Although efforts have been made to improvise this technology by improving RNA guide – design strategies, ribonucleoprotein delivery, protein engineering, using spatiotemporally controlled Cas9 or base editing to name a few. This technique is costly as it has high levels of technical requirements, so after patenting of the crop variety it won't be accessible by all types of income groups. Chew et al. [88] reported immunogenic effects of Cas9 protein in mice.

In India rules and regulations to control research, development, and use of genetically modified organisms (GMOs) and their products, including new gene technologies, were established in 1989 and the regulations which control GMOs are applied on CRISPR edited food as well. The body which regulates GMO and other related issues is known by the name The Food Safety and Standards Authority of India (FSSAI) and it defines genetically engineered (GE) or modified food as "any food or food ingredient composed or containing engineered or GMO obtained through modern biotechnology, or food and food ingredients produced from but not containing engineered or GMO obtained through modern biotechnology."

In USA, certain crops developed by CRISPR were excluded from regulatory oversight. The United States Department of Agriculture (USDA) declared in March 2018 that GE is equivalent to conventional breeding in some cases and hence are not categorized and regulated under GM Crops pertaining to the fact that no foreign DNA or transgene is inserted during editing and that the resulting change did not involve resistance to pesticides or herbicides.

In EU, the Court of Justice (on July 2018) ordered that gene-edited crops should be subject to the same stringent regulations as conventional GMO. However, conventional mutagenic techniques used in a number of applications and having a long safety record are exempted from this regulation. Canada has different regulations, and it considers that any gene editing technology that is subjected to creating a novel product requires additional regulatory control on allergenicity, toxicity, and its impacts on non-target organisms.

In New Zealand, Hazardous Substances and New Organisms Act 1996 (HSNO Act) controls the import, development, field testing, and release of GMOs. In 2016, the HSNO Act was amended and kept plant breeding by GE under same regulatory laws as the GMOs. It has most stringent laws for the regulation of GMOs due to which country has yet not commercialized any single GM crops also no GM meat or fresh produce is sold in the country.

In Japan the Ministry of Health, Labor, and Welfare (MHLW) has also recently amended their law. The genome edited crops stand distinct to that of GMOs as they do not contain any foreign DNA hence newer guidelines are needed to be designed for the crops to be well regulated to reach the market for end use also this remarkable feature makes them more socially acceptable than plants that GMOs. In the end we can conclude that every country has different laws regulating the gene edited crops and this technology has a long way to go.

14.8 CONCLUSION AND FUTURE PROSPECTS

Biofortification of staple crops can prove to be a cost-effective method to reduce malnutrition cases. Biofortication of food crops helps in accumulation of essential nutrients like minerals; Zn, Fe, Se, vitamins; provitamin A, vitamin C, etc., in human body. Biofortified staple foods can be obtained by two methods; selective breeding and genetic modification. In selective breeding plants are cross bred with the one having desired and superior traits to produce selected trait offspring. This approach to develop plants with desired characters is time consuming and tedious. Genome modification is a faster and robust technology to develop plants with the desired characters in less time. Also, it requires fewer generation than selective breeding method. Genetic modification has few disadvantages associated such as a lot of research and advance biotechnical tools are needed, and it has not yet got much of societal acceptance which is a major issue. CRISPR-Cas systems have been improved to enhance its preciseness and lower the off-target effects. Base editing and prime editing are improved forms of CRISPR/Cas system. In near future one can see promising results in the crop improvement by CRISPR/Cas9 tool to curb hidden hunger.

KEYWORDS

- **base editors**
- **biofortification**
- **CRISPR**
- **genome editing**
- **micronutrient deficiency**
- **TALENs CRISPR-Cas9**
- **zinc finger nuclease**

REFERENCES

1. Jacquier, A., & Dujon, B., (1985). An intron-encoded protein is active in a gene conversion process that spreads an intron into a mitochondrial gene. *Cell, 41*(2), 383–394.
2. Tingdong, L., Xinping, Y., Yuan, Y., Xiaomin, S., Xiawan, Z., Huawei, Z., Wenxia, D., et al., (2018). Domestication of wild tomato is accelerated by genome editing. *Nature Biotechnology, 36,* 1160–1163.
3. Srinivasan, C., & Dana, C., (2016). Origins of programmable nucleases for genome engineering. *J. Mol. Biol., 428*(5 Pt B), 963–989.
4. Zhao, X., Hongyi, L., Yang, Y., Weiqi, H., Mengyuan, H., & Min, W., (2020). Applications of genome editing technology in the targeted therapy of human diseases: Mechanisms, advances and prospects. *Signal Transduction and Targeted Therapy, 5,* Article number:1.

5. Chevalier, B. S., & Stoddard, B. L., (2001). Homing endonucleases: Structural and functional insight into the catalysts of intron/intein mobility. *Nucleic Acids Res., 29*(18), 3757–3774.
6. Schiffer, J. T., Aubert, M., Weber, N. D., Mintzer, E., Stone, D., & Jerome, K. R., (2012). Targeted DNA mutagenesis for the cure of chronic viral infections. *J. Virol., 86*(17), 8920–8936.
7. Weber, N. D., Aubert, M., Dang, C. H., Stone, D., & Jerome, K. R., (2014). DNA cleavage enzymes for treatment of persistent viral infections: Recent advances and the pathway forward. *Virology, 454, 455*, 353–361.
8. Zaslavskiy, M., Bertonati, C., Duchateau, P., Duclert, A., & Silva, G. H., (2014). Efficient design of meganucleases using a machine learning approach. *BMC Bioinformatics, 15*, 191.
9. McMurrough, T. A., Brown, C. M., Zhang, K., Hausner, G., Junop, M. S., Gloor, G. B., & Edgell, D. R., (2018). Active site residue identity regulates cleavage preference of LAGLIDADG homing endonucleases. *Nucleic Acids Res., 46*, 11990–12007.
10. Ochiai, H., & Yamamoto, T., (2017). Construction and evaluation of zinc finger nucleases. *Methods Mol. Biol., 1630*, 1–24.
11. Watanabe, T., Ochiai, H., Sakuma, T., Horch, H. W., Hamaguchi, N., Nakamura, T., Bando, T., et al., (2012). Non-transgenic genome modifications in a hemimetabolous insect using zinc-finger and TAL effector nucleases. *Nat. Commun., 3*, 1017.
12. Wood, A. J., Lo, T. W., Zeitler, B., Pickle, C. S., Ralston, E. J., Lee, A. H., Amora, R., et al., (2011). Targeted genome editing across species using ZFNs and TALENs. *Science, 333*(6040), 307.
13. Boch, J., & Bonas, U., (2010). Xanthomonas AvrBs3 family-type III effectors: Discovery and function. *Annu. Rev. Phytopathol., 48*, 419–436.
14. Boch, J., Scholze, H., Schornack, S., et al., (2009). Breaking the code of DNA binding specificity of TAL-type III effectors. *Science, 326*, 1509–1512.
15. Moscou, M. J., & Bogdanove, A. J., (2009). A simple cipher governs DNA recognition by TAL effectors. *Science, 326*, 1501.
16. Miller, J. C., Tan, S., Qiao, G., Barlow, K. A., Wang, J., Xia, D. F., Meng, X., et al., (2011). A TALE nuclease architecture for efficient genome editing. *Nat. Biotechnol., 29*(2), 143–148.
17. Cermak, T., Doyle, E. L., Christian, M., Wang, L., Zhang, Y., Schmidt, C., Baller, J. A., et al., (2011). Efficient design and assembly of custom TALEN and other TAL effector-based constructs for DNA targeting. *Nucleic Acids Res., 39*(12), e82.
18. Bedell, V. M., Wang, Y., Campbell, J. M., Poshusta, T. L., Starker, C. G., Krug, R. G. 2nd., Tan, W., et al., (2012). *In vivo* genome editing using a high-efficiency TALEN system. *Nature, 491*(7422), 114–118.
19. Carlson, D. F., Tan, W., Lillico, S. G., Stverakova, D., Proudfoot, C., Christian, M., Voytas, D. F., et al., (2012). Efficient TALEN-mediated gene knockout in livestock. *Proc. Natl. Acad. Sci. U S A., 109*(43), 17382–17387.
20. Keith, J., & Jeffry, D. S., (2013). TALENs: A widely applicable technology for targeted genome editing *Nat. Rev. Mol. Cell Biol., 14*(1), 49–55.
21. Lei, Y., Guo, X., Liu, Y., Cao, Y., Deng, Y., Chen, X., Cheng, C. H., et al., (2012). Efficient targeted gene disruption in *Xenopus* embryos using engineered transcription activator-like effector nucleases (TALENs). *Proc. Natl. Acad. Sci. U S A., 109*(43), 17484–17489.

22. Hockemeyer, D., Wang, H., Kiani, S., Lai, C. S., Gao, Q., Cassady, J. P., Cost, G. J., et al., (2011). Genetic engineering of human pluripotent cells using TALE nucleases. *Nat. Biotechnol., 29*(8), 731–734.
23. Li, T., Huang, S., Zhao, X., Wright, D. A., Carpenter, S., Spalding, M. H., Weeks, D. P., & Yang, B., (2011). Modularly assembled designer TAL effector nucleases for targeted gene knockout and gene replacement in eukaryotes. *Nucleic Acids Res., 39*(14), 6315–6325.
24. Li, T., Liu, B., Spalding, M. H., Weeks, D. P., & Yang, B., (2012). High-efficiency TALEN-based gene editing produces disease-resistant rice. *Nat. Biotechnol., 30*(5), 390–392.
25. Liu, J., Li, C., Yu, Z., Huang, P., Wu, H., Wei, C., Zhu, N., et al., (2012). Efficient and specific modifications of the *Drosophila* genome by means of an easy TALEN strategy. *Genet Genomics, 39*(5), 209–215.
26. Sander, J. D., Cade, L., Khayter, C., Reyon, D., Peterson, R. T., Joung, J. K., & Yeh, J. R., (2011). Targeted gene disruption in somatic zebrafish cells using engineered TALENs. *Nat. Biotechnol., 29*(8), 697, 698.
27. Shuhei, Y., Naoyuki, U., Hyoung, J. L., Masaru, N., Satoru, S., Tetsushi, S., Takashi, Y., et al., (2019). Efficient genome engineering using platinum TALEN in potato. *Plant Biotechnol. (Tokyo), 36*(3), 167–173.
28. Tesson, L., Usal, C., Ménoret, S., Leung, E., Niles, B. J., Remy, S., Santiago, Y., et al., (2011). Knockout rats generated by embryo microinjection of TALENs. *Nat Biotechnol., 29*(8), 695, 696.
29. Sakuma, T., Ochiai, H., Kaneko, T., Mashimo, T., Tokumasu, D., Sakane, Y., Suzuki, K., Miyamoto, T., Sakamoto, N., Matsuura, S., et al., (2013a). Repeating pattern of non-RVD variations in DNA-binding modules enhances TALEN activity. *Sci. Rep., 3*, 3379.
30. Ma, L., Zhu, F., Li, Z., Zhang, J., Li, X., Dong, J., et al., (2015). TALEN-based mutagenesis of lipoxygenase LOX3 enhances the storage tolerance of rice (*Oryza sativa*) seeds. *PLoS One, 10*, e0143877.
31. Jinek, M., Chylinski, K., Fonfara, I., Hauer, M., Doudna, J. A., & Charpentier, E., (2012). A programmable dual-RNA-guided DNA endonuclease in adaptive bacterial immunity. *Science, 337*(6096), 816–821.
32. Mojica, F. J., Díez-Villaseñor, C., Soria, E., & Juez, G., (2000). Biological significance of a family of regularly spaced repeats in the genomes of archaea, bacteria and mitochondria. *Mol. Microbiol., 36*, 244–246.
33. Li, J., et al., (2016). Gene replacements and insertions in rice by intron targeting using CRISPR-Cas9. *Nat. Plants, 2*, 16139.
34. Liu, Y., Han, J., Chen, Z., Wu, H., Dong, H., & Nie, G., (2017). Engineering cell signaling using tunable CRISPR–Cpf1-based transcription factors. *Nat. Commun., 8*, 2095.
35. Zetsche, B., Gootenberg, J. S., Abudayyeh, O. O., Slaymaker, I. M., Makarova, K. S., Essletzbichler, P., et al., (2015). Cpf1 is a single RNA-guided endonuclease of a class 2 CRISPR-Cas system. *Cell, 163*, 759–771.
36. Komor, A. C., Kim, Y. B., Packer, M. S., Zuris, J. A., & Liu, D. R., (2016). Programmable editing of a target base in genomic DNA without double stranded DNA cleavage. *Nature, 533*, 420–424.

37. Lu, Y., & Zhu, J. K., *(2017).* Precise editing of a target base in the rice genome using a modified CRISPR/Cas9 system. *Mol. Plant, 10*, 523–525.
38. Huang, Q. N., Shi, Y. F., Zhang, X. B., Song, L. X., Feng, B. H., Wang, H. M., et al., (*2016*). Single base substitution in OsCDC48 is responsible for premature senescence and death phenotype in rice. *J. Integr. Plant Biol., 58*, 12–28.
39. Yan, F., Kuang, Y., Ren, B., Wang, J., Zhang, D., Lin, H., et al., (*2018*). High-efficient A·T to G·C base editing by Cas9n-guided tRNA adenosine deaminase in rice. *Mol. Plant., 11*, 631–634.
40. *Micronutrient Fortification of Food: Technology and Quality Control*, (2016). Wayback Machine.
41. Mkambula, P., Mbuya, M. N. N., Rowe, L. A., Sablah, M., Friesen, V. M., Chadha, M., Osei, A. K., et al., (2020). The unfinished agenda for food fortification in low- and middle-income countries: Quantifying progress, gaps and potential opportunities. *Nutrients, 29*, 354.
42. Bouis, H. E., Christine, H., McClafferty, B., Meenakshi, J. V., & Wolfgang, H. P., (2011). Biofortification: A new tool to reduce micronutrient malnutrition. *Sage Journal,* S31–S40.
43. Sun, Y., Jiao, G., Liu, Z., Zhang, X., Li, J., & Guo, X., (2017). Generation of high-amylose rice through CRISPR/Cas9-mediated targeted mutagenesis of starch branching enzymes. *Front. Plant Sci., 8*, 298.
44. Chen, K., et al., (2019). CRISPR/Cas genome editing and precision plant breeding in agriculture. *Annu. Rev. Plant Biol., 70*, 667–697.
45. Tuncel, A., Corbin, K. R., Ahn-Jarvis, J., Harris, S., Hawkins, E., Smedley, M. A., Harwood, W., et al., (2019). Cas9-mediated mutagenesis of potato starch branching enzymes generates a range of tuber starch phenotypes. *Plant Biotechnology Journal*, pbi.13137.
46. Wang, H., Wu, Y., Zhang, Y., Yang, J., Fan, W., Zhang, H., Zhao, S., et al., (2019). CRISPR/Cas9-based mutagenesis of starch biosynthetic genes in sweet potato (*Ipomoea* Batatas) for the improvement of starch quality. *International Journal of Molecular Sciences, 20*(19), 4702.
47. Jingying, L., Guiai, J., Yongwei, S., Jun, C., Yingxin, Z., Lei, Y., Dong, J., et al., (2021). Modification of starch composition, structure and properties through editing of TaSBEIIa in both winter and spring wheat varieties by CRISPR/Cas9. *Plant Biotechnol J., 19*(5), 937–951.
48. Pioneer, (2016). *DuPont Announces Intentions to Commercialize First CRISPR-Cas Product.* Press Release. https://www.prweb.com/releases/dupont-pioneer-seed/crispr-cas-corn/prweb13349828.htm (accessed on 22 June 2023).
49. Andersson, M., Turesson, H., Nicolia, A., Fält, A. S., Samuelsson, M., & Hofvander, P., (2017). Efficient targeted multiallelic mutagenesis in tetraploid potato (*Solanum tuberosum*) by transient CRISPR-Cas9 expression in protoplasts. *Plant Cell Rep., 36*, 117–128.
50. Andersson, M., Turesson, H., Olsson, N., Fält, A. S., Ohlsson, P., Gonzalez, M. N., et al., (2018). Genome editing in potato via CRISPR-Cas9 ribonucleoprotein delivery. *Physiol. Plant, 164*, 378–384.
51. Zhang, S., Zhang, R., Gao, J., Song, G., Li, J., Li, W., Qi, Y., Li, Y., & Li, G., 2021. CRISPR/Cas9-mediated genome editing for wheat grain quality improvement. *Plant Biotechnology Journal, 19*(9), 1684.

52. Dong, L., Qi, X., Zhu, J., Liu, C., Zhang, X.,Cheng, B., Mao, L., & Xie, C., (2019). Supersweet and waxy: Meeting the diverse demands for specialty maize by genome editing. *Plant Biotechnol J., 17*(10), 1853–1855.
53. Gao, H., Gadlage, M. J., Lafitte, H. R., et al., (2020b). Superior field performance of waxy corn engineered using CRISPR–Cas9. *Nat. Biotechnol., 38*, 579–581.
54. Qi, X., Wu, H., Jiang, H., et al., (2020). Conversion of a normal maize hybrid into a waxy version using *in vivo* CRISPR/Cas9 targeted mutation activity. *Crop J., 8*, 440–448.
55. Che, J., Yamaji, N., & Ma, J. F. (2021). Role of a vacuolar iron transporter OsVIT2 in the distribution of iron to rice grains. *New Phytologist, 230*(3), 1049–1062.
56. Sheng-Kai, S., Xuejie, X., Zhong, T., Zhu, T., Xin-Yuan, H., Markus, W., Rüdiger, H., & Fang-Jie, Z., (2021). A molecular switch in sulfur metabolism to reduce arsenic and enrich selenium in rice grain. *Nature Communications, 12*, Article number: 1392.
57. Mohan, M. A. V., & Malireddy, K. R., (2021). CRISPR-Cas9 mediated mutation in GRAIN WIDTH and WEIGHT2 (GW2) locus improves aleurone layer and grain nutritional quality in rice. *Scientific Reports, 11*, Article number: 21941.
58. Ali, A., Paul, S., Gayen, G., Sarkar, S. N., Datta, K., & Datta, S. K., (2013). Development of low phytate rice by RNAi mediated seed-specific silencing of inositol 1, 3, 4, 5, 6-pentakisphosphate 2-kinase gene (IPK1). *PLoS One, 8*(7), Article e68161.
59. Khan, M. S. S., Basnet, R., Islam, S. A., & Shu, Q., (2019). Mutational analysis of OsPLDα1 reveals its involvement in phytic acid biosynthesis in rice grains. *J. Agric. Food Chem., 67*, 11436–11443.
60. Sashidhar, N., Harloff, H. J., Potgieter, L., & Jung, C., (2020). Gene editing of three BnITPK genes in tetraploid oilseed rape leads to significant reduction of phytic acid in seeds. *Plant Biotechnology Journal.*
61. Saira, I., Bilal, S., Nazia, R., Syed, A., Zafar, M., Kashif, N., & Muhammad, R. K., (2021). CRISPR/Cas9 mediated disruption of *Inositol Pentakisphosphate 2-Kinase 1* (*TaIPK1*) reduces phytic acid and improves iron and zinc accumulation in wheat grains. *Journal of Advanced Research.*
62. Xiaoyu, Y., Lei, C., Junxian, H., & Weichang, Y., (2017). Knocking out of carotenoid catabolic genes in rice fails to boost carotenoid accumulation, but reveals a mutation in strigolactone biosynthesis. *Plant Cell Rep.*, (10), 1533–1545.
63. Li, X., et al., (2018). Lycopene is enriched in tomato fruit by CRISPR/Cas9-mediated multiplex genome editing. *Front. Plant Sci., 9*, 559.
64. Akira, E., Hiroaki, S., Miho, T., Norihiko, M., & Seiichi, T., (2019). A novel approach to carotenoid accumulation in rice callus by mimicking the cauliflower orange mutation via genome editing. *Rice, 12*, Article number: 81.
65. Dong, O. X., et al., (2020). Marker-free carotenoid-enriched rice generated through targeted gene insertion using CRISPR-Cas9. *Nat. Commun., 11*, 1–10.
66. Paine, J. A., et al., (2005). Improving the nutritional value of golden rice through increased pro-vitamin A content. *Nat. Biotechnol., 23,* 482–487.
67. Kaur, N., et al., (2020). CRISPR/Cas9 directed editing of lycopene epsilon-cyclase modulates metabolic flux for βcarotene biosynthesis in banana fruit. *Metab. Eng., 59*, 76–86.
68. Nonaka, S., Arai, C., Takayama, M., Matsukura, C., & Ezura, H., (2017). Efficient increase of ɣ-aminobutyric acid (GABA) content in tomato fruits by targeted mutagenesis. *Scientific Reports, 7*(1).

69. Sánchez-León, S., Gil-Humanes, J., Ozuna, C. V., Giménez, M. J., Sousa, C., Voytas, D. F., & Barro, F., (2017). Low-gluten, non-transgenic wheat engineered with CRISPR/Cas9. *Plant Biotechnology Journal.*
70. Zhang, Y., Li, D., Zhang, D., Zhao, X., Cao, X., Dong, L., Liu, J., et al., (2020). Analysis of the functions of TaGW2 homoeologs in wheat grain weight and protein content traits. *The Plant Journal.*
71. Akama, K., Akter, N., Endo, H., et al., (2020). An in vivo targeted deletion of the calmodulin-binding domain from rice glutamate decarboxylase 3 (OsGAD3) increases gamma-aminobutyric acid content in grains. *Rice (n y), 13*(1), 20.
72. Camerlengo, F., Frittelli, A., Sparks, C., et al., (2020). CRISPR-Cas9 multiplex editing of the α-amylase/trypsin inhibitor genes to reduce allergen proteins in durum wheat. *Front Sustain Food Syst.*
73. Baysal, C., He, W., Drapal, M., et al., (2020). Inactivation of rice starch branching enzyme IIb triggers broad and unexpected changes in metabolism by transcriptional reprogramming. *Proc. Natl. Acad. Sci., 42*, 26503–26512.
74. Jiang, W. Z., Henry, I. M., Lynagh, P. G., Comai, L., Cahoon, E. B., & Weeks, D. P., (2017). Significant enhancement of fatty acid composition in seeds of the allohexaploid, Camelina sativa, using CRISPR/Cas9 gene editing. *Plant Biotechnology Journal, 15*(5), 648–657.
75. Ayako, O., Takumi, O., Chie, K., Kanako, K., Mizue, I., Jun, I., & Nobuya, K., (2018). CRISPR/Cas9-mediated genome editing of the fatty acid desaturase 2 gene in *Brassica napus*. *Plant Physiol. Biochem., 131*, 63–69.
76. Phat, T. D., Cuong, X. N., Hien, T. B., Ly, T. N. T., Gary, S., Jason, D. G., Zhanyuan, J. Z., & Minviluz, G. S., (2019). Demonstration of highly efficient dual gRNA CRISPR/Cas9 editing of the homeologous *GmFAD2–1A* and *GmFAD2–1B* genes to yield a high oleic, low linoleic and α-linolenic acid phenotype in soybean. *BMC Plant Biol., 19*, 311.
77. Zhai, Y., Yu, K., Cai, S., Hu, L., Amoo, O., Xu, L., Yang, Y., et al., (2019). Targeted mutagenesis of BnTT8 homologs controls yellow seed coat development for effective oil production in *Brassica napus* L. *Plant Biotechnology Journal.*
78. Ma, J., Sun, S., Whelan, J., & Shou, H., (2021). CRISPR/Cas9-mediated knockout of GmFATB1 significantly reduced the amount of saturated fatty acids in soybean seeds. *International Journal of Molecular Sciences, 22*(8), 3877.
79. Arao, T., & Ishikawa, S., (2006). Genotypic differences in cadmium concentration and distribution of soybean and rice. *JARQ: Japan Agricultural Research Quarterly; Biology.*
80. Zhang, L., Tang, L., Mao, B., Li, Y., Lv, Q., Chen, C., et al., (2017). Knockout of OsNramp5 using the CRISPR/Cas9 system produces low Cd-accumulating indica rice without compromising yield. *Sci. Rep., 7*, 14438.
81. Gago, C., Drosou, V., Paschalidis, K., Guerreiro, A., Miguel, G., Antunes, D., & Hilioti, Z., (2017). Targeted gene disruption coupled with metabolic screen approach to uncover the LEAFY COTYLEDON1-LIKE4(L1L4) function in tomato fruit metabolism. *Plant Cell Reports, 36*(7), 1065–1082.
82. Huawei, Z., Xiaomin, S., Xiang, J., Rong, F., Jinxing, L., Kunling, C., Daowen, W., & Caixia, G., (2018). Genome editing of upstream open reading frames enables translational control in plants. *Nature Biotechnology, 36*, 894–898.
83. Lemmon, Z. H., et al., (2018). Rapid improvement of domestication traits in an orphan crop by genome editing. *Nat. Plants, 4*, 766–770.

84. Zsögön, A., Čermák, T., Naves, E., et al., (2018). *De novo* domestication of wild tomato using genome editing. *Nat. Biotechnol., 36*, 1211–1216.
85. Waltz, E., (2016). Gene-edited CRISPR mushroom escapes US regulation. *Nature, 532*(7599), 293.
86. *News of This Trial Will Likely be Welcomed by the Food Industry': Europe's First CRISPR-Edited Wheat set to be Grown.* Article By: Oliver Morrison, Press Release. https://www.foodnavigator.com/Article/2021/09/03/News-of-this-trial-will-likely-be-welcomed-by-the-food-industry-Europe-s-first-CRISPR-edited-wheat-set-to-be-grown (accessed on 22 May 2023).
87. Zafar, K., Khalid, E. M. S., Gundra, S. R., Muhammad, Z. K., Imran, A., Radwa, K., Zahid, M., et al., (2020). Genome editing technologies for rice improvement: Progress, prospects, and safety concerns. *Front. Genome Ed.*
88. Chew, W. L., Tabebordbar, M., Cheng, J. K., Mali, P., Wu, E. Y., Ng, A. H., et al., (2016). A multifunctional AAV-CRISPR-Cas9 and its host response. *Nat. Methods, 13*, 868–874.
89. Hiroyuki, S., & Susumu, K. (1981). New restriction endonucleases from *Flavobacterium okeanokoites* (FokI) and *Micrococcus luteus* (MluI). *Gene,16*(1–3), 73–78.

CHAPTER 15

Algae and Fungi-Based Micronutrients Enrichment in Food

NIVEDITA SINGH[1] and SAIMA SOHRAB[2]

[1]*Phycology Laboratory, Department of Botany, University of Allahabad, Prayagraj, Uttar Pradesh, India*

[2]*Department of Botany, Ewing Christian College, University of Allahabad, Prayagraj, Uttar Pradesh, India*

ABSTRACT

Micronutrient malnutrition or deficiencies of iron, zinc, iodine, vitamins, etc., produce several negative impacts on the individuals' as well as societies' health system. Iron deficiency is found to be common among anemic children and women, leading to growth retardation and cognitive impairment. Vitamin A deficiency (VAD) causes loss of sight and also increases the threat of mortality and morbidity of expecting women while lack of vitamin D leads to deteriorating effects on bones. Zinc and iodine deficiency increases the susceptibility to diseases and retards the physical and mental well-being of individuals. An effective way to curb this inadequacy of micronutrients in plants and animals is biofortification, which is a process of upgrading of the dietary quality of food crops. Genetic and classical breeding approaches are augmented which are often found to be inapt for the rural population. Thus, it is more lucrative to circumvent the conventional fortification dependency and move to biofortification using agriculturally significant bacteria, fungi, and algae as a substitute and sustainable tool to increase micronutrients in crop plants. Mycorrhizal fungi provide tolerance against biotic and abiotic stress, promote the plant productivity and nutritional value by enriching the plants with N, P, Zn, Fe, Mn, and Cu and also serve as a

Phytoremediation and Biofortification: Strategies for Sustainable Environmental and Health Management.
Nand Kumar Singh, Shadma Afzal, & Tariq Aftab (Eds.)

natural fertilizer. Members of basidiomycetes (e.g., *Agaricus*) are capable of bioaccumulating metals of pharmacological and nutritional significance such as iron, zinc, selenium, and lithium. Algae biomass-based micronutrient bio-components also serve as a natural fertilizer and have been implemented in the biofortification of maize. Microalgae (such as *Spirulina, Chlorella,* and *Arthrospira*) and certain seaweeds when incorporated in food products enhance proteins, carbohydrates, essential minerals, polyunsaturated fatty acids (FAs), and vitamins in them. Thus, algae/fungi based biofortification should be encouraged to promote a healthy and nutritious diet feasible for all and more rigorous research is needed to promote the commercialization of bio-fortification for eradicating the problem of malnutrition in the world.

15.1 INTRODUCTION

Micronutrient deficiency is a severe health issue that affects people all over the world. Increased micronutrient density in staple meals appears to be the solution. Micronutrient directed fertilization can be utilized as a form of food biofortification, allowing micronutrient deficiencies to be overcome. Intensive soil exploitation produces a slew of issues, one of which is micronutrient depletion [1]. Micronutrient deficiency is most common among women and children in underdeveloped nations. Hidden hunger is a term for micronutrient malnutrition [2]. More than 2 billion individuals (more than 30% of the world's inhabitants) undergo from zinc insufficiency [3]. Growth stunting, susceptibility to viral illnesses, are all symptoms of Zn deficiency in humans [4]. Slow growth, anemia, reduced immunity, and nyctalopia are all symptoms of vitamin deficiency [5]. Plant nutrition depends heavily on micronutrients. Crops, livestock, and humans all require them in modest amounts [1]. They function as cofactors in a variety of metabolic processes. Trace elements are also required for effective implementation of respiration and photosynthesis functions [6]. Many plants are chiefly vulnerable to mark out element shortage in the soil for example the result in maize is a reduced crop production [7]. Micronutrients in grain, such as Zn, Cu, Mn, and Fe, have a direct impact on food quality [1]. Nonetheless, micronutrient fertilization strategies have received too little attention in recent years [5]. Biofortification, by definition, is a method of increasing nutrient substances in staple food crops by agricultural methods such as usual agronomic techniques, mutagenesis, breeding, and genetic engineering [8]. Use of micronutrient fertilizers appears in the direction to subsist the most straightforward method of biofortification [9]. Micronutrient fertilizers have been found to be beneficial in enhancing micronutrient source in wheat, rice, and maize grains, which are significant cereal crops due to their fiscal worth in farm animals feeding [10]. Agronomic zinc biofortification by fertilizer appliance is a balancing strategy to increasing grain Zn concentration even further.

Different fertilizers, including chelates and procedural salts of micronutrients, are available in the market. Chelates are rather expensive, yet they have an elevated bioavailability. Although Procedural salts are inexpensive fertilizers, micronutrients are rapidly eluted from the soil by water [11]. Furthermore, chelating compounds like EDTA are both hazardous and non-biodegradable [12]. Traditional micronutrient fertilizers are being replaced by new biocomponents with micronutrients derived from waste biomass [11].

Experts called for employing various ways for the elemental biofortification of crop plants to combat micronutrient deficiency and sustain elevated concentrations of important micronutrients. Agronomic techniques, genetic engineering, plant breeding, and the use of earth microbes (plant growth-promoting *Mycorrhiza* and *Rhizobacteria*) are some of the important tactics that might be used to increase the absorption of micronutrients (Mn, Fe, Cu, Se, and Zn) in plant parts which are suitable for eating. Every strategy has its own set of benefits and drawbacks [13].

The biofortification method aims to capitalize on all family members' constant daily ingestion of substantial amounts of food staples, particularly women and offspring, who are the most susceptible to micronutrient scarcity. This method wholly targets low-income households due to the prevalence of food staples in low-income households' diets. Recurring costs are modest after a one-time investment in generating self-fortifying seeds, and germplasm can be shared internationally. Plant breeding is so cost-effective because of the multiplier effect it has across time and distance. Even if management focus and intercontinental financing for micronutrient issues decrease, manufacture, and utilization of nutritionally enhanced varieties are very sustainable once in place. Biofortification is a viable method of addressing underfed populations in rural areas by distributing naturally prepared meals to those who are not privileged enough to use commercially enriched foods, which are further easily available in metropolitan zones. To be achieved, however, this will take decades, efficient policy formation by government bodies, and a huge venture into agricultural investigation and additional public and on-farm infrastructure is another pressing need [13]. Interdisciplinary contact between human nutrition experts and plant scientists has significant potential in conceiving treatments for a variety of nutritional deficits. Human nutritionists, for example, should be aware of the extent to which plant breeding can alter the vitamin and mineral concentration of specific foods, as well as the chemicals (for instance, prebiotics) that endorse and hinder its bioavailability. Plant breeders must be cognizant of both the historical impact of agricultural research on the use of nutrients (e.g., the bioavailability of trace minerals in current variety versus conventional kinds) and the future prospective of plant breeding for nourishment and health gains [14].

Plant growth-promoting microorganisms (PGPM) such as cyanobacteria, bacteria, fungi, and mycorrhiza, can alter nutrient accessibility in the soil through solubilization, sequestration, and oxidation/reduction processes, which are a new and promising option in this regard [15–17]. Microorganism-mediated biofortification is linked to their important function in nutrient cycling, which includes the breakdown and facilitation of the mineralization of crude matter along with the conversion of nutrients of inorganic nature into usable forms [18, 19]. Microbes can as well influence plant growth and nutrient uptake by releasing growth-stimulating or -inhibiting chemicals to change root composition and root system structural design [20–22]. Bacteria also act together with plant roots, causing the discharge of root exudates, which can affect soil qualities such as pH and nutritional standing, affecting the variety and commotion of plant-associated beneficial microorganisms (BMOs) [23]. In this chapter, fungi and algae-mediated biofortification of food is discussed.

15.2 FUNGI MEDIATED BIOFORTIFICATION

Mushrooms have long been considered as a fine meal, particularly for their restrained essence and flavor, in addition to as a culinary marvel by humans. There are about 2,000 different types of mushrooms, yet just 25 are commonly eaten and only a few are professionally farmed [1]. Since they have high functional and dietary value, mushrooms are frequently employed as nutraceutical foods. They've also gotten a lot of attention because of their trade and industrial value as well as their organoleptic and therapeutic features [2, 3]. *Agaricus bisporus* is the most extensively farmed fungus, followed by *Lentinus edodes*, *Flammulina velutipes,* and *Pleurotus*. The basic protein substance of mushrooms suitable for eating be typically elevated, although it differs widely moreover influenced by some factors such as the mushroom's species and phase of growth [5]. The free amino acid content of mushrooms is typically little and almost lacks sulfur-containing methionine and cysteine, however, they are high in threonine and valine, and it contributes to the principal flavor qualities [6]. Several vitamins are abundant in cultivated mushrooms, including folates, niacin, and riboflavin. Mushrooms contain more vitamin B_2 than most vegetables, making them a rich source of this vitamin [7]. Folic acid is the most bioavailable type of folate in mushrooms [8]. Vitamins B_1 and C, as well as traces of vitamin B_{12}, are found in modest amounts in cultivated mushrooms [7]. Total soluble sugars are in short supply in edible mushrooms, but oligosaccharides are plentiful [9]. The carbohydrate value of edible mushrooms ranges from one species to another and varies starting 35 to 70% by dry weight (DW). Mushrooms have a fatty acid (FA) content ranging from 2 to 8% [11–13]. Contrary to FAs, polyunsaturated FAs report for around 3/4th of the total FAs, whereas palmitic acid (saturated fatty acid) is reported to be the most abundant [10]. All the attributions and nutritional qualities described above make the mushroom one of the best sources to be applied for the biofortification of food.

In addition, arbuscular mycorrhizal fungi (AMF) application has been set up to be effective in increasing macronutrients (P, N, Ca, K) and micronutrients (B, Mn, Fe, Cu, Mo, Zn) range in crop plants like okra as well as pea crops. Other studies also suggest that AMF inoculation is important for improving crop efficiency, agronomic bio-fortification with respect to micronutrients, and eminence augmentation, as well as assembling plant nutrient desires proficiently and cost-effectively. Thus, if AMF is applied to the Himalayan production systems (which is a poor resource system) would result assist in harnessing nutritionally affluent farm turnout for Himalayan communities afflicting from malnutrition, particularly anemia and zinc insufficiency, as well as resource-poor Himalayan farmers who cannot meet the expense of costly outside inputs, such as phosphatic fertilizers [24].

15.2.1 ZINC BIOFORTIFICATION

Mitigating Zinc deficiency in dietary grains is a difficult task, and a new technique to improve Zn status must be developed. Biofortification is a novel way to address the predicament of Zinc deficiency in the human diet. The strategy has proven to be long-term, low-cost to execute, extremely effective, and widespread, particularly in the world's poorest places. Exogenous Zn sources such as $ZnSO_4$ are fixed in the soil in around 75% of cases. Fixation of Zn in earth with pH > 7.0 rises by way of escalating carbonate concentrations, making it unavailable. It can be reverted to accessible form with bacteria which solubilizes Zn by producing organic acids such as gluconic acid, a strong acid in the monocarboxylic group of acids that is easily biodegradable. The main anion in gluconic acid may have been a crucial factor in the process of solubilization of indissoluble Zn amalgam. It loosens the compactly attached residual structure of zinc and increases its accessibility. As a result of the increased amount of Zn soil utilized by plants, AMF are recognized to boost the nutrient demand of the host plant [25].

15.2.2 SELENIUM BIOFORTIFICATION

BMOs help to improve Se biofortification (Table 15.1). Plant-BMO-symbiosis promotes plant growth, micronutrient percipience, in addition to resilience to abiotic and biotic stressors. In bio-filmed biofertilizers, BMOs such as mycorrhizal fungi as endo- and ectomycorrhizal fungi), root endophytic fungi (REFs), and PGPRs are common. Angiosperm roots are colonized by AMF, while gymnosperm roots are colonized by ecto-mycorrhizal fungi. The majority of REFs have a diverse spectrum of plant hosts. Since their potential is headed for improvement of the nourishment intake of their host flora, AMF are utilized for Selenium biofortification. Consequently, the interaction between the plants and mycorrhizal fungus, phosphate absorption and transfer to host plants are improved [26–30]. Se is absorbed as selenate or selenite, which is metabolized in plants through sulfur assimilation route, resulting in the creation of SeCys, SeMet, and further Se isoforms of other metabolites of sulfur [31–35]. The sulfur transporters connecting mycorrhizal fungi and their host plants can transport Se to host plants and are involved in the buildup of selenium in food crops. For example, Selenium biofortification has been done on wheat plants and it has been seen that among different genotypes of wheat tested, the most Se-tolerant wheat genotype is 'Puelche' which is reported to accumulate the most, to the point where its Se accumulation has been connected to the highest amount of sulfate transporter named TaeSultr4.1 mRNA in the roots [87]. Other transporters like silicon transporters in rice [36] and tomato [37], phosphate/orthophosphate transporters in wheat [38], tomato [89], rice [39–41], and yeast *(Saccharomyces cerevisiae)* [43, 44], and monocarboxylates transporters in yeast (*S. cerevisiae*) [42], have been identified, also play a role in Se transport. Furthermore, it has been found that competing ions such as phosphate and sulfate alter the availability of selenate and selenite in plants [46, 47]. For example, conflict among phosphate and Selenium uptake results in decreased Se accumulation translocation coefficients and Se concentrations in wheat roots, stems, leaves, and spikes when phosphate fertilizers are functional to selenite-treated soil [48]. Sulfate and selenate uptake in *Astragalus species*, on the other hand, demonstrated that sulfur deficit increased Se amassing in equally roots as well as shoot tissue, while increased Se availability increased sulfate accumulation [49].

15.2.3 LITHIUM BIOFORTIFICATION

Bio-enrichment of mushrooms during cultivation is being investigated more and more as a possible technique for producing functional meals [50–52]. Li nutritional intake has been shown to be advantageous to health notably in the hindrance or amelioration of neurological and mental changes [53, 54]. Studies show that some mushroom species may efficiently acquire Li from overgrown substrates and store it in their fruiting bodies without affecting their development or appearance significantly. Li-enriched mushrooms' potential commercial application as a nutraceutical would necessitate cost-effective and efficient production in terms of produced biomass. Artificially increasing the content of various elements in the cultivation substrate can have a substantial impact on mushroom development and appearance, leading to higher cultivation costs, a loss of attraction to potential buyers, and a reduced market value [50]. Grown species grew a larger mycelium colony and were extra defiant to Li in the figure of acetate, resulting in higher biomass of fruiting bodies. The therapeutic qualities of *G. lucidum*, which have long been recognized [55, 56], may also promote the economic appliance of Li-fortified *G. lucidum*. The profitable preface of *G. lucidum's* Li-enriched version supposed not to be difficult, given to facilitate the market for *G. lucidum* is by now appeal over 2.5 billion dollars [57]. The product's safety ought to be assessed initial, if possible, using both *in vitro* and *in vivo* models; the product's target group of possible regulars must then be distinct, probably utilizing randomized trials on various populations. Overall, Li-rich mushrooms like *G. lucidum*, *P. ostreatus*, and *P. eryngii* are more likely to have a realistic relevance in prophylactics of temper disturbances than in psychiatric handling. Importantly, significantly increased Li consumption may help former drinkers and drug addicts stabilize their moods [53, 58]. Suicides, anxiousness, and homicides have all been connected to low serum Li levels [59], therefore such food crop could assure a sufficient dietary intake of Li for the goal of reducing the regularity of belligerence and cruelty. The interim suggested dietary intake for Li in adults has been set at 1 mg for a 70-kg mature to maintain physiological levels, despite the fact that tenfold larger doses have not been observed to cause any significant health effects [61]. Some writers have lately advised that cereal grain harvest be prepared with Li or that Li be supplemented to dietary supplements to meet the nutritional claim for Li [61]. When one assumes their esteem in many cuisines, their conventional utility in diverse ethinicity for fitness and illness deterrence, and their societal recognition as consumables, mushrooms may be an intriguing or even superior alternative in this aspect [62]. All three species included in this study could provide moderately enhanced dietary Li at concentrations significantly lower than those considered hazardous to humans [60]. On the other hand, the recent studies found rather high fluctuations in Li assessment in mushroom fruiting bodies, which could pose a problem in standardizing Li levels in ultimate food yield derived from such mushrooms. To assess the therapeutic prospective of Li-enriched mushrooms, more biomedical research is needed, counting investigational *in vitro* and *in vivo* prototype.

In conclusion, the current research looked into regardless 3 economically imperative mushroom variety could be used in Li biofortification to fabricate Li-enriched fruiting structures. In addition to evaluating Li accumulation, the impact on the mycelium settlement and the biomass of fruiting bodies was evaluated. *G. lucidum* showed the most promising results, together in conditions of growth and in conditions of Li amassing. Given its already proven medicinal status, the acceptance of this variety may be mainly intriguing in raising Li dietary consumption for mood regulation reasons. Mushrooms biofortified with Li may be useful in reducing aggressive behavior rates—a hypothesis that has to be experienced in randomized manage trials followed by security assessments.

15.2.4 COPPER, IRON, AND MANGANESE BIOFORTIFICATION

AMF has a strong favorable impact on crop plant Cu sustenance and, underneath some situations, moreover on Fe nutrition. For Mn, there was just a minor AM fungal-mediated advantage. There were no downbeat or neutral AM fungal mediated effects for Cu. Only 'herbs' showed a beneficial AM fungal mediated effect on crop plants. Taken together, AMF has a clear involvement in agricultural plant nutrition for copper and iron, while its effect when observed on Mn appears to be limited to herbs. Another significant step is to better understand the procedure of micronutrient uptake and its translocation by AMF to the related plant, as well as to assess AMF's contribution in crop micronutrient homeostasis. Because there are so many possible challenges and problems, only understanding the fundamental mechanisms will allow AMF to be used effectively as pre-harvest fortification agents. However, assessing the role of AM fungi as possible fortification agents in relation to other fortification approaches, such as agricultural and genetic biofortification is critical. The confirmation that AMF increases the bioavailability of micronutrients in edible plant portions, taking into account the influence of anti-nutrients [63, 64] and the aftermath of food dispensation and storage on micronutrient bioavailability, would be required to establish an affirmative AM fungal mediated effect on individual sustenance [66, 67]. To profit sustainably from AMF, future research must explain the causal mechanisms of plant-fungus nutrition balance and recognize added suitable and efficient agricultural executive strategies. Certain pressures on cultivation to provide for an increasing individual inhabitant, as well as the latent downbeat consequences of parts of worldwide modification on crop nutritional excellence, this is becoming an increasingly vital undertaking [68].

TABLE 15.1 List of Crop Plants and Mycorrhizal Fungi Used for Se Biofortification [63]

Crop Plants or Vegetables as Host Plants	Fungi Involved in Biofortification
Wheat	*Glomus claroideum* (AMF)
	Glomus versiform
	Funneliformis mosseae
Garlic	*Glomus fasciculatum*
	Glomus irtraradices
	Glomus mosseae
Asparagus officinalis	*Rhizophagus intraradices*
	Funneliformis mosseae
Corn plant	*Glomus mosseae*
Onion	*Rhizophagus intraradices*
	Trichoderma harzianum
Lettuce	*Rhizophagus intraradices*
	Funneliformis mosseae
Soybean	*Glomus mosseae*

Source: Ref. [63]

15.3 ALGAE MEDIATED BIOFORTIFICATION

15.3.1 INNOVATIVE NUTRITIONAL FORMULATION: BIOFORTIFICATION OF HENS EGGS WITH POLYUNSATURATED FATTY ACIDS BY SUPERCRITICAL MICROALGAL EXTRACT

There has recently been marketplace need for a new age group of purposeful foods along with pro-health effects. Consumers have become more aware of the link between diet and health. It is now more vital than ever to understand not just the calorie substance of food, although also that it has all of the crucial elements. This opened up fresh market for goods having higher concentrations of biologically vigorous chemicals [69]. Polyunsaturated fatty acids (PUFAs) are a model of such molecules required by consumers, owing to their well-being (brain plus vision development and defense, protection against cardiovascular illnesses) [70].

Many items, such as oils, mayonnaise, margarine, salad dressing, bakery foodstuffs, newborn formula, milk, farmed fish, meat products, and eggs, are reinforced with PUFA n–3 and PUFA n–6 (vary in the site of the initial twofold bond among carbon atoms) [72]. Many PUFA-fortified goods come from the poultry industry, and enriched eggs, in particular, are gaining popularity in the human diet. Sardine oil [73], linseed oil [76], menhaden oil [74], flaxseed oil [78], rapeseed/corn oils [75], and fish oil (FO) [76] are some of the PUFA-rich supplements used in the poultry feed (chicken/hens). The chief downside of eggs enhanced with FA from FO, for instance, is the possibility of oxidation and a reduction in PUFA n–3 content during preservation.

Further ordinary sources of PUFAs and antioxidants include aquatic algae (different species) [73] or microalgae (different species) [75]: *Chlorella* sp. [78], *Phaeodactylum tricornutum, Chlorella fusca, Isochrysis galbana* [79], *Schizochytrium* sps. [76, 80], *Nannochloropsis oculata* [74], and *Nannochloropsis gadit*. Some workers have already investigated the usage of microalgae (*Spirulina maxima*) derived biomass in the diet of laying hens (Figure 15.1) [82].

Typically, poultry diets are based on a regular feeding mixture made up of PUFA-rich basic materials [83]. The microalgae *Spirulina platensis* (Arthrospira) was used to extract active chemicals (including PUFAs). Commercially, this alga is grown in big outside ponds under controlled conditions, or it can be harvested directly from lakes. The current global production of *Spirulina* is expected to be around 3,000 metric tons per year. It was primarily utilized for food and specialty feeds in the last 30 years [85].

Spirulina sp. is a good source of protein (greater than egg or meat, beef, beer yeast, cheese, skimmed powdered milk, and fish) as well as minerals, vitamins, and pigments [85]. Spirulina consists of 6 to 13% lipids, half of that can be represented as whole FAs, according to the concentrate of lipids found in this microalga [86]. Spirulina platensis has a lipid component with a high quantity of PUFAs like gamma-linoleic acid, according to Zinnai et al. [87]. Many factors influence the constitution of PUFA in Spirulina sp., counting atmosphere, development phase [88], temperature, as well as the content of culture means [89].

The withdrawal of physiologically dynamic chemicals from *Spirulina* sp. using supercritical fluid extraction (SFE) was investigated. Mendes et al. [90] achieved the extraction of GLA from *Spirulina* using CO_2 and a polar co-solvent in the second variation (ethanol). The use of this solvent enhanced the withdrawal yield of lipids in the glycolipid fraction, including GLA. Another group of researchers, Andrich et al. [92] had also extracted the oils from *Spirulina platensis* which again was reported to have GLA.

According to the available research, concentrate of algae can be used as a straight resource of dietary FAs which might exist a viable option to presently accessible FA root for makers of PUFA n–3 and PUFA n–6-rich products [73].

15.3.2 NEW DIETARY FORMULATION

15.3.2.1 SPIRULINA AS BIOFORTIFICATION AGENT

Biofortification using *Spirulina platensis* Trace elements, often known as micronutrients, take part in a significant role in plant and human metabolism. Zinc dearth is a huge hazard to world agriculture and human health. Around 2 billion individuals suffer from zinc deficiency around the world today. The rise in zinc value can be linked to increased iron concentration. In dicot plants, there is a crosstalk between the iron and zinc uptake mechanisms. As a result, as the iron concentration of the soil rises, the zinc content of the plants rises as well. Besides, organic manures are enriched in microflora that can induce crops to absorb micronutrients, and the synergistic effect of *Spirulina* and organic manures increased zinc levels in the plant.

The effect of using post-extraction residues of *Spirulina platensis* loaded with Zn, Mn, and Cu by means of biosorption has been researched and their use as a micronutrient fertilizer for biofortification of maize grains along with micronutrients was investigated. The bioavailability of micronutrients was more advanced in bio-preparations than in reference fertilizer. Maize grains fertilized with *Spirulina* had the maximum quantity of micronutrients given to plants, indicating that biofortified maize grain was acquired. Micronutrient-fortified corn grains can be used as a main diet or else supply to prevent micronutrient deficiency. Micronutrient bio-components derived from *Spirulina* biomass can be used to create a valuable fertilizer rich in bioavailable micronutrients [93].

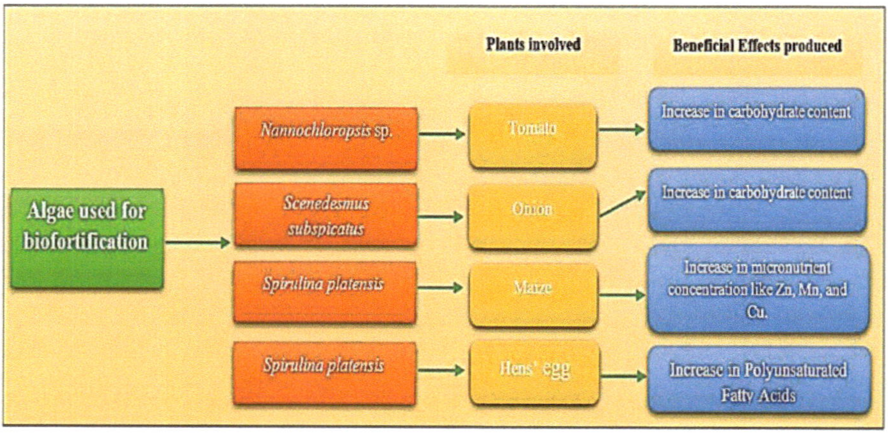

FIGURE 15.1 Biofortification of different food plants using algae.

15.3.2.2 ORGANIC ONION BIOFORTIFICATION VIA HUMIC ACID (HA) AND MICROALGAE

Biofertilizers are an intriguing choice in unconventional construction systems since they can improve food dietetic eminence and plant improvement though also contributing to environmental sustainability. Although the utilization of biofertilizers based on humic acid (HA) and microalgae has led to increased nutritional and biomolecule substances in plants, their collective appliance has yet to be investigated. Onions were studied to see how root raptness in *Scenedesmus subspicatus* (Sc) and HA solutions affected macro- and micronutrients, total sugars, reducing sugar, free total amino acids, total soluble solids, soluble proteins, and antioxidant activity. Seedling roots from two onion variety were immersed in solutions that contain microalgae and HA for one minute before being transplanted to an organic system field. Using microalgae with HA association increased the level of nitrogen, carbohydrates, and soluble proteins in onion bulbs while also increasing antioxidant activity [94, 95].

The microalgae *Nannochloropsis* sp. and *Scenedesmus subspicatus*, when applied to tomato and onion plants, respectively promoted an increase in carbohydrates [95].

- **Effects of biofortified diets containing algae and selenized yeast on nutrient utilization, gilthead seabream growth, and tissue composition (*Sparus aurata*):** Fish meal (FM) as well as FO have been replaced, along with attempts have been undertaken to develop usual, highly nutritious substitutes that can advance the nutritious content of filets in favor of human being and animal eating while also enhancing fish health. The study looked at the effects of biofortified foods that included microalgae (as a substitute for FM and FO), macroalgae (as a natural means of iodine), and selenized yeast (an organic means of selenium) on gilthead seabream growth, nutrient utilization, tissue structure, and gene expression. Microalgae are high-quality suppliers of protein and lipids in seabream muscle, according to the findings. Selenized yeast is a superior enrichment vehicle for boosting selenium values in fish. However, novel iodine supplementation procedures should be investigated [45].

15.4 CONCLUSION

Micronutrient deficiency, which endanger human wellbeing, are linked to all environmental factors, both directly and indirectly. Micronutrient deficit in plants is caused by factors such as low or high light and temperature, excessive rain, high CO_2, flooding, pathogenic illnesses, soil nature, and the opposed effects of micronutrients (Fe–Cu, Cu–Zn, Cu–Mn, Fe–Mn). Traditional plant breeding relies on manipulating plant genomes to develop crops and raise levels of certain key nutrients. Traditional breeding approaches were utilized in large research programs like HarvestPlus and BioCassava Plus to generate biofortified crops [65]. Biofortification is a novel way to fixing the challenges of mineral deficiency in the diet. Biofortification is a critical strategy for addressing the micronutrient shortage that is endangering human health. Plants and soil microbes live in a symbiotic relationship. As an effect of the augmented amount of soil utilization by plants, AMF are recognized to boost host plant nutrition. AMF, often known as a biofertilizer, are increasingly being used in breeding research around the world. The application of AMF in farming crops attempts to address micronutrient requirements in a cost-effective manner, while also conserving water and protecting the soil. AMF promotes root growth and nutrient uptake, including calcium (Ca), nitrogen (N), copper (Cu), manganese (Mn), phosphorus (P), sulfur (S), Fe, and zinc (Zn). Besides fungi, algae are also used in biofortification of different crops such as maize and onion as discussed above. Algal extracts are also widely used as a biofortifying agent as FAs are abundant in algae biomass. Micronutrient biocomponents generated from algal biomass can be utilized to make a valuable bioavailable micronutrient fertilizer. Application of different algae increased the carbohydrate, lipid, and protein content in the food crop. Hence, it can be concluded that utilization of microbes such as algae or fungi can be a better substitution as biofortifying agent to reduce the insufficiency of mineral nutrients in the edible crop.

KEYWORDS

- **algae mediated biofortification**
- **biofortification**
- **malnutrition**
- **microalgae**
- **micronutrients**
- **mycorrhiza**
- **polyunsaturated fatty acids**

REFERENCES

1. Valverde, M. E., Hernández-Pérez, T., & Paredes-López, O., (2015). Edible mushrooms: Improving human health and promoting quality life. *Int. J. Microbiol., 2015*, 376387.
2. Chang, S. T., & Miles, P. G., (2008). *Mushrooms: Cultivation, Nutritional Value, Medicinal Effect, and Environmental Impact* (2nd edn.). CRC Press: Boca Raton, FL, USA.
3. Ergönül, P. G., Akata, I., Kalyoncu, F., & Ergönül, B., (2013). Fatty acid compositions of six wild edible mushroom species. *Sci. World J., 2013*, 163964.
4. Guillamón, E., García-Lafuente, A., Lozano, M., D'Arrigo, M., Rostagno, M. A., Villares, A., & Martínez, J. A., (2010). Edible mushrooms: Role in the prevention of cardiovascular diseases. *Fitoterapia, 81*, 715–723.
5. Longvah, T., & Deosthale, Y. G., (1998). Composition and nutritional studies on edible wild mushroom from Northeast India. *Food Chem., 63*, 331–334.
6. Maga, J. A., (1981). Mushroom flavor. *J. Agric. Food Chem., 29*, 1–4.

7. Mattila, P., Konko, K., Euvola, M., Pihlava, J., Astola, J., & Vahteristo, L., (2001). Contents of vitamins, mineral elements and some phenolic compound in cultivated mushrooms. *J. Agric. Food Chem., 42*, 2449–2453.
8. Clifford, A. J., Heid, M. K., Peerson, J. M., & Bills, N. D., (1991). Bioavailability of food folates and evaluation of food matrix effects with a rat bioassay. *J. Nutr., 121*, 445–453.
9. Bano, Z., & Rajarathnam, S., (1988). Pleurotus mushrooms. Part II. Chemical composition, nutritional value, post-harvest physiology, preservation, and role as human food. *Crit. Rev. Food Sci. Nutr., 27*, 87–158.
10. Ribeiroa, B., Pinhoa, P. G., Andradea, P. B., Baptistab, P., & Valentao, P., (2009). Fatty acid composition of wild edible mushrooms species: A comparative study. *Microchem. J., 93*, 29–35.
11. Antunes, F., Marçal, S., Taofiq, O., Morais, A. M. M. B., Freitas, A. C., Ferreira, I. C. F. R., & Pintado, M., (2020). Valorization of mushroom by-products as a source of value-added compounds and potential applications. *Molecules, 25*, 2672.
12. Lim, S. H., Lee, Y. H., & Kang, H. W., (2013). Efficient recovery of lignocellulolytic enzymes of spent mushroom compost from oyster mushrooms, *Pleurotus* spp., and potential use in dye decolorization. *Mycobiology, 41*, 214–220.
13. Tuhy, T., Samoraj, M., Witkowska, Z., & Chojnacka, K., (2015). Biofortification of maize with micronutrients by Spirulina. *Open Chemistry, 13*, 1119–1126.
14. Howarth, E. B., Christine, H., McClafferty, B., Meenakshi, J. V., & Wolfgang, H., (2011). Pfeiffer biofortification: A new tool to reduce micronutrient malnutrition. *Food and Nutrition Bulletin, 32*(1). ©The United Nations University.
15. De Santiago, A., Quintero, J. M., Aviles, M., & Delgado, A., (2011). Effect of *Trichoderma asperellum* strain T34 and glucose addition on iron, copper, manganese, and zinc uptake by wheat grown on calcareous medium. *Plant and Soil, 342*, 97–104.
16. Khan, A. G., (2005). Role of soil microbes in the rhizospheres of plants growing on trace metal contaminated soils in phytoremediation. *Journal of Trace Elements in Medicine and Biology, 18*, 355–364.
17. Prasanna, R., Pattnayak, S., Sugitha, T. C. K., Nain, L., & Saxena, A. K., (2011). Development of cyanobacterium-based biofilms and their in vitro evaluation for agriculturally useful traits. *Folia Microbiologica, 56*, 49–58.
18. Rana, A., Joshi, M., Prasanna, R., Shivay, Y. S., & Nain, L., (2012a). Biofortification of wheat through inoculation of plant growth promoting rhizobacteria and cyanobacterial. *European Journal of Soil Biology, 50*, 118–126.
19. Rana, A., Saharan, B., Nain, L., Prasanna, R., & Shivay, Y. S., (2012b). Enhancing micronutrient uptake and yield of wheat through bacterial PGPR consortia. *Soil Science and Plant Nutrition, 58*, 573–582.
20. Ryu, C. M., Hu, C. H., Locy, R. D., & Kloepper, J. W., (2005). Study of mechanisms for plant growth promotion elicited by rhizobacteria in *Arabidopsis thaliana. Plant and Soil, 268*, 285–292.
21. Nain, L., Rana, A., Joshi, M., Shrikrishna, D. J., Kumar, D., Shivay, Y. S., Paul, S., & Prasanna, R., (2010). Evaluation of synergistic effects of bacterial and cyanobacterial strains as biofertilizers for wheat. *Plant and Soil, 331*(1), 217–230. DOI: 10.1007/s11104-009-0247-z.

22. Prasanna, R., Chaudhary, V., Gupta, V., Babu, S., Kumar, A., Shivay, Y. S., & Nain, L., (2013). Cyanobacteria mediated plant growth promotion and bioprotection against *Fusarium* wilt in tomato. *European Journal of Plant Pathology, 136*, 337–353.
23. Dakora, F. D., & Phillips, D. A., (2002). Root exudates as mediators of mineral acquisition in low-nutrient environments. *Plant Soil, 245*(1), 35–47.
24. Anil, K., Anil, K. C., & Vinod, K. S., (2017). Agronomic Bio-fortification and Quality Enhancement in Okra–Pea Cropping System Through Arbuscular Mycorrhizal Fungi at Varying Phosphorus and Irrigation Regimes in Himalayan Acid Alfisol. *Journal of Plant Nutrition, 40*(8), 1213–1229. DOI: 10.1080/01904167.2016.1267208.
25. Tran, B. T. T., Cavagnaro, T. R., & Watts-Williams, S. J., (2019). Arbuscular mycorrhizal fungal inoculation and soil zinc fertilization affect the productivity and the bioavailability of zinc and iron in durum wheat. *Mycorrhiza, 29*(5), 445–457. doi: 10.1007/s00572-019-00911-4.
26. Smith, S. E., & Read, D. J., (2008). Arbuscular mycorrhizaes. In: Smith, S. E., & Read, D. J., (eds.), *Mycorrhizal Symbiosis* (3rd edn., pp. 13–187). Academic Press: London, UK.
27. Smith, S. E., Jakobsen, I., Grønlund, M., & Smith, F. A., (2011). Roles of arbuscular mycorrhizas in plant phosphorus nutrition: Interactions between pathways of phosphorus uptake in arbuscular mycorrhizal roots have important implications for understanding and manipulating plant phosphorus acquisition. *Plant Physiol., 156*, 1050–1057.
28. Higo, M., Takahashi, Y., Gunji, K., & Isobe, K., (2018). How are arbuscular mycorrhizal associations related to maize growth performance during short-term cover crop rotation? *Sci. Food Agric., 98*, 1388–1396.
29. Higo, M., Tatewaki, Y., Gunji, K., Kaseda, A., & Isobe, K., (2019). Cover cropping can be a stronger determinant than host crop identity for arbuscular mycorrhizal fungal communities colonizing maize and soybean. *Peer J., 7*, e6403.
30. Higo, M., Tatewaki, Y., Iida, K., Yokota, K., & Isobe, K., (2020). Amplicon sequencing analysis of arbuscular mycorrhizal fungal communities colonizing maize roots in different cover cropping and tillage systems. *Sci. Rep., 10*, 6093.
31. Eah, P. S., (2015). Se in plants. In: Lüttge, U., & Beyschlag, W., (eds.), *Progress in Botany* (Vol. 76, pp. 93–107). Springer: Cham, Switzerland.
32. Ellis, D. R., & Salt, D. E., (2003). Plants, selenium and human health. *Curr. Opin. Plant Biol., 6*, 273–279.
33. Sors, T. G., Ellis, D. R., Na, G. N., Lahner, B., & Salt, D. E., (2005). Analysis of sulfur and selenium assimilation in *Astragalus* plants with varying capacities to accumulate selenium. *Plant J., 42*, 785–797.
34. Wessjohann, L. A., Schneider, A., Abbas, M., & Brandt, W., (2007). Selenium in chemistry and biochemistry in comparison to sulfur. *Biol. Chem., 388*, 997–1006.
35. Chauhan, R., Awasthi, S., Srivastava, S., Dwivedi, S., Pilon-Smits, E. A. H., Dhankher, O. P., & Tripathi, R. D., (2019). Understanding selenium metabolism in plants and its role as a beneficial element. *Crit. Rev. Environ. Sci. Technol., 49*, 1937–1958.
36. Zhao, X. Q., Mitani, N., Yamaji, N., Shen, R. F., & Ma, J. F., (2010). Involvement of silicon influx transporter OsNIP2;1 in selenite uptake in rice. *Plant Physiol., 153*, 1871–1877.
37. Wang, M., Yang, W., Zhou, F., Du, Z., Xue, M., Chen, T., & Liang, D., (2019). Effect of phosphate and silicate on selenite uptake and phloem-mediated transport in tomato (*Solanum lycopersicum* L.). *Environ. Sci. Pollut. Res., 26*, 20475–20484.

38. Li, H. F., Mcgrath, S. P., & Zhao, F. J., (2008). Selenium uptake, translocation and speciation in wheat supplied with selenate or selenite. *New Phytol., 178*, 92–102.
39. Chen, G. H., Yan, W., Yang, S. P., Wang, A., & Zhu, Y. L., (2015). Overexpression of rice phosphate transporter gene OsPT2 enhances tolerance to low phosphorus stress in soybean. *J. Agric. Sci. Technol., 17*, 469–482.
40. Zhang, L., Hu, B., Li, W., Chen, R., Deng, K., Li, H., Yu, F., et al., (2014). OsPT2, a phosphate transporter, is involved in the active uptake of selenite in rice. *New Phytol., 201*, 1183–1191.
41. Zhang, M., Wilson, L., Xing, G., Jiang, L., & Tang, S., (2020). Optimizing root architecture and increasing transporter gene expression are strategies to promote selenium uptake by high-se accumulating rice cultivar. *Plant Soil, 447*, 319–332.
42. McDermott, J. R., Rosen, B. P., & Liu, Z., (2010). Jen1p: A high affinity selenite transporter in yeast. *Mol. Biol. Cell, 21*, 3934–3941.
43. Lazard, M., Blanquet, S., Fisicaro, P., Labarraque, G., & Plateau, P., (2010). Uptake of selenite by *Saccharomyces cerevisiae* involves the high and low affinity orthophosphate transporters. *J. Biol. Chem., 285*, 32029–32037.
44. Pérez-Sampietro, M., Serra-Cardona, A., Canadell, D., Casas, C., Ario, J., & Herrero, E., (2016). The yeast Aft2 transcription factor determines selenite toxicity by controlling the low affinity phosphate transport system. *Sci. Rep., 6*, 32836.
45. Ferreira, M., Ribeiro, P. C., Ribeiro, L., Barata, M., Domingues, V. F., Sousa, S., Soares, C., et al., (2022). Biofortified diets containing algae and selenized yeast: Effects on growth performance, nutrient utilization, and tissue composition of gilthead seabream (*Sparus aurata*). *Front. Physiol., 12*, 812884. doi: 10.3389/fphys.2021.812884.
46. Hopper, J. L., & Parker, D. R., (1999). Plant availability of selenite and selenate as influenced by the competing ions phosphate and sulfate. *Plant Soil, 210*, 199–207.
47. Yang, X., Lu, Y., Zhong, J., Qian, Y., Zhao, Z., & Liu, X., (2017). The positive effect of sulfur on selenium detoxification under selenite condition in wheat. *Commun. Soil Sci. Plant Anal., 48*, 1564–1573.
48. Zhang, D., Dong, T., Ye, J., & Hou, Z., (2017). Selenium accumulation in wheat (*Triticum aestivum* L.) as affected by coapplication of either selenite or selenate with phosphorus. *Soil Sci. Plant Nutr., 63*, 37–44.
49. Cabannes, E., Buchner, P., Broadley, M. R., & Hawkesford, M. J., (2011). A comparison of sulfate and selenium accumulation in relation to the expression of sulfate transporter genes in astragalus species. *Plant Physiol., 157*, 2227–2239.
50. Rzymski, P., Mleczek, M., Niedzielski, P., Siwulski, M., & Ga̧secka, M., (2016). Potential of cultivated *Ganoderma lucidum* mushrooms for the production of supplements enriched with essential elements. *J. Food Sci., 81*, 587–592.
51. Rzymski, P., Mleczek, M., Niedzielski, P., Siwulski, M., & Ga̧secka, M., (2016). Cultivation of *Agaricus bisporus* enriched with selenium, zinc and copper. *J Sci Food Agric*. doi: 10.1002/jsfa.7816.
52. Gil-Ramírez, A., Soler-Rivas, C., Rodriguez-Casado, A., Ruiz-Rodríguez, R. G., & Marín, F. R., (2015). Effect of selenium-enriched *Agaricus bisporus* (higher *Basidiomycetes*) extracts, obtained by pressurized water extraction, on the expression of cholesterol homeostasis related genes by low-density array. *Int J Med Mushrooms., 17*, 105–116.

53. Schrauzer, G. N., & De Vroey, E., (1994). Effects of nutritional lithium supplementation on mood. A placebo-controlled study with former drug users. *Biol Trace Elem Res., 40*, 89–101.
54. Klemfuss, H., & Schrauzer, G. N., (1995). Effects of nutritional lithium deficiency on behavior in rats. *Biol Trace Elem Res., 48*, 131–139.
55. Bishop, K. S., Kao, C. H., Xu, Y., Glucina, M. P., Paterson, R. R., & Ferguson, L. R., (2015). From 2000 years of *Ganoderma lucidum* to recent developments in nutraceuticals. *Phytochemistry, 114*, 56–65.
56. Ma, H. T., Hsieh, J. F., & Chen, S. T., (2015). Anti-diabetic effects of *Ganoderma lucidum*. *Phytochemistry, 114*, 109.
57. Li, J., (2013). Zhang, J., Chen, H., Chen, X., Lan, J., & Liu, C., (2013). Complete mitochondrial genome of the medicinal mushroom *Ganoderma lucidum*. *PLoS One, 8*, e72038.
58. Sartori, H. E., (1986). Lithium orotate in the treatment of alcoholism and related conditions. *Alcohol., 3*, 97–100.
59. Młyniec, K., Davies, C. L., De Agüero, S. I. G., Pytka, K., Budziszewska, B., & Nowak, G., (2014). Essential elements in depression and anxiety. *Part I Pharmacol. Rep., 66*, 534–544.
60. Schrauzer, G. N., (2002). Lithium: Occurrence, dietary intakes, nutritional essentiality. *J. Am. Coll. Nutr., 21*, 14–21.
61. Goldstein, M. R., & Mascitelli, L., (2016). Is violence in part a lithium deficiency state? *Med. Hypotheses, 89*, 40–42.
62. Chang, S. T., & Wasser, S. P., (2012). The role of culinary-medicinal mushrooms on human welfare with a pyramid model for human health. *Int. J. Med. Mushrooms., 14*, 95–134.
63. Ryan, M. H., & Kirkegaard, J. A., (2012). The agronomic relevance of arbuscular mycorrhizas in the fertility of Australian extensive cropping systems. *Agriculture, Ecosystems and Environment, 163*, 37e53.
64. Subramanian, K. S., Balakrishnan, N., & Senthil, N., (2013). Mycorrhizal symbiosis to increase the grain micronutrient content in maize. *Australian Journal of Crop Science, 7*, 900e910.
65. Garg, M., Sharma, N., Sharma, S., Kapoor, P., Kumar, A., Chunduri, V., & Arora, P., (2018). Biofortified crops generated by breeding, agronomy, and transgenic approaches are improving lives of millions of people around the world. *Front. Nutr., 5*, 12. doi: 10.3389/fnut.2018.00012.
66. Toepfer, E. W., Slover, H. T., Hepburn, F. N., Polansky, M. M., Morris, E. R., Eheart, J. F., & Quackenbush, F. W., (1972). Nutrient composition of selected wheats and wheat products. 11. summary. *Cereal Chemistry, 49*, 173e186.
67. Lorenz, K., Loewe, R., Weadon, D., & Wolf, W., (2014). Natural levels of nutrients in commercially milled wheat flours. 3. Mineral analysis. *Cereal Chemistry, 57*, 65e69.
68. Myers, S. S., Zanobetti, A., Kloog, I., Huybers, P., Leakey, A. D. B., Bloom, A. J., Carlisle, E., et al., (2014). Increasing CO_2 threatens human nutrition. *Nature, 510*, 139e142.
69. Gul, K., Singh, A. K., & Jabeen, R., (2016). Nutraceuticals and functional foods: The foods for the future world. *Crit. Rev. Food Sci. Nutr., 56*, 2617–2627.

70. Zarate, R., El Jaber-Vazdekis, N., Tejera, N., Perez, J. A., & Rodriguez, C., (2017). Significance of long chain polyunsaturated fatty acids in human health. *Clin. Transl. Med., 6*, 25.
71. Simopoulos, A. P., (2000). Human requirement for n-3 polyunsaturated fatty acids. *Poult. Sci., 79*, 961–970.
72. Carrillo, S., Ríos, V. H., Calvo, C., Carranco, M. E., Casas, M., & Pérez-Gil, F., (2012). n-3 fatty acid content in eggs laid by hens fed with marine algae and sardine oil and stored at different times and temperatures. *J. Appl. Phycol., 24*, 593–599.
73. Herber, S. M., & Van, E. M. E., (1996). Dietary marine algae promotes efficient deposition of n-3 fatty acids for the production of enriched shell eggs. *Poult. Sci., 75*, 1501–1507.
74. Fredriksson, S., Elwinger, K., & Pickova, J., (2006). Fatty acid and carotenoid composition of egg yolk as an effect of microalgae addition to feed formula for laying hens. *Food Chem., 99*, 530–537.
75. Rizzi, L., Bochicchio, D., Bargellini, A., Parazza, P., & Simioli, M., (2009). Effects of dietary microalgae, other lipid sources, inorganic selenium and iodine on yolk n-3 fatty acid composition, selenium content and quality of eggs in laying hens. *J. Sci. Food Agricult., 89*, 1775–1781.
76. Gładkowski, W., Kiełbowicz, G., Chojnacka, A., Bobak, Ł., Spychaj, R., Dobrza nski, Z., Trziszka, T., & Wawrze ´nczyk, C., (2014). The effect of feed supplementation with dietary sources of n-3 polyunsaturated fatty acids, flaxseed and algae *Schizochytrium* sp., on their incorporation into lipid fractions of Japanese quail eggs. *Int. J. Food. Sci. Technol., 49*, 1876–1885.
77. Al-Harthi, M. A., & El-Deek, A. A., (2012). Effect of different dietary concentrations of brown marine algae *(Sargassum dentifebium)* prepared by different methods on plasma and yolk lipid profiles, yolk total carotene and lutein plus zeaxanthin of laying hens. *Ital. J. Anim. Sci., 11*, 347–353.
78. Grigorova, S., Surdjiiska, S., Banskalieva, V., & Dimitrov, G., (2006). The effect of biomass from green algae of Chlorella genus on the biochemical characteristics of table eggs. *J. Cent. Eur. Agricult., 7*, 111–116.
79. Lemahieu, C., Bruneel, C., Termote-Verhalle, R., Muylaert, K., Buyse, J., & Foubert, I., (2013). Impact of feed supplementation with different omega-3 rich microalgae species on enrichment of eggs of laying hens. *Food Chem., 141*, 4051–4059.
80. Šefer, D., Andonov, A., Šobaji´c, S., Markovi´c, R., Radulovi´c, S., Jaki´c-Dimi´c, D., & Petrujki´c, B., (2011). Effects of feeding laying hens diets supplemented with omega 3 fatty acids on the egg fatty acid profile. *Biotechnol. Animal. Husb., 27*, 679–686.
81. Bruneel, C., Lemahieu, C., Fraeye, I., Ryckebosch, E., Muylaert, K., Buyse, J., & Foubert, I., (2013). Impact of microalgal feed supplementation on omega-3 fatty acid enrichment of hen eggs. *J. Funct. Foods, 5*, 897–904.
82. Saeid, A., Chojnacka, K., Opali´nski, S., & Korczy´nski, M., (2016). Biomass of *Spirulina maxima* enriched by biosorption process as a new feed supplement for laying hens. *Algal Res., 19*, 342–347.
83. Van, E. M. E., (1997). Comparison of n-3 fatty acids sources in laying hen rations for improvement of whole egg nutritional quality: A review. *Br. J. Nutr., 78*, 61–69.
84. Michalak, I., & Chojnacka, K., (2015). Algae as production systems of bioactive compounds. *Eng. Life Sci., 15*, 160–176.

85. Koru, E., (2012). Earth food *Spirulina (Arthrospira)*: Production and quality standards. In: El-Samragy, Y., (ed.), *Food Additive*. InTech: London, UK.
86. Chaiklahan, R., Chirasuwan, N., Loha, V., & Bunnag, B., (2008). Lipid and fatty acids extraction from the cyanobacterium *Spirulina. Sci. Asia, 34*, 299–305.
87. Zinnai, A., Venturi, F., Sanmartin, C., & Andrich, G., (2012). A simplified method to estimate the time evolution of oil extraction from different substrates by supercritical CO_2. *Am. J. Analyt. Chem., 3*, 939–945.
88. Mühling, M., Belay, A., & Whitton, B. A., (2005). Variation in fatty acid composition of *Arthrospira (Spirulina)* strains. *J. Appl. Phycol., 17*, 137–146.
89. Colla, L. M., Bertolin, T. E., & Costa, J. A. V., (2004). Fatty acids profile of *Spirulina platensis* grown under different temperatures and nitrogen concentrations. *Zeitschrift Für Naturforschung C, 59*, 55–59.
90. Mendes, R. L., Reis, A. D., & Palavra, A. F., (2006). Supercritical CO_2 extraction of γ-linolenic acid and other lipids from *Arthrospira* (*Spirulina*) *maxima*: COMPARISON with organic solvent extraction. *Food Chem., 99*, 57–63.
91. Quihui, H., (1999). Supercritical carbon dioxide extraction of *Spirulina platensis* component and removing the stench. *J. Agricult. Food. Chem., 47*, 2705, 2706.
92. Andrich, G., Zinnai, A., Nesti, U., & Venturi, F., (2006). Supercritical fluid extraction of oil from microalga *Spirulina (Arthrospira) platensis. Acta Aliment., 35*, 195–203.
93. Łukasz, T., Mateusz, S., Zuzanna, W., & Katarzyna, C., (2015). Biofortification of maize with micronutrients by *Spirulina. Open Chemistry, 13*, 1119–1126. doi: 10.1515/chem-2015-0126.
94. Luiz, G. G., Gabriel, B. De. L., Átila, F. M., Gilda, M., & Christiane De, Q., (2021). Organic onion biofortification using microalgae and humic acid. *Research, Society and Development, 10*(13), e320101321432. doi: Http://dx.doi.org/10.33448/rsd-v10i13.21432.
95. Bettoni, M. M., Mogor, Á. F., Pauletti, V., Goicoechea, N., Aranjuelo, I., & Garmendia, I., Nutritional quality and yield of onion as affected by different application methods and doses of humic substances. *Journal of Food Composition and Analysis, 51*, 37–44. doi: 10.1016/j.jfca.2016.06.008.

CHAPTER 16

Nutritional Revolution: Biotechnological Strategies for Fortified Crop Produce

ELANGOVAN DHARANI,[1] SENTHILKUMAR SHRICHARAN,[1] BALAMURUGAN BALAJI,[2] and JESHIMA KHAN YASIN[3]

[1]*Division of Plant Physiology, ICAR–Indian Agricultural Research Institute, Pusa Campus, New Delhi, India*

[2]*Division of Molecular Biology and Biotechnology, ICAR–Indian Agricultural Research Institute, Pusa Campus, New Delhi, India*

[3]*Division of Genomic Resources, ICAR–National Bureau of Plant Genetic Resources, Pusa Campus, New Delhi, India*

ABSTRACT

Micronutrients are inevitably essential for housekeeping, enzymatic activities and homeostasis. The micronutrients comprise vitamins and minerals. A diet with well-balanced amount of vitamins and minerals can be difficult to maintain. But now due to the recent advancements in agricultural biotechnology, it is possible to produce nutritionally enriched food crops with enhanced bioavailability of micronutrients. Vegetatively propagated crops are more suitable for biofortification especially fruits as their bioavailability will be more due to lack of processing. It can be achieved by strategies like introducing the novel gene, overexpression of a target gene, down regulating the expression of interfering genes in biosynthesis pathways. To combat vitamin A deficiency (VAD), vitamin A enriched biofortified golden rice with two genes were developed. A strategic approach to overcome iron deficiency is to lower the phytate level and to increase the iron content. Ferritin was introduced with threefold increase of iron in the rice endosperm. Biofortification

Phytoremediation and Biofortification: Strategies for Sustainable Environmental and Health Management. Nand Kumar Singh, Shadma Afzal, & Tariq Aftab (Eds.)

can also be achieved through modern biotechnological approaches like transgenic, cis-genic, RNAi, and CRISPR-Cas based genome editing (GE). Biofortification promises for the nutritive improvement in crops. As there are lot of regulations in permitting these GMOs into circulation, cultivating all nutrition rich conventional crops on street roads can alleviate hidden hunger in few months. The conventional fast growing *Moringa*, papaya, amla, *Amaranthus*, *Sesbania*, and curry leaf can be grown as street plants with other fruit trees to achieve 'Nutritional Revolution' rather than searching for costly techniques. However, apart from SMART agriculture, having expertise in recent techniques will help during adverse conditions. Biofortified crops hold a great future in alleviating malnutrition of the global population.

16.1 INTRODUCTION

Of the three basic needs food tops the essentials for all the living organisms in this world. The food is the source of energy for our activities which supplements nutrients and minerals required for body function and survival. Lack of the essential micronutrients creates deficiency syndrome and in severe cases irreversible health effects. Most of our diet consciousness depends mainly on calories rather than regular supply of essential micronutrients. In the evolving industrial and urban lifestyle supply of daily nutrient requirements through the food we consume is difficult. Apart from lifestyle diseases, the COVID-19 pandemic threatened the food and nutritional security. Unemployment, vicious cycle of poverty loss of human resource and income has created more stress and divide in the society. During the year of 2020, about 761 million people were facing hunger and hidden hunger which is 118 million more than previous year and under nourishment increased from 8.4% to 9.9% [1]. In India it has been project to a very high rate as per the global hunger index; which is at present on par with 1950's index. Hence there is an urge to improve our food and nutritional security status especially the younger generation.

Fortification is the process of improving the quality of the food by enhancing the content of essential vitamins, minerals, and other micronutrients, i.e., Food fortification (or) industrial fortification increase the micronutrients through postharvest processing like iodized salts. These processed and fortified foods can reach only the affordable people and not to the one in hidden hunger and the rural population. Biofortification can make this achievable by reaching the grass root level to make the impossible as possible. By biofortification the nutrient content of the edible parts can be increased through agronomic practices, conventional breeding techniques and biotechnological approaches. Fortification can make the nutrient rich food to reach the global population within a shorter duration of time. Consumption of the biofortified staple food crops available in supply system can improve the nutrient status. Biofortification removes the need for post processing fortification to increase the nutrient content in the food. This is more feasible for the malnourishment hit rural areas. Several millions of people will be benefited even there is 5% increase in nutrient content of any food item. Apart from staple food supply of more than four types of food item in any meal will eventually green revolution alleviated starvation by increasing of food grain production facilitated by well managed supply system. Whereas only biofortifying and making the nutrient rich fruits and vegetables available to the common public only can result in 'Nutritional Revolution.'

16.2 MICRONUTRIENT DEFICIENCY IN HIDDEN HUNGER

Micronutrients are vitamins and minerals required in varying quantities for regular functioning of biological and biochemical processes of the human body. For vitamins and their deficiencies, the recommended daily amount (RDA) are given in Table 16.1. Poverty, improper diet, modern unhealthy lifestyle, poor bioavailability leads to micronutrient deficiency and hidden hunger. Infants, children, and pregnant women are more vulnerable to micronutrient deficiency as all developmental stages require these nutrients at higher amounts. Most common micronutrients, their sources, impacts due to their deficiencies are as in subsections.

16.2.1 IODINE

Iodine plays a vital role in metabolic rate regulation. Iodine is essential for thyroid hormone, i.e., triiodothyronine and thyronine. Iodine deficiency causes goiter (inflammation of thyroid gland), Hypothyroidism, etc. Iodine deficiency during pregnancy, affects neurodevelopment of the infant. Iodine rich sources are sea foods, eggs, dairy products, iodized salt, etc. Nearly 20–60% of the world population and iodine deficiency affects half of Europe's population.

16.2.2 VITAMIN A

Vitamin A is one of the fat-soluble vitamins that is important for immune system, normal vision and in Reproduction. Vitamin A is found in fish liver oils, spinach, lettuce, carrots, tomato, and other foods. Vitamin A deficiency (VAD) leads to night blindness, dry eyes, xeropthalmia, infertility, blindness, skin irritation, etc. Infants and pregnant women are at high risk and almost around 40% of them suffer with VAD. VAD was reported in terms of serum retinol concentration. It was characterized as serum retinol levels of less than 0.7 mol/L, which indicated moderate to severe deficiency [2].

16.2.3 IRON

Iron is commonly known for its oxygen carrier function and as a part of hemoglobin, which carries oxygen throughout our body through blood stream. Insufficient intake of iron causes anemia. Common symptoms of iron deficiency are pale skin, extreme fatigue, headache shortness of breath and inflammation of tongue, etc. Moringa leaf, curry leaf, Date palm, carrots, spinach, broccoli, sweet potatoes, apricots, and peaches are rich in iron content. Iron deficiency affects about 2 billion people all over the world.

16.2.4 ZINC

Zinc is a vital mineral that our body requires for growth, immune function, DNA synthesis, protein synthesis, gene expression, etc. Zinc is essential co-factor for immune cell function and cell signaling related activities. In that case, deficiency of zinc can lead to ineffective immune responses, skin lesions, hair loss, anorexia, persistent diarrhea and mental retardation. Zinc content is higher in meat, shellfish, legumes, seeds, nuts, dairy products and eggs.

16.2.5 CALCIUM

Calcium plays a major role in quantity as well as quality functions of the body. Calcium is essential for bone health; virtually all of the calcium in the human body is found in bones and teeth. Calcium deficiency has profound impacts on bones that result in rickets and osteoporosis. Fatigue, muscle cramps, numbness, osteopenia, abnormal heart rhythm are some of the signs of low-level calcium in the body. Dairy products including milk, cheese, yogurt, papayas, oranges, spinach, Moringa leaf and soybean, etc., are rich in calcium.

16.2.6 FOLATE

Folate is a form of Vit B9, highly essential during pregnancy, infancy, and adolescence. Some of the folate rich sources are beef liver, beans, spinach, yeast, eggs, mushroom raw mango, moringa leaf, and oranges. Deficiency in folate often cause ulcer in mouth, fatigue, sore tongue, depression, vision problem, etc. Synthetic form of folate is folic acid, which is fortified in food products for commercial purpose. When folate intake is reduced, it increases the risk of birth complication and neural tube defect in infants.

TABLE 16.1 Vitamins and Their Deficiency [2]

Vitamins	RDA	Deficiency	Solubility in Water
Vitamin A	900 μg	Night blindness	No
Vitamin D	15 μg	Rickets	No
Vitamin E	15 mg	Sterility	No
Vitamin K	120 μg	Blood clotting	No
Thiamine (B_1)	1.2 mg	Beriberi	Highly
Riboflavin (B_2)	1.3 mg	Glossitis	Slightly
Niacin (B_3)	16 mg	Pellagra	Yes
Pantothenic acid (B_5)	5 mg	Burning feet	Quite stable
Pyridoxine (B_6)	1.3 mg	Anemia	Yes
Biotin (B_7)	30 μg	Paralysis and hair falling	Slightly
Folic acid (B_9)	400 μg	Anemia	Yes
Cobalamin (B_{12})	2.4 μg	Pernicious anemia	Yes
Vitamin C	90 mg	Scurvy	Unstable

Source: Ref. [2]

16.3 BIOTECHNOLOGICAL APPROACHES FOR CROP IMPROVEMENT

Crop nutrient content is affected by many factors. A crop may synthesize vitamins and minerals in required quantities for their function or they may accumulate the excess synthesized vitamins in sinks. They may lack the genes to produce proteins and enzymes that help in accumulating in economic part of a crop plant which is edible. Using biotechnological approaches and strategies it is possible to improve the vitamin and mineral contents in commercial produce of the plant. The biofortified crop varieties released by ICAR [3] is listed in Table 16.2. Protein engineering involves the modification of proteins by changing the amino acids sequence without loss in biological function through site directed mutagenesis. Adding or altering desirable amino acids in polypeptides is done using site guided mutagenesis. This addition or alteration of amino acids improve the stability without loss in biological activity and less prone to degradation comparing native proteins. Enzymes involved in pathways of vitamins and minerals can be edited by this site directed mutagenesis. These engineered enzymes enhance the synthesis and accumulation of vitamins and minerals result in improved nutrients of the edible part. There may be certain specific enzymes which are inhibitory and downregulate the pathways of vitamins and nutrients which need editing.

The expression of gene coding for enzymes can be silenced by gene silencing biotechnology such as antisense-RNA and RNA interference (RNAi). RNAi is the method of interfering the gene expression post transcriptionally with the help of short dsRNA that binds to the target mRNA preventing its translation. It has become a promising tool in biotechnological aspects for crop improvement. Diverse methodologies were developed in pest control, diseases resistance and crop improvement for enhanced nutritional levels. The RNAi acts through dicer or DCL1 to 4 cleaving the dsRNA into 21 to 24nt long smaller functional fragment known as small interfering RNA (siRNA). This siRNA forms complex machinery called RNA induced silencing complex (RISC) in association with group of proteins. The sense and antisense strands of siRNA are identified by siRNA in RISC. The antisense RNA in RISC complex binds to complementary mRNA strand and directs the mRNA for decapping and deadenylation to further degradation. The messenger RNA that codes for the polypeptide gets cleaved by degradation pathway as a result protein synthesis is inhibited. Another well-known technology to enhance nutritional values of crops is antisense RNA technology. Antisense RNA technology is small, non-coding RNA, which is complementary to mRNA. It is generally used to stop the expression of a target gene by regulating its expressions. It has been reported that antisense RNA technology improved protein quality. Antisense RNA is complementary to mRNA, it hybridized with mRNA and form mRNA – antisense RNA duplex resulting in gene silencing by blocking its translation. Zeins are proteins that are spacio-temporally regulated and expressed in and during seed development. It serves as a source of free amino acids. By knocking off the expression of 22-KD maize zein storage proteins, RNA silencing has been successfully used to develop high lysine maize variants in maize [4].

TABLE 16.2 Biofortified Crop Varieties Released by ICAR, New Delhi

Crop	Variety Released	Content Biofortified	Method Involved	Year of Release
Rice	CR Dhan 310	Protein	Pure line	2016
	DRR Dhan 45	Zinc	Pure line	2016
Wheat	WB 02	Zinc and Iron	Pure line	2017
	HPBW 01	Zinc and Iron	Pure line	2017
Maize	Pusa Vivek QPMP	Provitamin A, Lysine, and Tryptophan	Hybrid	2017
	Pusa HM 4 improved	Tryptophan and Lysine	Hybrid	2017
	Pusa HM 8 improved	Tryptophan and Lysine	Hybrid	2017
	Pusa HM 9 improved	Tryptophan and Lysine	Hybrid	2017
Pearl millet	HHB 299	Iron and Zinc	Hybrid	2017
	AHB 1200	Iron	Hybrid	2017
Lentil	Pusa Ageti Masoor	Iron	Pure line	2017
Mustard	Pusa Mustard 30	Erucic acid	Pure line	2013
	Pusa double zero Mustard 31	Erucic acid and Glucosinolates	Pure line	2016
Cauliflower	Pusa Beta Kesari 1	β Carotene	Pure line	2015
Potato	Bhu Sona	β Carotene	Pure line	2017
Sweet potato	Bhu Krishna	Anthocyanin	Pure line	2017
Pomegranate	Solapur Lal	Iron, Zinc, and Vitamin C	Hybrid	2017

Source: Ref. [3]

16.4 VITAMIN A ENRICHED GOLDEN RICE

Retinol known as Vitamin A, is one among the fat-soluble vitamins. Vitamin A has a vital functions like proper immune responses, good vision, etc. Vitamin A is found in foods such as papaya, sweet potato, eggs, cheese, milk, yogurt, and Moringa leaf. It is also obtained by including a good source of β-carotene, a precursor molecule in daily diet. The RDA intake of vitamin A for adult men is 700 μg and for adult women is 600 μg, but vitamin A consumption is significantly low as compared to the RDA due to a lack of a well-balanced diet. Rice (Oryza sativa) is the primary food, hence it is produced and consumed in greater quantities. The edible part of the rice grain is its endosperm. Rice grains are processed to remove the outer glumes, husk, and aleurone layer. The polished white rice grain is rich in carbohydrates but lacks vitamins. It is also low in dietary fiber content. Rice endosperm produces geranylgeranyl diphosphate (GGPP) a precursor of β-carotene [5]. The large section of the population doesn't have access to vitamin A rich sources like papaya, mango, and green leafy vegetables. Due to which they suffer with vision related problems. Hence, to combat VAD, normal rice was transgenically biofortified to produce grains with high beta carotene. Production of β-carotene from GGPP requires certain enzymes. Phytoene synthase and lycopene β-cyclase from daffodils and carotene desaturase from a bacterium *Erwinia* was used to enhance vit A in rice. The biosynthesis pathway involves the conversion of GGPP into phytoene by the enzyme phytoene synthase and the enzyme carotene desaturase converts phytoene into lycopene. Lycopene β-cyclase produces β-carotene from lycopene. The overview of β carotene synthesis is given in Figure 16.1. The genes transferred by genetic engineering in rice endosperm enhanced vitamin A content. As a result, "Golden rice" was developed. Consequently, crop improvement techniques can be employed in biofortification of other rice varieties, in combating VAD diseases globally [6].

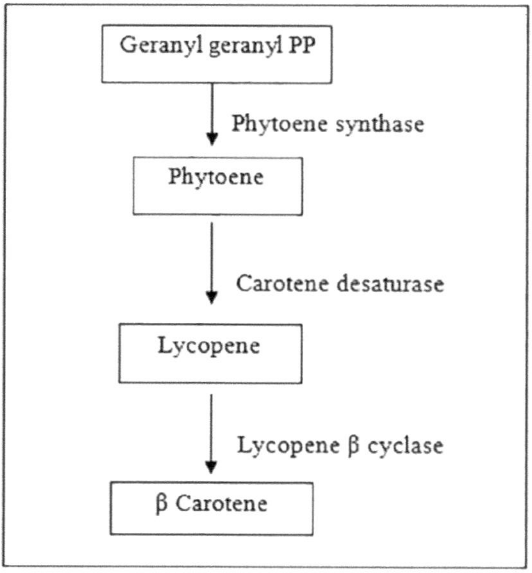

FIGURE 16.1 An overview of β-carotene synthesis.

16.5 ZINC BIOFORTIFICATION IN PLANTS

Zinc is another important micronutrient next to iron. Zinc is very important for both plants and human metabolism. In plants zinc is responsible for activation of many enzymes involved in protein synthesis and for the biosynthesis of auxin. In human system zinc catalyzes enzymes responsible for gene expression and protein synthesis. The ultimate aim of zinc biofortification is to increase the zinc bioavailability. In this process of developing a new plant variety, molecular breeding approaches are highly preferred. Manipulation of Zn content of cereal grain may be less straightforward than Fe [7]. Ozturk et al. [8] reported the strong correlations between protein content, Fe and Zn content. Over expression of the gene is one of the approaches used in expression of NA synthase (NAS) gene under 35S enhancer elements showed 2–3-fold zinc content increase in rice. Similarly, a transgenic rice was developed by transferring and expressing nicotianamine (NA) synthase gene *Hv*NAS1 gene from barley under rice actin-1 promoter. This also resulted in 2–3-fold increase in zinc content in rice grain. Masuda et al. [9] reported that overexpressing the iron(II) NA transporter OsYSL2 in rice endosperm boosted the accumulation of the iron storage protein ferritin and improved iron translocation. The yield remained similar to the original rice variety, but the transgenic rice line showed higher levels of iron and zinc.

16.6 BIOFORTIFICATION OF RICE TO IMPROVE IRON STATUS

Humans relay on plants to include iron in diet from fruits, vegetables, and other greens. Iron deficiency and anemia are now becoming threatening and major global health issue. An anemic mother gives birth to an underdeveloped anemic child. Biotechnological methods which are using to boosting the iron content of plants and their edible parts is more promising and long-term. The commonly used strategy to increase iron content is introduction of FERRITIN gene (gene encoding iron storage protein) from *Phaseolus vulgaris* especially in endosperm which increase iron content up to two-fold. On the other way if we enhance the production of *Phaseolus* and made it available to the common public the task will be simpler. But reaching the unreachable roots is a serious concern. In soybean (*Glycine max* (L). *Merr*), The endosperm specific Glutelin B-1 promoter (pOsGluB-1) was used to express FER (GmFER), which resulted in a 2.1-fold rise in endosperm Fe content [10]. The Fe concentration in unpolished grains increased when the common bean (*P. vulgaris* L.) FER (PvFER) was expressed under the same promoter [11]. Cystine residues act as a major enhancer for iron absorption, so overexpression of cysteine rich metallothionein like protein, increases iron absorption. Native metallothioneins (MTs) are also reported in crops like pigeon pea to have impact in quality of grains [12]. Another strategy is to boost the expression of the NAS gene to increase Fe transport inside the plant body. Nicotinamine (NA) is a metal cation chelator that is biosynthesized from s-adenosyl methionine by NAS [13]. Nicotinamine is responsible for internal transport of Fe and other metals in plants. Overexpression of NAS gene increases the iron content in rice. Furthermore, overexpression of the barley NAS gene, HvNAS1, resulted in higher Fe and Zn concentrations in tobacco plant leaves, flowers, and seeds [14].

16.7 BIOFORTIFICATION OF TAPIOCA

Tapioca or cassava is one of the most suitable crops for biofortification. Tapioca is cultivated around the globe in about 95 countries and is consumed by 800 million people. Tapioca grows well in arid condition and highly tolerant to drought. Tapioca is rich in calories but lack micronutrient content. Tapioca is under rigorous crop improvement for micronutrient content in of tubers. Tapioca germplasm, lack enough variability for micronutrient accumulations. Hence it is not possible to select through conventional breeding techniques to improve micronutrients content of Tapioca. Transgenic approaches are in place to be adopted for tapioca such as upregulating genes from other donors. Iron and Zinc content can be increased by over expression of certain transporter genes from *A. thaliana*. The gene VIT1 which is the vacuolar iron transporter (VIT) gene of *A. thaliana* introduced in tapioca. Overexpression of this gene with organ specific patatin promoter in Tapioca result in seven-fold increase in accumulation of iron content in tubers comparing non transgenic plant. Co expression of IRT1 (iron regulated transporter gene) and FER1 (Feretin gene) from *A. thaliana* in Tapioca result in 18 times higher iron and 10 times higher zinc content in roots comparing non transgenic plant [15]. In addition to iron and zinc biofortification, β-carotene levels in the roots of tapioca can be increased by transgenes by enzymes deoxy-d-xylulose-5-phosphate synthase (DXS) and bacterial phytoene synthase (crtB) with patatin type 1 promoter increase the β-carotene content about 15% on comparing non transgenic plants [16].

16.8 BIOFORTIFICATION OF BANANA

Banana is an important fruit crop with 97.5 million tons of production worldwide. β-Carotene is a provitamin A carotenoid, that human body readily converts it into vitamin A. Due to its parthenocarpic nature and triploid nature, the genetic improvement in banana is a challenging task [17]. CRISPR/Cas9 based genome editing (GE) approach was used to develop cavendish banana rich in β-carotene. In lycopene epsilon-cyclase (LCYε) gene ORF fifth exon was targeted. The capacity of the developed guide-RNA to create indels in the LCY gene at the A genome of cv. Rasthali (AAB) was also tested for targeting specificity. In comparison to untreated plants, metabolic analysis of the fruit pulp of chosen altered lines revealed a 6-fold increase in carotene concentration (24 g/g) [17]. The Queensland University of Technology (QUT) in Australia began biofortification research on bananas in order to generate pro vitamin A (carotene), tocopherol, and iron-rich variants, as well as disease-resistant kinds resistant to the Banana Bunchy Top Virus (BBTV) and Fusarium Wilt [18].

16.9 SOILLESS CULTURE

Positive stress includes moderate salinity/nutritional stress can be applied accurately by soilless systems. This can be achieved by maintaining the concentration and composition of nutrient solution such as cationic and anionic composition of single ions which prevents the accumulation of nitrates, i.e., antinutrient compounds and on other hand improves the nutritional value of vegetables [19–22]. In the present highly polluted condition, apart from water and air, soil is also polluted. Enhancing the expression of any transporter can have direct impact on ion exchange of any plant and its parts. Hence, higher mineral containing plants may have the exchanged pollutant by its ion exchange capacity. Applicability of compost for use in soil less culture is studied. Efficiency and aerobic degradation of kitchen waste with *Bacillus subtilis* to produce compost has been reported [23]. This compost can be exploited to use as a substrate for soil less culture systems. Soilless cultivation can help biofortify edible plant parts with needed and/or helpful micronutrients for human health without exposing people to heavy metals or pollution toxicity. Biofortification with essential or beneficial micronutrients could be a good way to get iodine(I), selenium (Se), zinc (Zn), and silicon into the human diet (Si) [24].

16.10 SPEED BREEDING AND PROTECTED CULTIVATION

Speed breeding is a developing technique that enables plant breeders to accelerate plant breeding cycles. Researchers at University of Queensland coined the term speed breeding. Under this method plants are grown under control environmental conditions, including day and night temperature, availability of light spectrum and intensity, photoperiod duration [25]. Speed breeding and other methods of protected cultivation can ensure the supply of nutrient rich crop produce even in agroclimatic zones not suitable for such crop cultivation [26]. Plants are exposed for long day and short-day conditions based on the crop. Under these conditions, wheat plants grown for six generations in three years glass house. Generally, a crop variety developed through conventional breeding program takes 8 to 10 years, but the speed breeding reduces the breeding cycle and generation time. Watson et al. [27] reported that speed breeding can accelerate generation time of major crops such as wheat, barley, pea, chickpea, and canola. The key elements as described by Fiker & Degefu [28] are it saves time, resources, enhances genetic gain, rate of variety replacement and improve productivity. Speed breeding enhances the photosynthetic rate which results in rapid growth of crops [29]. The new area of speed breeding needs proper research so that it assures food and nutritional security.

16.11 SYSTEMS BIOLOGY

Major impacts in omics platforms and modern sequencing technologies generated huge biological data. With the help of these enormous genomic resources system biology came into place to study effectively. In Systems Biology, biological data are incorporated to develop simulations and computational modeling by creating algorithms, data structure, data visualization and communication tools. Systems biology approaches are integrative and predictive, hold immense potential in understanding of molecular mechanism of agriculturally important complex traits linked to agricultural productivity. This has led to identification of some key genes [4, 30] and proteins involved in series of pathways involved in nutrition and anti-nutritional factors [31, 32]. Systems biology approaches are used to reduce yield barriers for enhancing the nutritional and productivity and values of food through integration of multi-omics data for modeling, simulation, and network analysis, This help in genetic manipulation and molecular breeding of key candidate genes in crop plants [4, 30, 33, 34].

16.12 GENOME EDITING (GE)

There are modern GE tools such as CRISPR CAS system. We can make precise cut and edit at specific DNA sites. This makes us possible to render the DNA of particular proteins that downregulating the pathway of vitamin metabolism. We can also edit the gene to synthesis new protein products that help in upregulating vitamin metabolic pathways. In rice gene associated with quality and plant height are demonstrated by GE [35]. There is another technology TALENs which can make targeted modification in DNA and cleave sequences at desired sites. So, it is aimed to enhance nutritive value of crops through recent advancements in field of biotechnology. In recent years GE tools have made the ability to carry out sequence specific modifications in the genome of organisms. The GE tools now commonly used are clustered regulated interspaced short palindromic repeats (CRISPR) – CRISPR associated protein 9 (Cas9), transcription activator like effector nucleases (TALENs), zinc finger nucleases (ZFNs). TALENs and CRISPR/Cas9 systems has been used to knock out *ZM*IPK gene involved in phytate biosynthesis in maize, to reduce phytate level [36]. Sanchez-Leon et al. [37] used CRISPR/Cas 9 technology to reduce the amount of α gliadins in seed kernel, developing bread and durum wheat with reduced immunoreactivity for gluten sensitive consumers. Cereal grains with high amylose content and resistant starch (RS) provides health benefits. Sun et al. [38] developed rice with high amylose content through CRISPR/Cas9 mediated targeted mutagenesis in SBEIIb gene. Trans fats produced by hydrogenation develops health risk on consumption. So, soybean lines were developed to produce poly-unsaturated fatty acids (PUFA) from oleic acids. TALENs editing method was adopted to induce mutation in FAD 2-1A and FAD2-1B, two fatty acid (FA) desaturase 2 genes [39]. Thus, GE has demonstrated their potential application in the area of qualitative trait improvement in crops.

16.13 CONCLUSION

Current crop improvement approaches focus on the improvement of quantitative and qualitative traits of a plant. Role of biotechnology is vital in crop improvement. Many useful traits like high yielding, disease, and pest resistance, abiotic stress tolerance, nutritional improvement can be done through molecular approaches. The use of biotechnology to create genetically modified organisms (GMOs) has a potential to design foods with biofortification special attributes [40]. The development of such crops can be achieved by selective breeding, speed breeding or biotechnological approaches [41].

Biofortification advancements through biotechnological strategies must meet the ultimate aim by ensuring that the benefits pass till the poor under nourished society and the negative consequences are mitigated. In the developing world most of the rural population is neither accessible nor affordable to nutritious food. So, nutritive improvement of staple food crops can be a good option to meet the nutritive requirement at affordable cost. Besides the advantages of transgenic biofortified crops, the concern regarding the impact and risk to the human and environmental health. Recent concepts of the use of synthetic metaboloms could also greatly enhance the efficiency and increase the outcome of metabolic engineering [42]. Despite many debates on the safety issues raised, Philippines is the first country in approving golden rice. The procedure of acceptance and release of GMOs differ from country. Every country may have their own regulatory procedures. By replacing ornamental garden and street plants with fast-growing nutrition-loaded crops, and incorporating fruit trees, we have the potential to initiate a "Nutritional Revolution" without resorting to costly techniques. This practice coupled with the principles of SMART agriculture, can effectively address hidden hunger and enhance nutrition accessibility within communities.

KEYWORDS

- **biofortification**
- **genome editing**
- **malnutrition**
- **micronutrient deficiency**
- **molecular approaches**
- **soilless culture**
- **zinc biofortification**

REFERENCES

1. The State of Food Security and Nutrition in the World (SOFI), (2021). *Food and Agricultural Organization of the United Nations.* https://www.fao.org/3/cb4474en/online/cb4474en.html (accessed on 22 May 2023).
2. Mason, J., Bailes, A., Beda-Andourou, M., Copeland, N., Curtis, T., Deitchler, M., Foster, L., et al., (2005). Recent trends in malnutrition in developing regions: Vitamin A deficiency, anemia, iodine deficiency, and child underweight. *Food and Nutrition Bulletin, 26*(1), 59–108. https://doi.org/10.1177/156482650502600107.

3. Yadava, D. K., Choudhury, P. R., Firoz, H., & Dinesh, K., (2017). *Biofortified Varieties: Sustainable Way to Alleviate Malnutrition*. Indian Council of Agricultural Research, New Delhi. https://icar.org.in/files/BiofortifiedEnglish_.pdf (accessed on 22 May 2023).
4. Singh, R. K., Krishnamachari, A., & Sharaff, M., (2020). Challenges of small RNA technology. In: *Plant Small RNA* (pp. 545–565). *Academic Press*.
5. Buu, M., (2003). Golden rice: Genetically modified to reduce vitamin a deficiency, benefit or hazard? *Nutrition Bytes, 9*(2). Retrieved from: https://escholarship.org/uc/item/2h01f05c (accessed on 22 May 2023).
6. Beyer, P., (2010). Golden rice and 'golden' crops for human nutrition. *New Biotechnology, 27*(5), 478–481. https://doi.org/10.1016/j.nbt.2010.05.010.
7. Brinch-Pedersen, H., Borg, S., Tauris, B., & Holm, P. B., (2007). Molecular genetic approaches to increasing mineral availability and vitamin content of cereals. *Journal of Cereal Science, 46*(3), 308–326.
8. Ozturk, L., Yazici, M. A., Yucel, C., Torun, A., Cekic, C., Bagci, A., & Cakmak, I., (2006). Concentration and localization of zinc during seed development and germination in wheat. *Physiologia Plantarum, 128*(1), 144–152.
9. Masuda, H., Ishimaru, Y., Aung, M. S., Kobayashi, T., Kakei, Y., Takahashi, M., & Nishizawa, N. K., (2012). Iron biofortification in rice by the introduction of multiple genes involved in iron nutrition. *Scientific Reports, 2*(1), 1–7.
10. Goto, F., Yoshihara, T., Shigemoto, N., Toki, S., & Takaiwa, F., (1999). Iron fortification of rice seed by the soybean ferritin gene. *Nature Biotechnology, 17*(3), 282–286.
11. Lucca, P., Hurrell, R., & Potrykus, I., (2001). Approaches to improving the bioavailability and level of iron in rice seeds. *Journal of the Science of Food and Agriculture, 81*, 828–834. doi: 10.1002/jsfa.886.
12. Chaudhary, S., Singh, A. K., & Yasin, J. K., (2017). Metallothioneins: Protein structure prediction and sequence analyses in *Cajanus cajan*: Protein structure prediction and sequence analyses in *Cajanus cajan*. *Journal of AgriSearch, 4*(4), 276–279.
13. Higuchi, K., Suzuki, K., Nakanishi, H., Yamaguchi, H., Nishizawa, N. K., & Mori, S., (1999). Cloning of nicotianamine synthase genes, novel genes involved in the biosynthesis of phytosiderophores. *Plant Physiology, 119*(2), 471–480. https://doi.org/10.1104/pp.119.2.471.
14. Takahashi, M., Terada, Y., Nakai, I., Nakanishi, H., Yoshimura, E., Mori, S., & Nishizawa, N. K., (2003). Role of nicotianamine in the intracellular delivery of metals and plant reproductive development. *The Plant Cell, 15*(6), 1263–1280. http://www.jstor.org/stable/3871982 (accessed on 22 May 2023).
15. Narayanan, N., Beyene, G., Chauhan, R. D., Gaitán-Solis, E., Grusak, M. A., Taylor, N., & Anderson, P., (2015). Overexpression of *Arabidopsis* VIT1 increases accumulation of iron in cassava roots and stems. *Plant Science: An International Journal of Experimental Plant Biology, 240*, 170–181. https://doi.org/10.1016/j.plantsci.2015.09.007.
16. Beyene, G., Solomon, F. R., Chauhan, R. D., Gaitán-Solis, E., Narayanan, N., Gehan, J., Siritunga, D., et al., (2018). Provitamin A biofortification of cassava enhances shelf life but reduces dry matter content of storage roots due to altered carbon partitioning into starch. *Plant Biotechnology Journal, 16*(6), 1186–1200. https://doi.org/10.1111/pbi.12862.
17. Kaur, N., Alok, A., Shivani, Kumar, P., Kaur, N., Awasthi, P., Chaturvedi, S., et al., (2020). CRISPR/Cas9 directed editing of lycopene epsilon-cyclase modulates metabolic

flux for β-carotene biosynthesis in banana fruit. *Metabolic Engineering, 59*, 76–86. https://doi.org/10.1016/j.ymben.2020.01.008.

18. Prasad, B. V. G., Mohanta, S., Rahaman, S., & Bareily, P., (2015). Bio-fortification in horticultural crops. *J. Agric. Eng. Food Technol., 2*(2), 95–99.
19. Colla, G., Kim, H. J., Kyriacou, M. C., & Rouphael, Y., (2018). Nitrate in fruits and vegetables. *Sci. Hortic., 237*, 221–238. doi: 10.1016/j.scienta.2018.04.016.
20. Rouphael, Y., & Kyriacou, M. C., (2018). Quality and safety of fresh fruits and vegetables at harvest. *Sci. Horti., 239*, 78–79. doi: 10.1016/j.scienta.2018.05.006.
21. Rouphael, Y., Kyriacou, M. C., Petropoulos, S. A., De Pascale, S., & Colla, G., (2018). Improving vegetable quality in controlled environments. *Sci. Hort., 234*, 275–289. doi: 10.1016/j.scienta.2018.02.033.
22. Rouphael, Y., Petropoulos, S. A., Cardarelli, M., & Colla, G., (2018). Salinity as eustressor for enhancing quality of vegetables. *Sci. Hortic., 234*, 361–369. doi: 10.1016/j.scienta.2018.02.048.
23. Nivedhitha, V. S., Neyavaruni, G., Nirmala, R., & Shricharan, S., (2020). Isolation of *Bacillus subtilis* and its efficiency in aerobic degradation of kitchen waste. *National Conference on Recent Trends in Chemical and Environmental Biotechnology (RTCEB) Conference Proceedings* (pp. 137–140). Annamalai University, Tamil Nadu.
24. White, P. J., & Broadley, M. R., (2005). Biofortifying crops with essential mineral elements. *Trends Plant Sci., 10*, 586–593. doi: 10.1016/j.tplants.2005.10.001.
25. Samantara, K., Bohra, A., Mohapatra, S. R., Prihatini, R., Asibe, F., Singh, L., Reyes, V. P., et al., (2022). Breeding more crops in less time: A perspective on speed breeding. *Biology, 11*, 275. doi.org/10.3390/biology11020275.
26. Ramya, K. T., Fiyaz, R. A., & Yasin, J. K., (2013). SMART agriculture for nutritional security. *Current Science, 105*(11), 1458.
27. Watson, A., Ghosh, S., Williams, M. J., Cuddy, W. S., Simmonds, J., Rey, M. D., & Hickey, L. T., (2018). Speed breeding is a powerful tool to accelerate crop research and breeding. *Nature Plants, 4*(1), 23–29.
28. Fikre, A., & Degefu, T., (2019). *A Guide to Accelerated Breeding Cycle in Chickpea to Enhance Rate of Gain.* Technical Manual No.1/2019. http://oar.icrisat.org/11288/1/Speed%20Breeding%20Manual.pdf (accessed on 22 May 2023).
29. Singh, H., & Janeja, H. S., (2021). Speed breeding a ray of hope for the future generation in terms of food security: A review. *Plant Archives, 21*(1), 155–158.
30. Yasin, J. K., Mishra, B. K., Pillai, M. A., Verma, N., Wani, S. H., Elansary, H. O., & Chinnusamy, V., (2020). Genome wide in-silico miRNA and target network prediction from stress responsive horsegram (*Macrotyloma uniflorum*) accessions. *Scientific Reports, 10*(1), 1–23.
31. Lone, J. K., Lekha, M. A., Bharadwaj, R. P., Ali, F., Pillai, M. A., Wani, S. H., Yasin, J. K., & Chandrashekharaiah, K. S., (2021). Multimeric association of purified novel Bowman-Birk inhibitor from the medicinal forage legume *Mucuna pruriens* (L.) DC. *Front. Plant Sci., 12,* 772046. doi: 10.3389/fpls.2021.772046.
32. Pathak, R. K., & Singh, D. B., (2020). Systems biology approaches for food and health. In: *Advances in Agri-Food Biotechnology* (pp. 409–426). Springer, Singapore.
33. Shabir, H. W., Vijayan, R., Choudhary, M., Kumar, A., Zaid, A., Singh, V., Kumar, P., & Yasin, J. K., (2021). Nitrogen use efficiency (NUE): Elucidated mechanisms, mapped genes and gene networks in maize (*Zea mays L.*). *Physiol. Mol. Biol. Plants, 27*(3), 2875–2891. https://doi.org/10.1007/s12298-021-01113-z.

34. Kumar, A., Pathak, R. K., Gupta, S. M., Gaur, V. S., & Pandey, D., (2015). Systems biology for smart crops and agricultural innovation: Filling the gaps between genotype and phenotype for complex traits linked with robust agricultural productivity and sustainability. *Omics: A Journal of Integrative Biology, 19*(10), 581–601.
35. Andrew-Peter-Leon, M. T., Selvaraj, R., Kumar, K. K., Muthamilarasan, M., Yasin, J. K., & Pillai, M. A., (2021). Loss of function of *OsFBX267* and *OsGA20ox2* in rice promotes early maturing and semi-dwarfism in γ-irradiated IWP and genome-edited pusa basmati-1. *Front. Plant Sci., 12*, 714066. doi: 10.3389/fpls.2021.714066.
36. Liang, Z., Zhang, K., Chen, K., & Gao, C., (2014). Targeted mutagenesis in *Zea mays* using TALENs and the CRISPR/Cas system. *Journal of Genetics and Genomics, 41*(2), 63–68.
37. Sánchez-León, S., Gil-Humanes, J., Ozuna, C. V., Giménez, M. J., Sousa, C., Voytas, D. F., & Barro, F., (2018). Low-gluten, nontransgenic wheat engineered with CRISPR/Cas9. *Plant Biotechnology Journal, 16*(4), 902–910.
38. Sun, Y., Jiao, G., Liu, Z., Zhang, X., Li, J., Guo, X., & Xia, L., (2017). Generation of high-amylose rice through CRISPR/Cas9-mediated targeted mutagenesis of starch branching enzymes. *Frontiers in Plant Science, 8*, 298.
39. Haun, W., Coffman, A., Clasen, B. M., Demorest, Z. L., Lowy, A., Ray, E., & Zhang, F., (2014). Improved soybean oil quality by targeted mutagenesis of the fatty acid desaturase 2 gene family. *Plant Biotechnology Journal, 12*(7), 934–940.
40. Johns, T., & Eyzaguirre, P. B., (2007). Biofortification, biodiversity and diet: A search for complementary applications against poverty and malnutrition. *Food Policy, 32*, 1–24.
41. Garg, M., Sharma, N., Sharma, S., Kapoor, P., Kumar, A., Chunduri, V., & Arora, P., (2018). Biofortified crops generated by breeding, agronomy, and transgenic approaches are improving lives of millions of people around the world. *Frontiers in Nutrition, 5*, 12. https://doi.org/10.3389/fnut.2018.00012.
42. Singleton, C., Howard, T. P., & Smirnoff, N., (2014). Synthetic metabolons for metabolic engineering. *Journal of Experimental Botany, 65*(8), 1947–1954.

Index

C

D

G

H

M

Q

R